IF FOUND, please notify and arrange return to owner. This handbook is important for the owner/pilot's responsibilities as an instrument pilot. Thank you.

Pilot's Name: ______________________________

Address: ______________________________

City State Zip Code

Telephone: (____) ____________ E-mail: ______________

Instrument Pilot Flight Maneuvers and Practical Test Prep is one of four related Gleim books that are cross-referenced and interdependent. Please obtain and use the following books as a package designed to prepare you for your instrument rating:

Instrument Pilot Flight Maneuvers and Practical Test Prep
Instrument Pilot FAA Written Exam
Aviation Weather and Weather Services
Pilot Handbook

If you purchased this book as part of Gleim's Instrument Pilot Kit, *Pilot Handbook* is not included since it is part of our Private Pilot Kit. If you do not have Gleim's *Pilot Handbook*, you should purchase it as a reference. All of the aviation books (and software) by Gleim are listed on the order form provided at the back of this book. Thank you for choosing Gleim.

Gleim Publications, Inc.
P.O. Box 12848, University Station
Gainesville, Florida 32604
(352) 375-0772 / (800) 87-GLEIM / (800) 874-5346
FAX: (352) 375-6940
Internet: www.gleim.com
E-mail: admin@gleim.com

Orders must be prepaid. Prices are subject to change without notice. We ship latest editions. Shipping and handling charges are added to all orders. Library and company orders may be on account. Add applicable sales tax to shipments within Florida. All payments must be in U.S. funds and payable on a U.S. bank. Please write or call for prices and availability of all foreign country shipments. Books will usually be shipped the day after your order is received. Allow 10 days for delivery in the United States. Please contact us if you do not receive your shipment within 2 weeks.

Gleim Publications, Inc. guarantees the immediate refund of all resalable texts returned within 30 days. The guaranteed refund applies only to books purchased direct from Gleim Publications, Inc. Shipping and handling charges are nonrefundable.

This book contains a systematic discussion and explanation of the FAA's Instrument Rating Practical Test Standards (Airplane, Single-Engine Land), which will assist you in (1) preparing for and (2) successfully completing your Instrument Rating FAA Practical Test!

REVIEWERS AND CONTRIBUTORS

Karen A. Hom, B.A., University of Florida, is our book production coordinator. Ms. Hom coordinated the production staff and provided assistance throughout the project.

Barry A. Jones, ATP, CFII, MEI, B.S. in Air Commerce/Flight Technology, Florida Institute of Technology, is our aviation project manager and also a charter pilot and flight instructor with Gulf Atlantic Airways in Gainesville, FL. Mr. Jones assembled the text, added new material, incorporated numerous revisions, and provided technical assistance throughout the project.

Travis A. Moore, M.B.A., University of Florida, assisted in the production process.

James A. Neinas, CFII, MEI, B.S. in Computer Science, University of Florida, is our aviation department assistant and a flight instructor with Gulf Atlantic Airways in Gainesville, FL. Mr. Neinas assisted Barry Jones throughout the project.

Nancy Raughley, B.A., Tift College, is our editor. Ms. Raughley reviewed the manuscript, revised it for readability, and assisted in all phases of production.

John F. Rebstock, B.S., School of Accounting, University of Florida, reviewed portions of the text.

Jan M. Strickland is our production assistant. She reviewed the final manuscript and prepared the page layout.

The many FAA employees who helped, in person or by telephone, primarily in Gainesville, FL; Orlando, FL; Oklahoma City, OK; and Washington, DC.

A PERSONAL THANKS

This manual would not have been possible without the extraordinary efforts and dedication of Jim Collis and Terry Hall, who typed the entire manuscript and all revisions, and prepared the camera-ready pages.

The author also appreciates the proofreading and production assistance of Matt Danner, Svetlana Dzyubenko, Robyn Fitzpatrick, Erin Foster, Rob Gallardo, Kirk Khan, Walter Mansfield, Jose Martinez, Jessica Medina, Kristin Morgan, Jaime Phillips, and Steve Rice.

Finally, I appreciate the encouragement, support, and tolerance of my family throughout this project.

Groundwood Paper and Highlighters -- This book is printed on high quality groundwood paper. It is lightweight and easy-to-recycle. We recommend that you purchase a highlighter specifically designed to be non-bleed-through (e.g., Avery *Glidestick*™) at your local office supply store.

GLEIM FLIGHT TRAINING SERIES

INSTRUMENT PILOT

FLIGHT MANEUVERS AND PRACTICAL TEST PREP

THIRD EDITION

by Irvin N. Gleim, Ph.D., CFII

with the assistance of
Barry A. Jones, ATP, CFII, MEI

ABOUT THE AUTHOR

Irvin N. Gleim earned his private pilot certificate in 1965 at the Institute of Aviation at the University of Illinois, where he subsequently received his Ph.D. He is a commercial pilot and flight instructor (instrument) with multiengine and seaplane ratings, and is a member of the Aircraft Owners and Pilots Association, American Bonanza Society, Civil Air Patrol, Experimental Aircraft Association, and Seaplane Pilots Association. He is author of flight maneuvers and practical test prep books for the private, instrument, commercial, and flight instructor certificates/ratings, and study guides for the private/recreational, instrument, commercial, flight/ground instructor, fundamentals of instructing, airline transport pilot, and flight engineer FAA knowledge tests. Two additional pilot training books are *Pilot Handbook* and *Aviation Weather and Weather Services*.

Dr. Gleim has also written articles for professional accounting and business law journals, and is the author of widely used review manuals for the CIA (Certified Internal Auditor) exam, the CMA (Certified Management Accountant) exam, the CPA (Certified Public Accountant) exam, and the EA exam (IRS Enrolled Agent). He is Professor Emeritus, Fisher School of Accounting at the University of Florida, and is a CIA, CMA, CFM, and CPA.

Gleim Publications, Inc.
P.O. Box 12848
University Station
Gainesville, Florida 32604
(352) 375-0772
(800) 87-GLEIM or (800) 874-5346
FAX: (352) 375-6940
Internet: www.gleim.com
E-mail: admin@gleim.com

ISSN 1085-3960
ISBN 1-58194-079-3
Third Printing: September 2001

This is the third printing of the third edition of ***Instrument Pilot Flight Maneuvers and Practical Test Prep***.
Please e-mail update@gleim.com with IPFM 3-3 in the subject or text. You will receive our current update as a reply.

EXAMPLE:

To: update@gleim.com
From: your e-mail address
Subject: IPFM 3-3

CAUTION: This book is an academic presentation for training purposes only. Under **NO** circumstances can it be used as a substitute for your *Pilot's Operating Handbook* or FAA-approved *Airplane Flight Manual*. **You must fly and operate your airplane in accordance with your *Pilot's Operating Handbook* or FAA-approved *Airplane Flight Manual*.**

The author is indebted to Jeppesen-Sanderson for permission to reprint various charts, legends, and textual material in Chapter 3 and elsewhere in this book. These are reprinted for academic illustration/training purposes only and are not for navigational uses. They have also provided a copy of their "Chart Training Guide" which is enclosed. See Chapter 3.

HELP !!

This is the Third Edition designed specifically for pilots or flight instructors seeking to add the instrument rating to their certificates. Please send any corrections and suggestions for subsequent editions to the author, c/o Gleim Publications, Inc. The last page in this book has been reserved for you to make comments and suggestions. It can be torn out and mailed to us.

NOTE: UPDATES

Send e-mail to update@gleim.com as described at the top right of this page, and visit our Internet site for the latest updates and information on all of our products. To continue providing our customers with first-rate service, we request that questions about our books and software be sent to us via mail, e-mail, or fax. The appropriate staff member will give each question thorough consideration and a prompt response. Questions concerning orders, prices, shipments, or payments will be handled via telephone by our competent and courteous customer service staff.

TABLE OF CONTENTS

THIRD PRINTING (09/01) CHANGES

1. Throughout the book, “National Ocean Service (NOS)” has been changed to “National Aeronautical Charting Office (NACO)” to reflect the recent restructuring of government agencies which produce aeronautical and marine charts.

2. Chapter IV, Flight by Reference to Instruments. Four flight instrument figures were corrected by reversing the direction of the arrow that appears above the heading indicator (pp. 149, 178, 187, and 188).

3. Appendix C, FAA Flight Instructor-Instrument Practical Test Standards. This appendix has been updated to reflect the latest FAA practical test standards for CFII applicants.

PREFACE

This book will facilitate your instrument flight training and prepare you to pass your INSTRUMENT RATING FAA PRACTICAL TEST. In addition, this book will assist you and your flight instructor in planning and organizing your flight training.

The instrument rating practical test is a rigorous test of both concept knowledge and motor skills. This book explains all of the knowledge that your instructor and FAA examiner will expect you to demonstrate and discuss with him/her. Previously, instrument rating candidates had only the FAA PTS "reprints" to study. Now you have PTSs followed by a thorough explanation of each task and a step-by-step description of each flight maneuver. Thus, through careful organization and presentation, we will decrease your preparation time, effort, and frustration, **and** increase your knowledge and understanding.

To save you time, money, and frustration, we have listed some of the common errors made by pilots in executing each flight maneuver or operation. You will be aware of *what not to do*. We all learn by our mistakes, but our *common error* list provides you with an opportunity to learn from the mistakes of others.

Most books create additional work for the user. In contrast, *Instrument Pilot Flight Maneuvers and Practical Test Prep* facilitates your effort; i.e., it is easy to use. The outline format, numerous illustrations and diagrams, type styles, indentions, and line spacing are designed to improve readability. Concepts are often presented as phrases rather than complete sentences.

Relatedly, our outline format frequently has an "a" without a "b" or a "1" without a "2." While this violates some journalistic *rules of style*, it is consistent with your cognitive processes. This book was designed, written, and formatted to facilitate your learning and understanding. Another similar counterproductive "rule" is *not to write in your books*. I urge you to mark up this book to facilitate your learning and understanding.

I am confident this book will facilitate speedy completion of your flight training and practical test. I also wish you the very best in subsequent flying, and in obtaining additional ratings and certificates. If you have *not* passed your instrument rating FAA knowledge test and do *not* have *Instrument Pilot FAA Written Exam* (another book with a red cover) or *FAA Test Prep* software, please order today. Almost everything you need to pass the FAA's knowledge and practical tests for the instrument rating is available from Gleim in our new Instrument Pilot Kit. If your FBO, flight school, or aviation bookstore is out of stock, call (800) 87-GLEIM, or (800) 874-5346.

I encourage your suggestions, comments, and corrections for future printings and editions. The last page of this book has been designed to help you note corrections and suggestions throughout your preparation process. Please use it, tear it out, and mail it to me. Thank you.

Enjoy Flying -- Safely!

Irvin N. Gleim

September 2001

PART I
GENERAL INFORMATION

Part I (Chapters 1 through 4) of this book provides general information to assist you in obtaining your instrument rating:

Part II consists of Chapters I through VIII, which provide an extensive explanation of each of the 26 tasks required of those taking the instrument rating FAA practical test in a single-engine airplane. Part II is followed by Appendix A, FAA Instrument Rating Practical Test Standards, which is a reprint of all 26 tasks in one location.

Instrument Pilot Flight Maneuvers and Practical Test Prep is one book in a series of six books for obtaining your instrument rating. The six books are

1. *Instrument Pilot Flight Maneuvers and Practical Test Prep*
2. *Instrument Pilot Syllabus*
3. *Instrument Pilot FAA Written Exam*
4. *Aviation Weather and Weather Services*
5. *Pilot Handbook*
6. *FAR/AIM*

Gleim's Instrument Pilot Kit contains all of the above books except *Pilot Handbook* and *FAR/AIM*, since these books were included in Gleim's Private Pilot Kit. They are available at your local FBO, flight school, bookstore, etc., or call 800-87-GLEIM (see order form at back of book).

Instrument Pilot Syllabus is a step-by-step syllabus of ground and flight training lesson plans for your instrument rating training.

Instrument Pilot FAA Written Exam contains all of the FAA's airplane-related questions and organizes them into logical topics called modules. The book consists of 91 modules, which are grouped into 11 chapters. Each chapter begins with a brief, user-friendly outline of what you need to know, and answer explanations are provided next to each question. This book will transfer knowledge to you and give you the confidence to do well on the FAA instrument rating knowledge test.

Aviation Weather and Weather Services combines all of the information from the FAA's *Aviation Weather* (AC 00-6A), *Aviation Weather Services* (AC 00-45E), and numerous other FAA publications into one easy-to-understand book. It will help you study all aspects of aviation weather and provide you with a single reference book.

Pilot Handbook is a complete text and reference for all pilots. Aerodynamics, airplane systems and instruments, and navigation systems are among the topics explained.

Gleim's *FAR/AIM* is an easy-to-read reference book containing all of the Federal Aviation Regulations (FARs) applicable to general aviation flying, plus the full text of the FAA's *Aeronautical Information Manual (AIM)*.

RECAP OF REQUIREMENTS TO OBTAIN AN INSTRUMENT RATING

1. Hold at least a private pilot certificate.
2. Be able to read, write, and converse fluently in English (certificates with operating limitations may be available for medically related deficiencies).
3. Hold a current FAA medical certificate.
4. Receive and log ground training OR complete a home-study course by studying *Instrument Pilot FAA Written Exam* (and the related Gleim *FAA Test Prep* software), *Pilot Handbook*, and *Aviation Weather and Weather Services*. Subjects include
   a. Airplane Instruments
   b. Airports and Air Traffic Control
   c. Aviation Weather
   d. Federal Aviation Regulations
   e. Navigation
   f. Aeromedical Factors and Aeronautical Decision Making
   g. Flight Operations
   h. Instrument Approaches
   i. IFR En Route
5. Pass the FAA instrument rating knowledge test with a score of 70% or better.
6. Accumulate flight experience (FAR 61.65).
   a. 50 hr. of cross-country flight time as pilot in command (PIC), of which at least 10 hr. must be in airplanes
      1) The 50 hr. includes solo cross-country time as a student pilot, which is logged as pilot-in-command time.
      2) Each cross-country must have a landing at an airport that was at least a straight-line distance of more than 50 NM from the original departure point.

   NOTE: Part 141 does not require you to have 50 hr. of cross-country flight time as PIC.

   b. 40 hr. of simulated or actual instrument time (of which up to 20 hr. with an instructor may be in an approved flight simulator or flight training device), including
      1) 15 hr. of instrument flight training from a CFII
      2) 3 hr. of instrument training from a CFII in preparation for the practical test within 60 days preceding the practical test
      3) Cross-country flight procedures that include at least one cross-country flight in an airplane that is performed under IFR and consists of
         a) A distance of at least 250 NM along airways or ATC-directed routing
         b) An instrument approach at each airport
         c) Three different kinds of approaches with the use of navigation systems

   NOTE: Part 141 requires 35 hr., instead of 40 hr., of instrument time.

7. Demonstrate flight proficiency (FAR 61.65). You must receive and log training, and obtain a logbook endorsement (sign-off) by your CFII on the following areas of operations:
   a. *Preflight preparation*
   b. *Preflight procedures*
   c. *Air traffic control clearances and procedures*
   d. *Flight by reference to instruments*
   e. *Navigation systems*
   f. *Instrument approach procedures*
   g. *Emergency operations*
   h. *Postflight procedures*
8. Successfully complete a practical test which will be given as a final exam by an FAA inspector or designated pilot examiner. The practical test will be conducted as specified in Part II of this book.

CHAPTER ONE
THE INSTRUMENT RATING

1.1 WHY GET AN INSTRUMENT RATING?

Achieving an instrument rating is not only a fun pursuit but also a worthwhile accomplishment. It provides the satisfaction of knowing that you have gone a step beyond the private pilot certificate and have demonstrated the skill and determination to achieve a higher and safer set of standards. For those not planning a career as a professional pilot, the expense can be justified several ways, including by

1. **Increased Skill:** From the standpoint of training, instrument flying is a logical extension of visual flying. You learn to use the instruments and navigation equipment to develop a precision impossible to achieve by visual and other sensory references alone. You become better able to utilize the full potential of your airplane.
2. **Increased Safety:** The more precise and thorough your piloting skills and knowledge are, the safer you will be. Safety should be the foremost concern of all pilots. By flying to higher standards (closer tolerances), you will increase the safety of every flight.
3. **Lowered Insurance Premiums:** A major factor in the cost of aviation insurance is the skill level of the pilot. Insurance companies do not send someone to fly with each client/customer. The level of pilot proficiency is assumed on a basis of the number of hours flown and the certificates and ratings held. By gaining your instrument rating, you are elevating yourself into a safer category, which will usually lower the cost of your insurance.
4. **Challenge:** Perhaps you want an instrument rating for the same basic reason you learned to fly in the first place -- because you like flying. Earn the rating, not because you might need it sometime, but because it represents achievement and provides training you will use continually and build upon as long as you fly.
5. **Confidence:** An instrument rating will give you increased confidence in your pilot capabilities. You will no longer be intimidated by marginal weather.
6. **Convenience:** You will not be totally weather dependent as you are as a noninstrument-rated pilot. An instrument rating substantially increases the "usability" of your flying convenience.

For those considering a career as a pilot, the instrument rating is required (except in some highly specialized commercial operations). A professional pilot career, whether in flight instruction, charter, corporate, or airline, is appealing to many beginning pilots. For some, the love of flying is strong enough for them to make a career of aviation.

1.2 WHAT IS AN INSTRUMENT RATING?

A. An instrument rating permits you to fly by reference to instruments only. It provides you with the capability to fly in conditions below basic VFR weather minimums. Your instrument rating will better enable you to use all the equipment in your airplane and have greater flexibility in various weather conditions.

B. An instrument rating is added to your private or commercial pilot certificate. A new certificate is sent to you by the FAA upon satisfactory completion of your training program, an FAA knowledge test, and an FAA practical test. A sample private pilot certificate with an instrument rating is reproduced below.

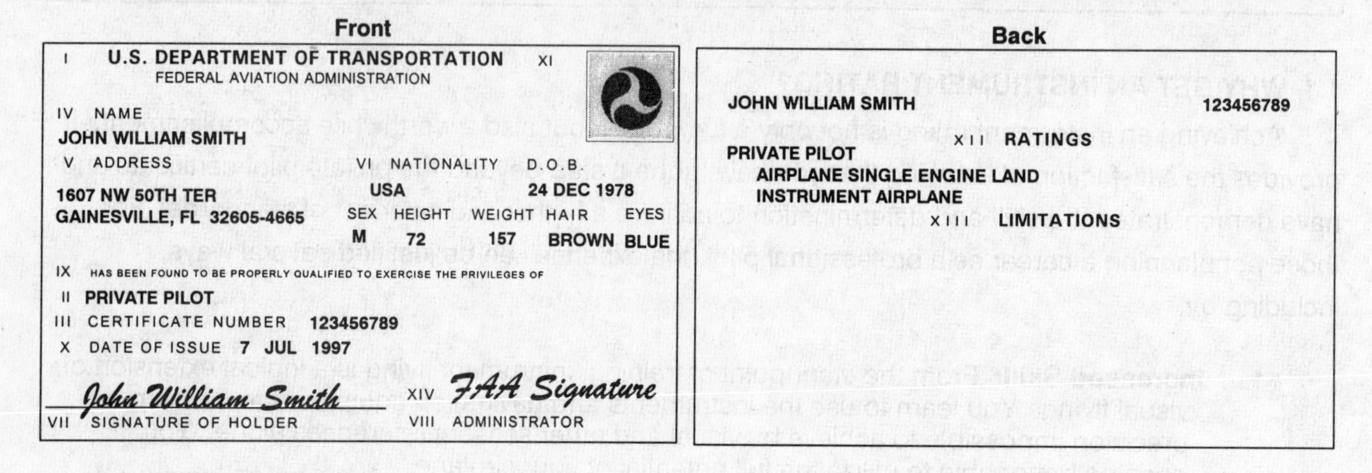

Front

I U.S. DEPARTMENT OF TRANSPORTATION XI
FEDERAL AVIATION ADMINISTRATION

IV NAME
JOHN WILLIAM SMITH
V ADDRESS VI NATIONALITY D.O.B.
1607 NW 50TH TER USA 24 DEC 1978
GAINESVILLE, FL 32605-4665 SEX HEIGHT WEIGHT HAIR EYES
M 72 157 BROWN BLUE
IX HAS BEEN FOUND TO BE PROPERLY QUALIFIED TO EXERCISE THE PRIVILEGES OF
II PRIVATE PILOT
III CERTIFICATE NUMBER 123456789
X DATE OF ISSUE 7 JUL 1997

John William Smith XIV *FAA Signature*
VII SIGNATURE OF HOLDER VIII ADMINISTRATOR

Back

JOHN WILLIAM SMITH 123456789
XII RATINGS
PRIVATE PILOT
AIRPLANE SINGLE ENGINE LAND
INSTRUMENT AIRPLANE
XIII LIMITATIONS

1.3 HOW TO GET STARTED

A. **Talk to several people who have recently attained their instrument rating.** Visit your local airport and ask for the names of several people who have just completed their instrument rating. When you locate one person, (s)he can usually refer you to another. How did they do it?

1. Flight training: Airplane? CFII? What period of time? Cost? How structured was the program?
   a. You can search for a CFII in your area with Gleim's online CFI Directory at http://www.gleim.com/Aviation/cfidirectory.html.
2. Ask for their advice. How would they do it differently? What do they suggest to you?
3. What difficulties did they encounter?
4. Did they get experience in actual IFR conditions?

B. **Talk to several CFIIs.** Tell them you are considering an instrument rating. Evaluate each as a prospective instructor.

1. What do they recommend?
2. Do they use Gleim's *Instrument Pilot Syllabus*?
3. Ask to see their flight syllabus. How structured is it?
4. What are their projected costs?
5. What is the rental cost for their training aircraft and simulators, solo and dual?

6. Ask for the names and phone numbers of several persons who recently attained the instrument rating under their direction. Ask the questions suggested on the previous page.
7. Is the flight instructor willing to do some training (late in the syllabus) in actual IFR conditions?
8. Does the flight instructor's schedule and the schedule of available aircraft fit your schedule?
9. Where will they recommend that you take your practical test? What is its estimated cost?

C. Once you have made a preliminary choice of flight instructor and/or FBO, **sit down with your CFII and plan a schedule of flight instruction.**

1. When and how often you will fly
2. When you will take the FAA instrument rating knowledge test
3. When you should plan to take your practical test
4. When and how payments will be made for your instruction
5. Review, revise, and update the total cost to obtain your instrument rating (see below).

D. **Prepare a tentative written time budget and a written expenditure budget.**

Hours PIC*: ____ hours × $______	$______
Hours Dual: ____ hours × $______	$______
Simulator: ____ hours × $______	$______
FAA instrument rating knowledge test	$______
Practical test (examiner)	$______
Practical test (airplane)	$______
This book**	$ 18.95
Gleim's *Instrument Pilot Syllabus***	$ 14.95
Gleim's *Instrument Pilot FAA Written Exam***	$ 18.95
Gleim's *FAA Test Prep* software (CD for Windows 95, 98, or NT 4.0 or higher)**	$ 59.95
Gleim's *Aviation Weather and Weather Services***	$ 22.95
Gleim's *Pilot Handbook*	$ 13.95
Other books:	
Gleim's *FAR/AIM*	$ 15.95
Terminal Procedures Publications (NOS or JEPP)	$______
Low altitude en route chart(s) (NOS or JEPP)	$______
Information manual for your airplane	$______
TOTAL	$______

* You may practice your IFR maneuvers with a qualified safety pilot (who is not a CFII) aboard.
** Included in Gleim's Instrument Pilot Kit.

E. **Consider purchasing an airplane (yourself, or through joint ownership) or joining a flying club.** Frequently, shared expenses through joint ownership can significantly reduce the cost of flying.

1. Inquire about local flying clubs.
2. Call a member and learn about the club's services, costs, etc.

1.4 LOOKING AHEAD: USING YOUR INSTRUMENT RATING

A. As you begin your instrument training, it is imperative that you recognize and understand the importance of knowing in what type of weather you can and cannot fly. An instrument rating does NOT enable you to fly in thunderstorms or icing. Thus, actual IFR experience with an experienced CFII is essential, mainly to learn when NOT to fly.

B. Conversely, you must obtain experience in actual IFR weather conditions so you have the confidence to use your instrument rating. Perhaps you can ride along, right seat or even back seat, with a skilled instrument-rated pilot in actual IFR conditions and help with charts and/or navigation. Even if you cannot record this time in your pilot logbook, it will be very valuable experience.

C. Avoid both extremes: being afraid to enter IMC (instrument meteorological conditions, i.e., clouds, fog, etc.) or having the fatal misconception that you can fly in any weather, once instrument-rated.

1.5 INSTRUMENT RELATED FARs

A. Study Chapter 4, Federal Aviation Regulations, in the *Instrument Pilot FAA Written Exam* book.

1. Remember to purchase Gleim's *FAR/AIM* as your official FAR and *AIM* reference. Its revised formatting and larger type make it easier to read, and it has better indexes. In addition, an automatic e-mail update service is provided for *FAR/AIM*.

B. **The required instruments and equipment for flight under IFR are:**

1. All equipment required for day and (if the aircraft is to be operated under IFR at night) night VFR flight.
2. Two-way radio communications and navigation systems appropriate to the facilities to be used.
3. A gyroscopic rate-of-turn indicator (turn coordinator or turn-and-slip indicator).
   a. A slip-skid indicator (i.e., a ball) is also required.
4. A gyroscopic attitude indicator.
5. A gyroscopic heading indicator.
6. A sensitive altimeter (i.e., adjustable for variations in barometric pressure).
7. A clock with a sweep second hand or a digital display.
8. A generator or alternator of adequate output.
9. At or above 24,000 feet MSL, Distance Measuring Equipment (DME).

END OF CHAPTER

CHAPTER TWO
OPTIMIZING YOUR FLIGHT AND GROUND TRAINING

The purpose of this chapter is to help you get the most out of your ground and flight training. They should support each other: Ground training should facilitate your flight training and vice versa. While your immediate objective is to pass your practical test, your long-range goal is to become a very safe and proficient pilot. Thus, you have to work hard to be able to **do your best.** No one can ask for more!

2.1 PART 61 vs. PART 141 vs. PART 142 FLIGHT TRAINING PROGRAMS

A. The program laid out in FAR Part 61 is available to anyone in conjunction with a flight instruction program taught by any CFII.

   1. A Part 61 instrument rating course requires a minimum of 40 hr. of instrument flight time, including a minimum of 15 hr. of instruction by a CFII.

      a. Most Part 61 schools include 40 hr. of instrument instruction in their instrument rating courses.

B. An alternative is an FAR Part 141 training program, which is a program conducted by an FAA-approved flight school. Part 141 flight schools are more highly regulated and require physical facility inspection, approval of ground school and flight training syllabi, etc., by the FAA.

   1. The Part 141 instrument rating course will include a minimum of 30 hr. of ground training and a minimum of 35 hr. of instrument flight training, all of which must be conducted at a Part 141 school.

      a. Unlike Part 61, Part 141 does not require 50 hr. of cross-country flight time as pilot in command.

   2. Gleim Publications has developed a Part 141 syllabus for flight schools. The syllabus can also be used as a Part 61 syllabus! Please bring this syllabus to the attention of your CFII. We welcome comments and suggestions from your and/or your CFII.

C. Part 142 explains the certification and operation of aviation training centers, which can provide an alternative means to accomplish the training required by Part 61.

   1. Under Part 142, a maximum of 30 hr. of instrument training from an authorized instructor may be performed in a flight simulator or flight training device.

2.2 GROUND TRAINING

A. First and foremost: Ground training is extremely important to facilitate flight training. Each preflight and postflight discussion is as important as the actual flight training of each flight lesson!

1. Unfortunately, most students and some CFIIs incorrectly overemphasize the in-airplane portion of a flight lesson.
   a. The airplane, all of its operating systems, ATC, other traffic, etc., are major distractions from the actual flight maneuver and the aerodynamic theory/factors underlying the maneuver.
   b. This is not to diminish the importance of dealing with operating systems, ATC, other traffic, etc.
2. Note that the effort and results are those of the student. Instructors are responsible for directing student effort so that the results are maximized.
3. Formal ground school to support instrument flight training generally does **not** exist except at aeronautical universities and some Part 141 programs. Most community college, adult education, and FBO ground schools are directed toward the FAA knowledge tests.

B. Again, **the effort and results are those of the student.** Prepare for each flight lesson so you know exactly what is going to happen and why. The more you prepare, the better you will do, both in execution of maneuvers and in acquisition of knowledge.

1. At the end of each flight lesson, find out exactly what is planned for the next flight lesson.
2. At home, begin by reviewing everything that occurred during the last flight lesson -- preflight briefing, flight, and postflight briefing. Make notes on follow-up questions and discussion with your CFII to occur at the beginning of the next preflight briefing.
3. Study all new flight maneuvers scheduled for the next flight lesson and review flight maneuvers that warrant additional practice (refer to the appropriate chapters in Part II of this book). Make notes on follow-up questions and discussion with your CFII to occur at the beginning of the next preflight briefing.
4. Before each flight, your CFII will sit down with you for a preflight briefing. Begin with a review of the last flight lesson. Then focus on the current flight lesson. Go over each maneuver to be executed, including maneuvers to be reviewed from previous flight lessons.
5. During each flight lesson, be diligent about safety (continuously ask your instructor to check traffic when you are "under the hood"). During maneuvers, compare your actual experience with your expectations (based on your prior knowledge from completing your Flight Maneuver Analysis Sheet).
6. Your postflight briefing should begin with a self-critique by you, followed by evaluation by your CFII. Ask questions until you are satisfied that you have expert knowledge. Finally, develop a clear understanding of the time and the maneuvers to be covered in your next flight lesson.

2.3 FLIGHT TRAINING

A. Once in the airplane, the FAA recommends that your CFII use the "telling and doing" technique:

1. Instructor tells, instructor does.
2. Student tells, instructor does.
3. Student tells, student does.
4. Student does, instructor evaluates.

B. Each attribute of the maneuver should be discussed before, during, and after execution of the maneuver.

C. Additional student home study of flight maneuvers needs to be **integrated** with in-airplane training to make in-airplane training both more effective and more efficient. Effectiveness refers to learning as much as possible (i.e., getting pilot skills "down pat" so as to be a safe and proficient pilot). Efficiency refers to learning as much as possible in a reasonable amount of time.

2.4 SIMULATOR TRAINING

A. **Flight simulator** means a device that

1. Is a full-size cockpit replica of a specific type of aircraft, or make, model, and series of aircraft
2. Includes the hardware and software necessary to represent the aircraft in ground and flight operations
3. Uses a force cueing system that provides cues at least equivalent to those cues provided by a 3° freedom of motion system
4. Uses a visual system that provides at least a 45° horizontal field of view and a 30° vertical field of view simultaneously for each pilot
5. Has been evaluated, qualified, and approved by the FAA

B. **Flight training device** means a device that

1. Is a full-size replica of the instruments, equipment, panels, and controls of an aircraft in an open flight deck area or in an enclosed cockpit, including the hardware and software for the systems installed, that is necessary to simulate the aircraft in ground and flight operations
2. Need not have a force (motion) cueing or visual system
3. Has been evaluated, qualified, and approved by the FAA

C. **Personal computer-based aviation training device (PCATD)** means a device that

1. Meets minimum acceptable design criteria
2. Functionally provides a training platform for at least the procedural aspects of flight relating to an instrument training course
3. Has been qualified and approved by the FAA

D. A simulator can be an extremely important teaching tool in your instrument training. It can be used to teach the instrument scan, visualization of position, holding, and the approach procedures far more efficiently and cost-effectively than an airplane. Most simulators can be stopped in mid-flight to allow discussion of problems and can be reset so that difficult portions of a procedure can be practiced repeatedly.

E. Simulator time must be logged with an authorized instructor.

1. Part 61 allows
   a. A maximum of 30 hr. in a flight simulator or flight training device operated by a Part 142 training center
   b. A maximum of 20 hr. in a flight simulator or flight training device if the training was not by a Part 142 training center
   c. A maximum of 10 hr., of the time listed in a. or b. above, using a PCATD
2. Part 141 allows
   a. A maximum of 17.5 hr. in a flight simulator
   b. A maximum of 14 hr. in a flight training device
   c. A maximum of 10 hr., of the time listed in 2.a. or 2.b. above, using a PCATD

F. A flight simulator or flight training device may be authorized for use on the practical test, but a PCATD is not authorized.

2.5 VIEW-LIMITING DEVICES

A. In order to simulate instrument meteorological conditions (IMC), you will use an easily removable device (e.g., a hood, an extended visor cap, or foggles) which will limit your vision to the instrument panel.

1. You should purchase your own view-limiting device or have access to the same one for all of your training and practical test.
2. Most are like hats; each feels and weighs a little different.
3. Purchase or obtain a view-limiting device after you have talked to your CFII and other pilots and have tried various types in flight.

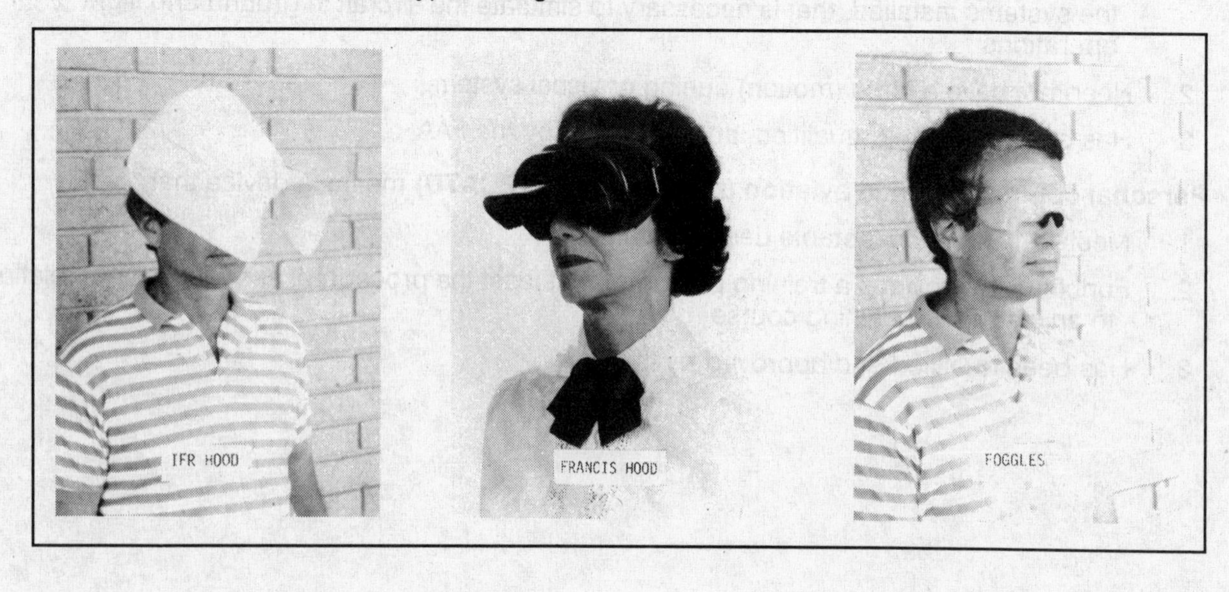

B. These view-limiting devices obviously require acclimation. You should spend a few minutes in "your" airplane with "your" device on before you meet your CFII for each flight. This added familiarity with (1) the view-limiting device and (2) the location of the instruments and their appearance will make it much easier for you to concentrate on flight maneuvers once in the air.

C. You must train yourself not to cheat with your view-limiting device on. It is usually possible to glimpse to the side and just over the instrument panel. **Do not do it!** You will not have this option in clouds or fog. Besides, these glimpses could distract you from the instruments. Some pilots get ill from vertigo caused by these distractions.

1. You may find it necessary to use pieces of cardboard fitted around the bottom of the windshield or on the side windows to prevent you from cheating.
   a. The major problem with this "fix" is that it impedes traffic surveillance.
   b. Another solution is to train at night when cheating will reveal much less information and not be worthwhile.

D. You will find flying in IMC easier than with an artificially limited view because your peripheral vision will not be restricted. Familiarity with the flight instruments, controls, and cockpit will facilitate your ability to concentrate on IFR flight.

1. Later in your flight training, you and your CFII should train in IMC whenever possible.
2. Flying in IMC is much different from flying with a view-limiting device.
   a. Your view of the cockpit is unrestricted, but there is usually no outside reference of movement.
   b. This can be quite unnerving in your first few IMC experiences. Thus, it is an excellent learning experience for you.

E. Always remember that, when you are using a view-limiting device, traffic surveillance is the RESPONSIBILITY of your CFII or safety pilot. Ask your CFII or safety pilot to clear right or left before making a turn. Continually remind your CFII or safety pilot of this responsibility by asking, "Do you have traffic?" Remove your device to find traffic reported by ATC and not found by your CFII or safety pilot. Remember, your first concern is safety with emphasis on other aircraft.

1. FAR 91.109 requires a safety pilot to occupy the other control seat when you are operating in simulated instrument conditions. You must enter the name of your safety pilot in your log book when you record simulated instrument conditions.
2. Your safety pilot must possess a current medical certificate and at least a private pilot certificate with category and class ratings appropriate to the airplane you are flying.

2.6 FLIGHT MANEUVER ANALYSIS SHEET (FMAS)

A. We have developed a method of analyzing and studying flight maneuvers that incorporates 10 variables:

1. Maneuver
2. Objective
3. Flight path
4. Power setting(s)
5. Altitude(s)
6. Airspeed(s)
7. Control forces
8. Time(s)
9. Traffic considerations
10. Completion standards

B. A copy of an FMAS (front and back) appears on pages 13 and 14 for your convenience. When you reproduce the form for your own use, photocopy on the front and back of a single sheet of paper to make the form more convenient. The front side contains space for analysis of the above variables. The back side contains space for

1. Make- and model-specific information
    a. Weight
    b. Airspeeds
    c. Fuel
    d. Center of gravity
    e. Performance data
2. Flight instrument review of maneuver
    a. Attitude Indicator . . . . AI
    b. Airspeed Indicator . . . . ASI
    c. Turn Coordinator . . . . TC
    d. Heading Indicator . . . . HI
    e. Vertical Speed Indicator . . . . . . VSI
    f. Altimeter . . . . ALT
3. Common errors

C. You should prepare/study/review FMAS for each maneuver you intend to perform before each flight lesson. Photocopy the form (front and back) on single sheets of paper. Changes, amplifications, and other notes should be added subsequently. Blank sheets of paper should be attached (stapled) to the FMAS including self-evaluations, “to do” items, questions for your CFII, etc., for your home study during your flight instruction program. FMAS are also very useful to prepare for the practical test.

1. A major benefit of the FMAS is preflight lesson preparation. It serves as a means to discuss maneuvers with your CFII before and after flight. It emphasizes preflight planning, make and model knowledge, and flight instruments.
2. Also, the FMAS helps you, in general, to focus on the operating characteristics of your airplane, including weight and balance. Weight and balance, which includes fuel, should be carefully reviewed prior to each flight.

CFI ______________
Student ______________
Date ______________

GLEIM'S
FLIGHT MANEUVER ANALYSIS SHEET

1. **MANEUVER** ______________
2. **OBJECTIVES/PURPOSE** ______________
3. **FLIGHT PATH (visual maneuvers)**
4. **POWER SETTINGS** 5. **ALT** 6. **A/S**

MP	RPM	SEGMENT OF MANEUVER	ALT	A/S
____	____	a. ______________	____	____
____	____	b. ______________	____	____
____	____	c. ______________	____	____

Pencil in expected indication on each of 6 flight instruments on reverse side.

7. **CONTROL FORCES**
   a. ______________
   b. ______________
   c. ______________
8. **TIME(S), TIMING** ______________
9. **TRAFFIC CONSIDERATIONS** **CLEARING TURNS REQUIRED** ____
10. **COMPLETION STANDARDS/ATC CONSIDERATIONS** ______________

AIRPLANE MAKE/MODEL ____________________

WEIGHT

Gross ____
Empty ____
Pilot/Pasngrs ____
Baggage ____
Fuel (gal x 6) ____

CENTER OF GRAVITY

Fore Limit ____
Aft Limit ____
Current CG ____

AIRSPEEDS

V_{S0} ____
V_{S1} ____
V_X ____
V_Y ____
V_A ____
V_{NO} ____
V_{NE} ____
V_{FE} ____
V_{LO} ____
V_R ____

ASI AI ALT
TC HI VSI

PRIMARY vs. SECONDARY INSTRUMENTS
(IFR maneuvers) -- instruments: AI, ASI, ALT, TC, HI, VSI, RPM and/or MP
(most relevant to instrument instruction)

	PITCH	BANK	POWER
ENTRY			
primary	____	____	____
secondary	____	____	____
ESTABLISHED			
primary	____	____	____
secondary	____	____	____

FUEL

Capacity L ____ gal R ____ gal
Current Estimate L ____ gal R ____ gal
Endurance (Hr.) ____
Fuel Flow -- Cruise (GPH) ____

PERFORMANCE DATA

	Airspeed	Power* MP	RPM
Takeoff Rotation	____	____	____
Climbout	____	____	____
Cruise Climb	____	____	____
Cruise Level	____	____	____
Cruise Descent	____	____	____
Approach**	____	____	____
Approach to Land (Visual)	____	____	____
Landing Flare	____	____	____

** If you do not have a constant-speed propeller, ignore manifold pressure (MP).*
*** Approach speed is for holding and performing instrument approaches.*

COMMON ERRORS

____
____
____
____

2.7 GLEIM'S INSTRUMENT PILOT SYLLABUS

A. Our syllabus consists of a flight training syllabus and a ground training syllabus, and is designed to meet the requirements of both Part 141 and Part 61. The ground and flight training may be done together as an integrated course of instruction, or each may be done separately. If done separately, the ground syllabus may be conducted as a home-study course or as a formal ground school.

1. This syllabus was constructed using the building-block progression of learning, in which the student is required to perform each simple task correctly before a more complex task is introduced. This method will promote the formation of correct habit patterns from the beginning.

B. **Flight Training Syllabus**

1. The Part 141 flight training syllabus contains 28 lessons, which are divided into three stages. It is recommended that each lesson be completed in segmental order. However, the syllabus is flexible enough to meet individual student needs, the particular training environment, or Part 61 requirements (except the PIC cross-country time).

2. A listing of the 28 lessons is presented in the table below.

Lesson	Topic	Lesson	Topic
	Stage One		**Stage Two**
1	Attitude Instrument Flying	14	VOR Holding
2	Instrument Takeoff, Steep Turns, and Airspeed Changes	15	GPS and ADF Holding
3	Rate Climbs/Descents, Timed Turns, and Magnetic Compass turns	16	Localizer Holding
4	Partial Panel Flying	17	DME and Intersection Holding
5	Attitude Instrument Flying (review)	18	VOR Instrument Approach
6	Partial Panel Flying (review)	19	GPS and NDB Instrument Approaches
7	Basic Instrument Flight Patterns	20	Localizer Instrument Approach
8	VOR/VORTAC Procedures	21	ILS Instrument Approach
9	VOR Time/Distance to Station and DME Arcs	22	Instrument Approaches (review)
10	GPS and ADF Procedures	23	Stage Two Check
11	GPS Procedures and ADF Time/Distance to Station		**Stage Three**
12	Tracking the Localizer	24	Cross-Country and Emergency Procedures
13	Stage One Check	25	Cross-Country Procedures
		26	Cross-Country Procedures
		27	Maneuvers Review
		28	Stage Three Check

3. The syllabus is reprinted in Appendix B, which begins on page 353.

C. **Ground Training**

1. The ground training syllabus contains 11 lessons, which are divided into three stages.
   a. The ground training can be conducted concurrently with the flight training, with the ground lessons completed in order as outlined in the lesson matrix.
   b. Ground training may also be conducted as part of a formal ground school or as a home-study program.

2. Each ground lesson involves studying the appropriate chapter in this book or Gleim's *Pilot Handbook*. After each chapter is completed, you need to answer the questions in the appropriate chapter in Gleim's *Instrument Pilot FAA Written Exam* book and review incorrect responses with your instructor.

END OF CHAPTER

CHAPTER THREE
IFR AERONAUTICAL CHARTS

This chapter provides an overview of the various IFR charts used for en route and terminal operations. There are three major providers of IFR charts:

1. **National Aeronautical Charting Office (NACO)** is a part of the FAA, and it produces charts of the United States and foreign areas.
2. **Jeppesen Sanderson, Inc. (JEPP)** is a private competitor to the NACO chart service. It is the most widely used chart service, and the graphic presentation of its charts differs from that of NACO in a number of respects. The information used by JEPP in developing its charts is furnished by the FAA. Thank you, Jeppesen, for providing a copy of your Chart Training Guide to be included with each Gleim *Instrument Pilot Flight Maneuvers and Practical Test Prep* book.
3. **Air Chart Systems** provides a chart service based on NACO charts. The concept is to sell IFR charts annually and then provide cumulative updates in narrative and graphic form.

Four basic IFR aeronautical charts are pertinent to IFR operations:

1. **En route** charts are maps about the size of sectional charts, but each en route chart covers an area of the U.S. equal to several sectional charts. Each gives navigational aids, IFR airports, airways, and the like, but little or no VFR information is included.
   a. En route low-altitude charts cover altitudes up to, but not including, 18,000 ft. MSL.
   b. En route high-altitude charts cover the airspace from 18,000 ft. MSL to 60,000 ft. MSL.
   c. Area charts, which show congested terminal areas (e.g., Miami, Atlanta, Denver) on a large scale, are also available.
2. **IAP** (instrument approach procedure) charts diagram the flight procedures to descend for landing while under IFR. Each chart usually covers an area 10 to 25 NM around the particular airport.
3. **STAR** (standard terminal arrival) charts are preplanned IFR arrival procedures in graphic and/or textual form. A STAR chart provides transition from an en route facility or airway to an outer fix or an instrument approach fix/arrival waypoint in the terminal area.
4. **DP** (instrument departure procedure) charts are preplanned IFR departure procedures in graphic and/or textual form. A DP provides a pilot with a way to depart the airport and transition to the en route structure safely.

JEPPESEN CHART TRAINING GUIDE

As a sales tool, Jeppesen Sanderson produces a training guide on a 17" x 34" sheet containing 14 panels on each side to illustrate Jeppesen charts as they compare with NACO charts. A copy of this guide was inserted inside the back cover of this book. If you do NOT have a copy, please e-mail jchart@gleim.com with your mailing address and we will send you a copy.

Panel Layout of JEPP Training Guide

1	2	3	4	5	6	7
8	9	10	11	12	13	14

Front Panels (Terminal Charts)

1. Front (cover) panel
2. Jeppesen ILS Rwy 30R, Terps, CA
3. Jeppesen Airport Chart, Terps, CA
4. Jeppesen approach chart information
5. NACO approach chart information
6. NACO ILS RWY 30R, Terps, CA
7. Jeppesen Terps Class B Airspace Area Chart
8. Jeppesen Winki One Arrival Chart
9. Jeppesen Thor One Departure Chart
10. Jeppesen VOR or GPS Rwy 24, Terps, CA
11. Jeppesen Radar-1 ASR Rwys 6, 12L, 24, 30R, Terps, CA
12. Jeppesen VOR DME RNAV Rwy 12L, Terps, CA
13. Jeppesen GPS Rwy 30R, Terps, CA
14. Jeppesen Airport Qualification Chart, Sitka, Alaska

Back Panels (En Route Charts)

1. Jeppesen Communications Tabulations
2. Jeppesen Low Altitude En Route Chart (left side)
3. Jeppesen Low Altitude En Route Chart (right side)
4. Jeppesen en route chart information
5. NACO en route chart information
6. NACO Low Altitude En Route Chart (left side)
7. NACO Low Altitude En Route Chart (right side)
8. Airport and Facility Listing
9. Jeppesen High Altitude En Route Chart (left side)
10. Jeppesen High Altitude En Route Chart (right side)
11. Jeppesen Area Chart (left side)
12. Jeppesen Area Chart (right side)
13. Jeppesen Area Navigation En Route Chart (left side)
14. Jeppesen Area Navigation En Route Chart (right side)

3.1 NACO CHARTS

A. NACO en route charts are revised every 56 days. These charts are printed back to back, similar to the VFR sectional charts.

B. IAPs, STARs, and DPs are all depicted in the U.S. *Terminal Procedures Publication*. These charts are published in 20 loose-leaf (with four holes punched at top) or perfect-bound volumes covering the conterminous U.S., Puerto Rico, and the Virgin Islands.

1. Each volume is revised every 56 days.
2. Changes to procedures are reflected in a Change Notice volume, issued on a 28-day midcycle. These changes are in the form of a new chart.

C. En route charts and *Terminal Procedures Publication* are available on a single-issue basis or by subscription from

FAA Distribution Division
AVN-530
National Aeronautical Charting Office
Riverdale, MD 20737-1199
Telephone: (800) 638-8972 (Toll free within U.S.)

1. NACO charts are also available from the Aircraft Owners and Pilots Association (AOPA; call 800-4CHARTS), FBOs, and other authorized NACO chart agents.

3.2 JEPP CHARTS

A. JEPP en route charts are printed back to back, similar to the NACO chart, but are printed on a special lightweight chart paper.

1. JEPP does not issue a new chart every 56 days, as NACO does, but issues changes in its NOTAM section.
   a. JEPP will issue a new chart when there are enough changes to warrant it.
   b. Before any flight, you must review the NOTAM section for any changes that may affect your flight.

B. Terminal procedures (i.e., IAPs, STARs, and DPs) are organized alphabetically by city within a given state. DPs and STARs are located immediately preceding the airports to which they pertain.

1. New charts are issued only when there is a change in a terminal procedure.

C. JEPP offers a variety of subscription services for various geographic coverage areas and pilot needs. Each comes with en route chart(s) and terminal procedures, which are kept in a 7-ring notebook.

1. The *Standard Airway Manual* is revised by biweekly updates, which you must organize.
2. *Airway Manual Express* is issued every 56 days with a new complete issue of en route chart(s) and terminal procedures, thus eliminating the time required to organize revisions.
3. For more information and current prices, call (800) 621-JEPP.

3.3 AIR CHART SYSTEMS

A. Air Chart Systems reproduces the NACO en route charts for the entire U.S. in one spiral-bound atlas, which is printed annually. The charts are updated in narrative and graphic form by use of a cumulative NOTAM system.

1. A cumulative NOTAM system means that you will be sent a regularly scheduled update of changes (i.e., NOTAMs), which will include both current and all past NOTAMs. Before a flight, you should use the most recent update to make changes on your charts.
    a. Thus, you will need to update your charts only for the time a flight is planned and for the area where you will be flying.

B. Air Chart Systems provides the NACO *Terminal Procedures Publication*, which is issued every May and updated by use of the cumulative NOTAM system.

C. For a free brochure that lists the geographic coverage areas and prices, call (800) 338-7221 or visit the Air Chart Systems web site at http://www.airchart.com.

3.4 EN ROUTE LOW-ALTITUDE CHARTS

A. IFR en route low-altitude charts provide aeronautical information for navigation under IFR below 18,000 ft. MSL.

B. Below are 10 major items that are presented on both the NACO and the JEPP en route charts. Think about each item and its use.

1. Airways (VHF and L/MF)
2. Airway distances
3. Minimum en route altitude (MEA)
4. Minimum obstruction clearance altitude (MOCA)
5. Reporting points
6. VHF NAVAID (position, frequency, identification, and geographic coordinates)
7. L/MF NAVAID (position, frequency, and identification)
8. Selected airports
9. Special-use airspace. Note the JEPP chart includes the limits of the airspace, while NACO has information on the margin.
10. Military training routes (MTR). Note that MTRs are not depicted on the JEPP chart.

C. The Jeppesen Chart Training Guide is provided with this book so you can become familiar with, and compare the differences between, the NACO and JEPP en route low-altitude charts.

1. You must be able to interpret the information on the en route chart (NACO or JEPP) that you plan to use. If you do not know the meaning of a symbol, you must look it up in the appropriate (NACO or JEPP) legend to learn its meaning.
    a. The Chart Training Guide lists 46 features on the NACO charts and 47 features on the JEPP charts.
    b. Consult panels 4 and 5 (as shown on page 18) on the back of your Chart Training Guide.
    c. Locate the corresponding numbers on each chart and note how the NACO and JEPP information differs in presentation.

3.5 INSTRUMENT APPROACH PROCEDURE (IAP) CHARTS

A. IAP charts portray the aeronautical data that are required to execute instrument approaches to airports. Each procedure is designated for use with a specific electronic navigational aid, such as ILS, VOR, NDB, RNAV, GPS, etc.

1. Not all IAPs are published in chart form. Radar IAPs are established where requirements and facilities exist but are printed in tabular form in the NACO *Terminal Procedures Publication*. On JEPP charts, radar IAPs are shown in chart form for each airport.

B. Below are 10 items that are presented on the NACO and JEPP IAP charts. Think about each item and its use.

1. Approach procedure identification
2. Plan view
3. Arrival communication frequencies
4. Primary approach NAVAID
5. Profile view
6. Glide slope intercept
7. Final approach fix (FAF)
8. Missed approach procedure
9. Minimums section
10. Airport diagram. Note that JEPP issues a separate airport chart for each airport.

C. A Jeppesen Chart Training Guide is provided with this book so you can become familiar with and compare the differences between the NACO and JEPP IAP charts.

1. You must be able to interpret the information on the IAP chart (NACO or JEPP) that you plan to use. If you do not know the meaning of a symbol, you must look it up in the appropriate (NACO or JEPP) legend to learn its meaning.

   a. The Chart Training Guide lists 29 features for the JEPP IAP chart and 51 features for the NACO chart.

      1) JEPP charts separate the IAP charts from the airport charts. NACO combines the IAP and airport charts on one page. There are 20 features for the JEPP airport chart.

      2) The NACO does provide full-page diagrams for some airports. These diagrams are designed to assist you at locations with complex runway/taxiway configurations.

   b. Consult panels 4 and 5 (as shown on page 18) on the front of your Chart Training Guide.

   c. Locate the corresponding numbers on each chart and note how the NACO and JEPP information differs in presentation.

3.6 STANDARD TERMINAL ARRIVAL (STAR) CHARTS

A. STAR charts are designed to expedite ATC arrival procedures and to facilitate transition between en route and instrument approach operations. Each STAR chart presents you with a preplanned IFR ATC arrival procedure in graphic and textual form. Each STAR procedure is presented as a separate chart and may serve a single airport or more than one airport in a given geographic location.

1. These charts are located in the front of the NACO *Terminal Procedures Publication* alphabetically by the STAR name. Each chart may apply to several airports.
2. The JEPP STAR chart(s) is(are) located prior to the IAPs for a particular airport. If a STAR procedure applies to more than one airport, a separate chart will be printed, with specific procedures included, for each airport.

B. An NACO STAR chart is reproduced on page 23, and the same JEPP STAR chart is reproduced on page 24. These are presented to you so you can become familiar with, and compare the differences between, the two charts.

1. You must be able to interpret the information on a STAR chart. If you do not know the meaning of a symbol, you must look it up in the appropriate (NACO or JEPP) legend to learn its meaning.

C. Below are five items that are presented on the NACO and JEPP STAR charts. Locate the corresponding number on each chart on the next two pages, and note how the NACO and JEPP information is presented.

1. STAR procedure identification
2. Transition route
3. Arrival route
4. MEA
5. STAR narrative

NACO STAR CHART

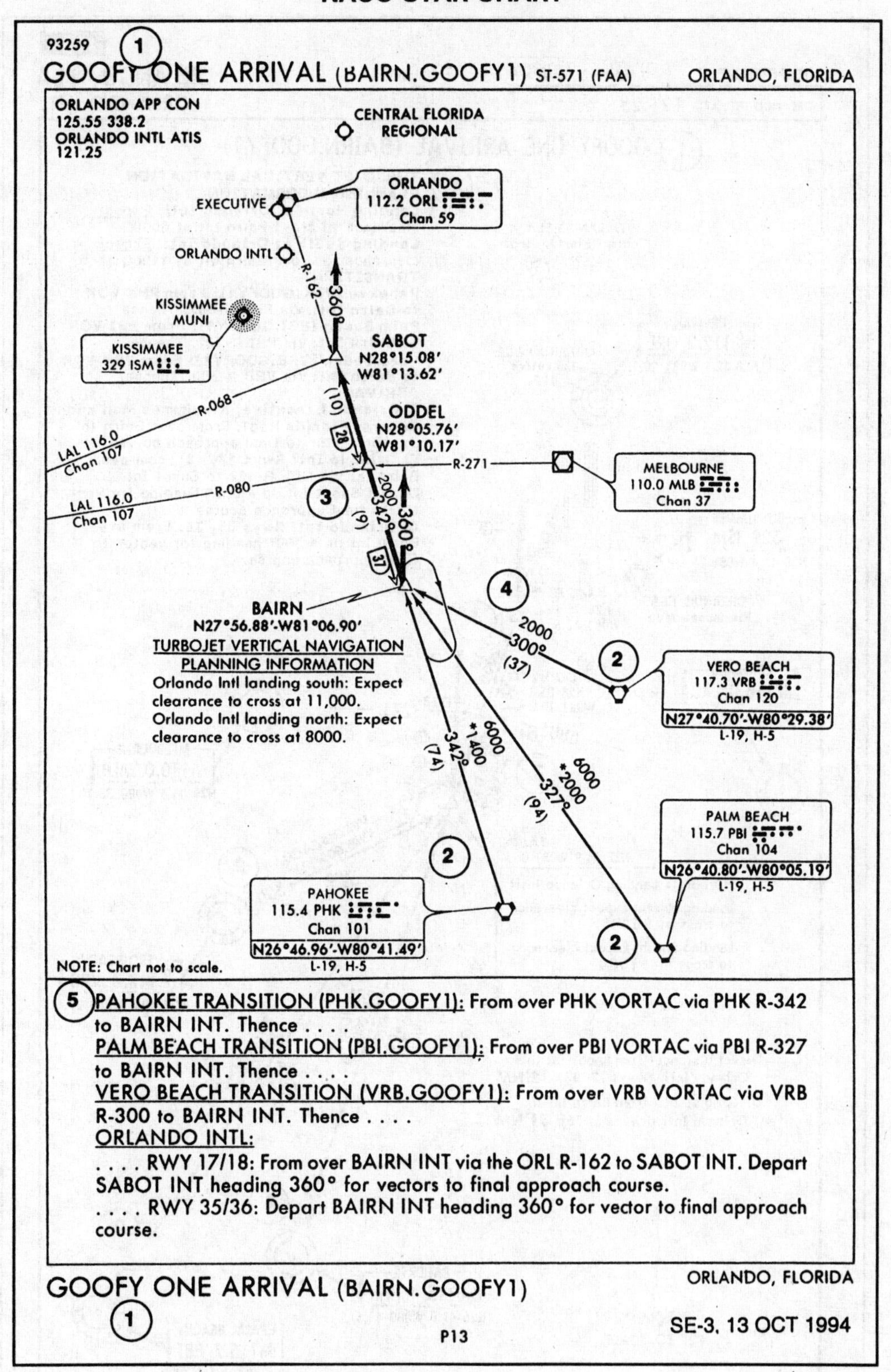

NOT FOR NAVIGATION

JEPPESEN STAR CHART

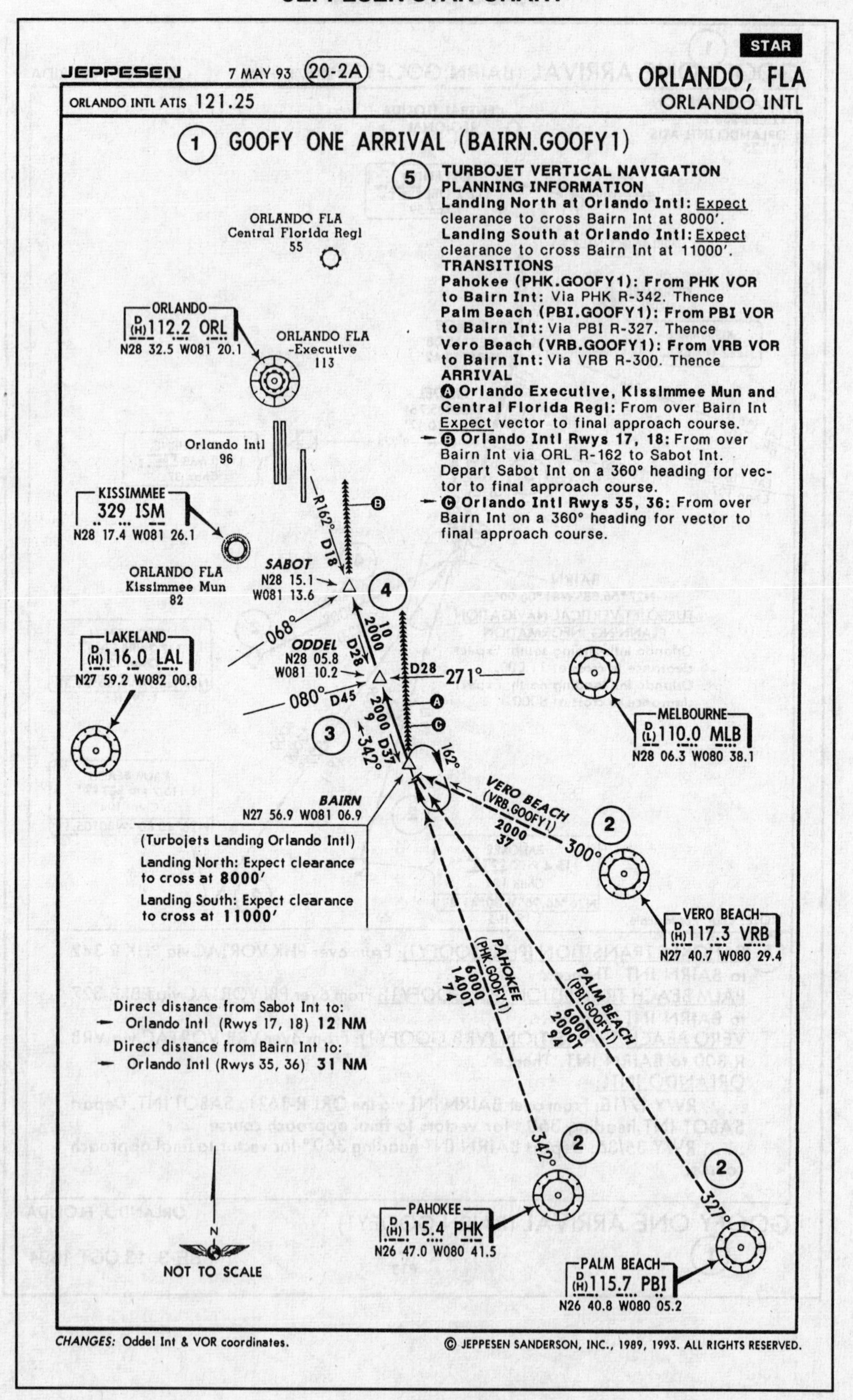

NOT FOR NAVIGATION

3.7 INSTRUMENT DEPARTURE PROCEDURE (DP) CHARTS

A. DP charts are designed to expedite clearance delivery and to facilitate transition between takeoff and en route operations. They provide you with departure routing clearance.

1. A DP is established primarily to provide obstacle clearance protection to aircraft operating in instrument meteorological conditions (IMC).

   a. A secondary reason a DP is established is to increase efficiency and reduce delays at busy airports.

B. NACO produces DPs in textual and/or graphical format. All DPs are listed in Section C of the NACO *Terminal Procedures Publications*.

1. If the DP is textual, it will be described in Section C.

   a. If the DP is graphical, it will be listed with a page reference in Section C.

2. Graphical DPs developed solely for obstacle clearance will also have the term "(OBSTACLE)" printed on the chart.

C. JEPP DPs are presented in graphical form and are located ahead of the IAP chart(s) for a particular airport.

D. An NACO DP chart is reproduced on page 26, and the same JEPP DP chart is reproduced on page 27. These are presented to you so you can become familiar with, and compare the differences between, the two charts.

1. You must be able to interpret the information on the DP chart. If you do not know the meaning of a symbol, you must look it up in the appropriate (NACO or JEPP) legend to learn its meaning.

E. Below are five items that are presented on the NACO and JEPP DP charts. Locate the corresponding number on each chart on the next two pages, and note how the NACO and JEPP information is presented.

1. DP identification
2. Departure route
3. NAVAID facility
4. DP narrative
5. Rate-of-climb table (not shown). Note that in NACO you need to refer to the rate-of-climb table located in the legend, while JEPP includes a table on the chart when a minimum rate of climb is required.

NACO DP CHART

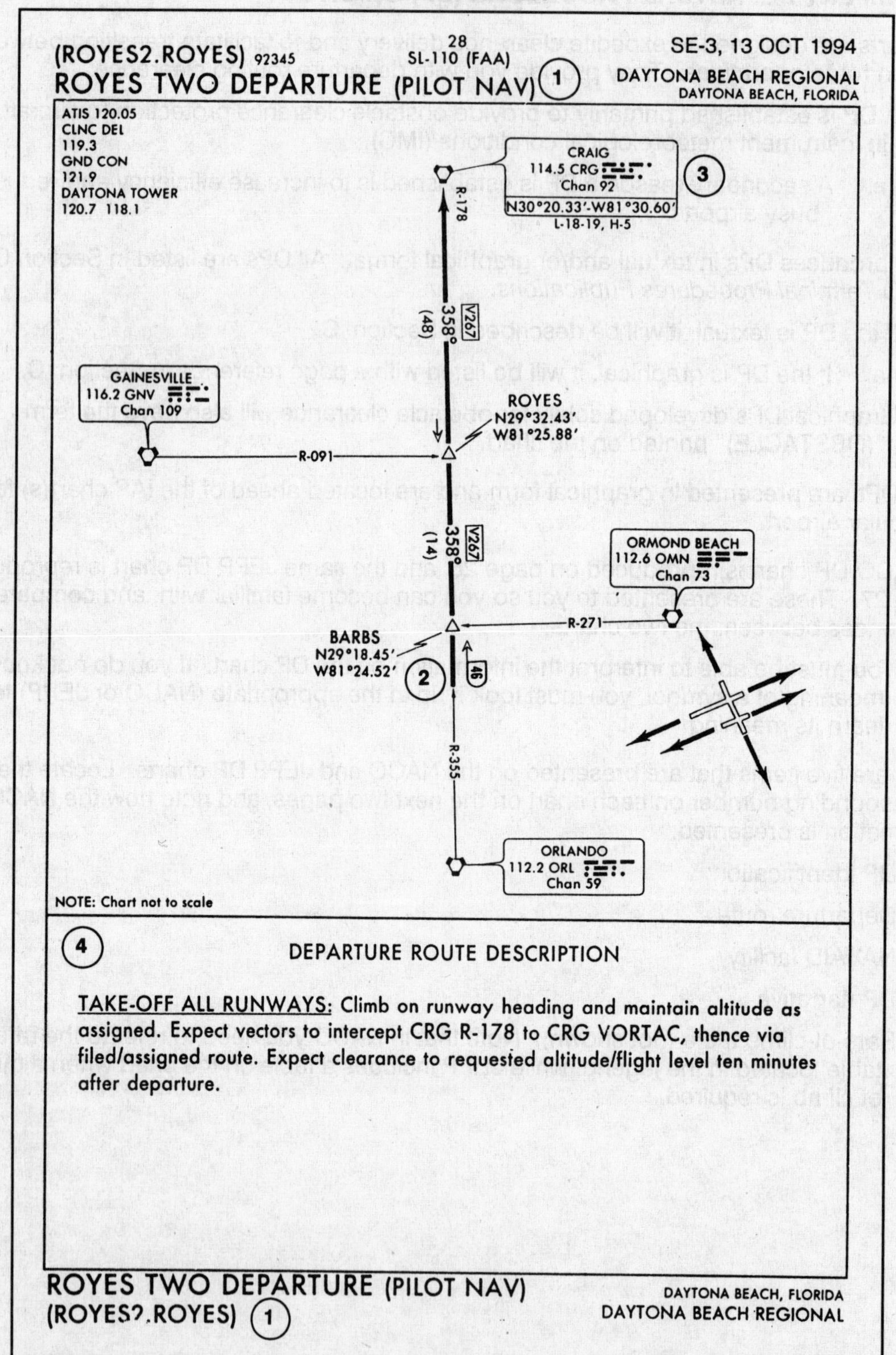

NOT FOR NAVIGATION

JEPP DP CHART

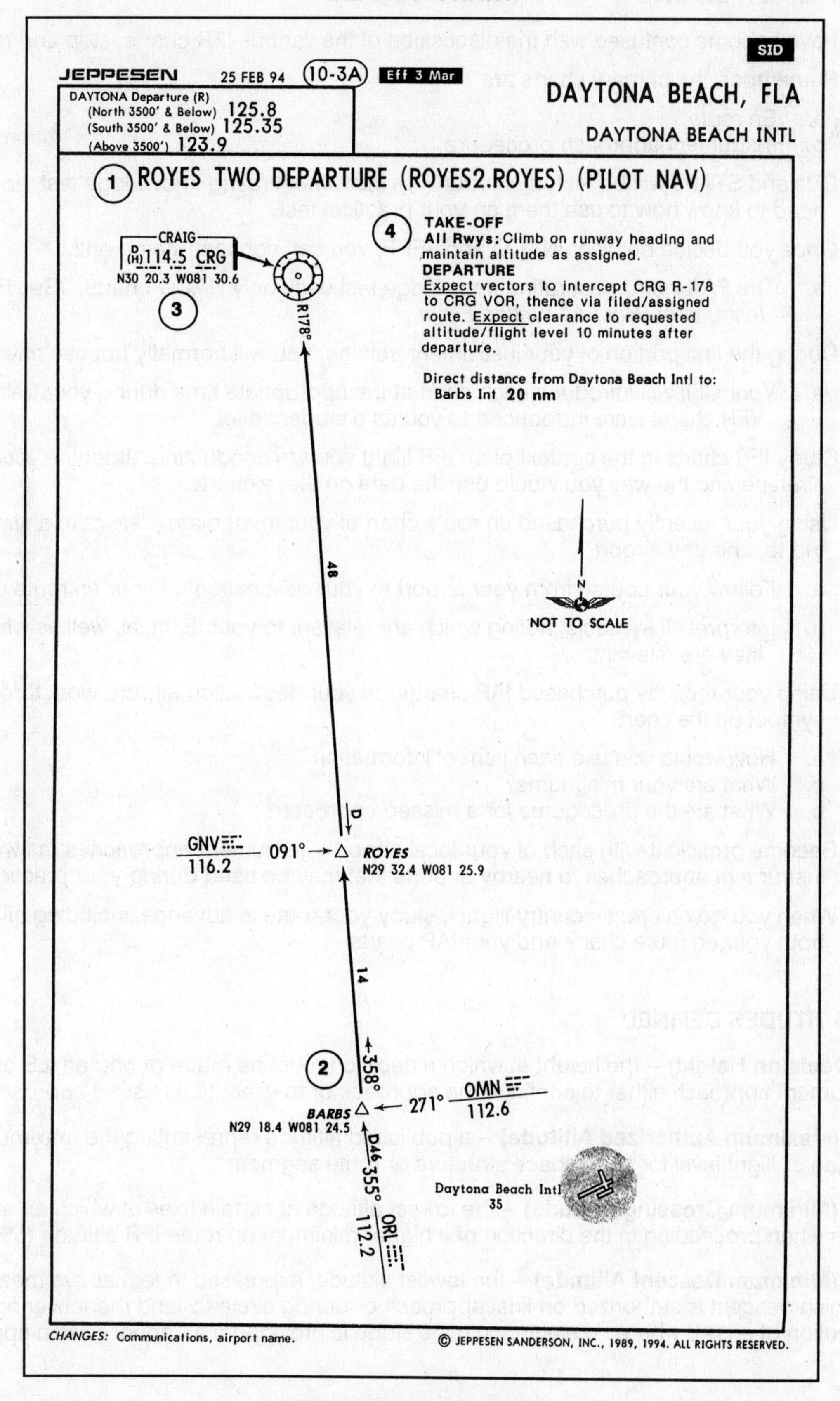

NOT FOR NAVIGATION

3.8 RECAP OF IFR CHARTS

A. If you have become confused with the discussion of the various IFR charts, stop and regroup.

1. Remember, the primary charts are
   a. En route
   b. Instrument approach procedure
2. DPs and STARs will be tested on your FAA instrument rating knowledge test, and you also need to know how to use them on your practical test.
3. Once you decide between NACO and JEPP, you can concentrate on one.
   a. The FAA instrument rating knowledge test uses only NACO charts. (See Gleim's *Instrument Pilot FAA Written Exam*.)
4. During the first portion of your instrument training, you will normally not use any IFR charts.
   a. Your CFII will introduce each chart at the appropriate time during your training, just as VFR charts were introduced to you as a student pilot.
5. Study IFR charts in the context of an IFR flight you are conducting. Imagine yourself in your airplane and the way you would use the data on these charts.
6. Using your recently purchased en route chart of your immediate area, take an imaginary trip to a nearby airport.
   a. Follow your course from your airport to your destination on your en route chart.
   b. Interpret all symbols, noting which are relevant to your flight, as well as why and how they are relevant.
7. Using your recently purchased IAP chart(s) of your destination airport, work through each symbol on the chart.
   a. How would you use each item of information?
   b. What are your minimums?
   c. What are the procedures for a missed approach?
8. Become proficient with each of your local airport's instrument approaches (as well as the instrument approaches to nearby airports that may be used during your practical test).
9. When you go on cross-country flights, study your route in advance, including all symbols of both your en route charts and your IAP charts.

3.9 IFR ALTITUDES DEFINED

A. **DH (Decision Height)** -- the height at which a decision must be made during an ILS or PAR instrument approach either to continue the approach or to execute a missed approach

B. **MAA (Maximum Authorized Altitude)** -- a published altitude representing the maximum usable altitude or flight level for an airspace structure or route segment

C. **MCA (Minimum Crossing Altitude)** -- the lowest altitude at certain fixes at which an aircraft must cross when proceeding in the direction of a higher minimum en route IFR altitude (MEA)

D. **MDA (Minimum Descent Altitude)** -- the lowest altitude, expressed in feet above mean sea level, to which descent is authorized on final approach or during circle-to-land maneuvering in execution of an IAP where no electronic glide slope is provided (i.e., nonprecision approach)

E. **MEA (Minimum En Route IFR Altitude)** -- the lowest published altitude between radio fixes that assures acceptable navigational signal coverage along the entire route segment and meets obstacle clearance requirements between those fixes

F. **MHA (Minimum Holding Altitude)** -- the lowest altitude prescribed for a holding pattern that assures navigational signal coverage and communications and meets obstacle clearance requirements

G. **MOCA (Minimum Obstruction Clearance Altitude)** -- the lowest published altitude in effect between radio fixes on VOR airways, off-airway routes, or route segments that meets obstacle clearance requirements for the entire route segment and that assures acceptable navigational signal coverage only within 25 SM (22 NM) of a VOR

H. **MRA (Minimum Reception Altitude)** -- the lowest altitude at which an intersection can be determined

I. **MSA (Minimum Safe/Sector Altitude)** -- altitudes depicted on IAP charts that provide at least 1,000 ft. of obstacle clearance normally within a 25-NM radius of the navigation facility for the approach (VOR or NDB); the runway or airport waypoint for an RNAV approach; or the missed approach waypoint for a GPS approach. MSA is for emergency use only and does not necessarily assure acceptable navigational signal coverage.

J. **MVA (Minimum Vectoring Altitude)** -- the lowest MSL altitude at which an IFR aircraft will be vectored by a radar controller, except as otherwise authorized for radar approaches, departures, and missed approaches. The altitude meets IFR obstacle clearance criteria.

K. **OROCA (Off-Route Obstruction Clearance Altitude)** -- an off-route altitude that provides obstruction clearance with a 1,000-ft. buffer in nonmountainous terrain areas and a 2,000-ft. buffer in designated mountainous areas within the U.S. This altitude may not provide signal coverage from ground-based navigational aids, ATC radar, or communications coverage.

3.10 COMMON ERRORS WITH IFR CHARTS

A. **Not having the necessary charts with you.** As part of your cockpit check, you should make sure all required charts (en route, approach, DPs, STARs) are accounted for and available.

B. **Not maintaining current charts.** Approaches, airways, intersections, and airports change. ATC is going to expect you to have the most current charts and will provide clearances based upon them.

C. **For flights near the edge of a chart, not carrying the charts for nearby airports in the adjoining chart.** The nearest alternate may be just on the adjoining chart. Carry charts for all possible diversions.

D. **Not maintaining an organized cockpit.** Charts have to be folded and unfolded. DPs and STARs, as well as approach plates, have to be taken out and returned. If chart usage is not done in a systematic way and kept organized, charts will be misplaced, folded inside-out, torn, etc. There is too much to do while flying, especially under IFR, to contend with unorganized, misplaced, or torn charts.

E. **Not being conversant with all the figures and notations on the charts.** It is surprising how many pilots are limited to the most basic information from their charts. Learn and understand all the information that is available from each type of chart. As you study charts for trip planning and while en route, decipher all available information. As a review, reread the legend periodically.

END OF CHAPTER

CHAPTER FOUR
YOUR FAA PRACTICAL (FLIGHT) TEST

After all the training, studying, and preparing, the final step to receive your instrument rating is the FAA practical test. It requires that you exhibit to your examiner your previously gained knowledge and that you demonstrate that you are a proficient and safe instrument pilot.

Your practical test will be similar to your private pilot practical test. It is merely repeating to an examiner flight maneuvers that are familiar and well practiced. Conscientious flight instructors do not send applicants to an examiner until the applicant can pass the practical test on an average day; i.e., an exceptional flight will not be needed. Theoretically, the only way to fail would be to commit an error beyond the scope of what your CFII expects.

Most applicants pass the instrument rating practical test on the first attempt. The vast majority of those having trouble will succeed on the second attempt. This high pass rate is due to the high quality of flight instruction and the fact that most examiners test on a human level, not a NASA shuttle pilot level. The FAA Instrument Rating Practical Test Standards are reprinted in Chapters I through VIII and again in their entirety in Appendix A. Study Chapters I through VIII carefully so that you know exactly what will be expected of you. Your goal is to exceed each requirement. Surpassing the requirements will ensure that even a slight mistake will fall within the limits, especially if you recognize your error and explain it to your examiner.

As you proceed with your flight training, you and your instructor should plan ahead and schedule your practical test. Several weeks before your practical test is scheduled, contact one or two individuals who took the instrument rating practical test with your examiner. Ask each person to explain the routine, length, emphasis, maneuvers, and any peculiarities (i.e., surprises). This is a very important step because, like all people, examiners are unique. One particular facet of the practical test may be tremendously important to one examiner, while another examiner may emphasize an entirely different area. By gaining this information beforehand, you can focus on the areas of apparent concern to your examiner. Also, knowing what to expect will relieve some of the apprehension and tension about your practical test.

When you schedule your practical test, ask your examiner for the cross-country flight you should plan for on the day of your test. Task I.B., Cross-Country Flight Planning (beginning on page 58), states that you are to present to your examiner a preplanned cross-country flight, which was previously assigned. However, some examiners may want to wait until the day of your test to assign you a cross-country flight.

4.1 FAA PRACTICAL TEST STANDARDS

A. The intent of the FAA is to structure and standardize practical tests by specifying required tasks and acceptable performance levels to FAA inspectors and FAA-designated pilot examiners. These tasks (procedures and maneuvers) listed in the PTS are mandatory on each practical test unless specified otherwise.

B. The 24 tasks for the instrument rating (single-engine airplane) are listed below in eight areas of operation as organized by the FAA.

1. The four tasks that can be completed away from the airplane are indicated below as "oral" and are termed "knowledge only" tasks by the FAA.
2. The 20 tasks that are usually completed in the airplane are indicated "flight" and are termed "knowledge and skill" tasks by the FAA.
3. Your examiner is required to test you on all 24 tasks.

NOTE: In the PTS format, the FAA has done away with reference to "oral tests" and "flight tests." Now the emphasis is that all tasks require oral examining about the applicant's knowledge. Nonetheless, we feel it is useful to separate the "knowledge only" tasks from the "knowledge and skill" tasks.

C. This book is based on the FAA's *Instrument Rating Practical Test Standards* (FAA-S-8081-4C), dated October 1998, with Change 1 (December 1998) and Change 2 (March 1999). E-mail update@gleim.com with IPFM 3-2 in the subject line to determine if new PTSs have been released or if there are any updates to this book. See page iv.

4.2 FORMAT OF PTS TASKS

A. Each of the FAA's 24 instrument rating tasks listed on the opposite page is presented in a shaded box in Chapters I through VIII, similar to Task I.A. reproduced below.

> **I.A. TASK: WEATHER INFORMATION**
>
> REFERENCES: 14 CFR Part 61; AC 00-6, AC 00-45; AIM
>
> **NOTE:** Where current weather reports, forecasts, or other pertinent information is not available, this information will be simulated by the examiner in a manner which will adequately measure the applicant's competence.
>
> **Objective.** To determine that the applicant:
>
> 1. Exhibits adequate knowledge of the elements related to aviation weather information by obtaining, reading, and analyzing the applicable items, such as --
>   a. Weather reports and forecasts.
>   b. Pilot and radar reports.
>   c. Surface analysis charts.
>   d. Radar summary charts.
>   e. Significant weather prognostics.
>   f. Winds and temperatures aloft.
>   g. Freezing level charts.
>   h. Stability charts.
>   i. Severe weather outlook charts.
>   j. Tables and conversion graphs.
>   k. SIGMETs and AIRMETs.
>   l. ATIS reports.
> 2. Correctly analyzes the assembled weather information pertaining to the proposed route of flight and destination airport, and determines whether an alternate airport is required, and if required, whether the selected alternate airport meets the regulatory requirement.

1. The task number is followed by the title.
2. The reference list identifies the FAA publication(s) that describe the task.
   a. Our discussion of each task is based on the FAA reference list. Note, however, that we will refer you to *Pilot Handbook* or *Aviation Weather and Weather Services* for further discussion of specific topics.
   b. A listing of the FAA references used in the PTS is on page 46.
3. Next the task has "**Objective**. To determine that the applicant . . .," followed by a number of "Exhibits knowledge . . ." of aviation concepts and "Demonstrates . . . " various maneuvers.

B. Each task in this book is followed by the following general format:

> A. General information
>   1. The FAA's objective and/or rationale for this task
>   2. A list of Gleim's *Pilot Handbook* or *Aviation Weather and Weather Services* chapters and/or modules that provide additional discussion of the task, as appropriate
>   3. Any general discussion relevant to the task
>
> B. Comprehensive discussion of each concept or item listed in the FAA's task
>
> C. Common errors for each of the flight maneuvers, i.e., tasks appearing in Chapters II through VIII, relative to knowledge and skill tasks. Chapter I contains "knowledge only" tasks.

4.3 AIRPLANE AND EQUIPMENT REQUIREMENTS

A. You are required to provide an appropriate and airworthy airplane for the practical test. Your airplane must be equipped for, and its operating limitations must not prohibit, the areas of operations required on the practical test.

1. Flight instruments are those required for controlling the airplane without outside references (FAR 91.205).
2. The required radio equipment is that necessary for communications with ATC and the performance of two nonprecision approaches using two different systems (e.g., VOR, LOC, NDB, LDA, SDF, or GPS) and an ILS (glide slope, localizer, and marker beacon) approach.
   a. EXAMPLE: You may do a VOR and a LOC approach, but you may not do a VOR/DME and a VOR approach.
3. All navigation equipment used during the practical test must be IFR approved.
4. Your airplane is not required to have an ADF.

4.4 WHAT TO TAKE TO YOUR PRACTICAL TEST

A. The following checklist from the FAA's Instrument Rating Practical Test Standards should be reviewed with your instructor both 1 week before and 1 day before your scheduled practical test:

1. Acceptable airplane with dual controls
   a. Aircraft documents
      1) Airworthiness certificate
      2) Registration certificate
      3) Operating limitations
   b. Aircraft maintenance records
      1) Logbook record of airworthiness inspections and AD compliance
   c. *Pilot's Operating Handbook* (FAA-approved *Airplane Flight Manual*)
2. Personal Equipment
   a. View-limiting device (foggles, hood)
   b. Current IFR aeronautical charts
      1) En route
      2) Terminal procedures (IAPs, DPs, STARs)
   c. Computer and plotter
   d. Flight plan form
   e. Flight logs
   f. Current *AIM*, *A/FD*, and appropriate publications
3. Personal Records
   a. Identification -- photo/signature ID
   b. Pilot certificate
   c. Current medical certificate
   d. Completed application for an airman certificate and/or rating (FAA Form 8710-1)
   e. Airman Knowledge Test Report
   f. Logbook with instructor's endorsement for your instrument rating practical test
   g. Notice of Disapproval of Application (only if you previously failed your practical test)
   h. Approved school graduation certificate (if applicable)
   i. Examiner's fee (if applicable)

4.5 PRACTICAL TEST APPLICATION FORM

A. Prior to your practical test, your instructor will assist you in completing FAA Form 8710-1 (which appears on pages 37 and 38) and will sign the top of the back side of the form.

1. An explanation on how to complete the form is attached to the original, and we have reproduced it on page 36.
   a. The form is not largely self-explanatory.
   b. For example, the FAA wants dates shown as 02-14-92, **not** 02/14/92.
2. Do not go to your practical test without FAA Form 8710-1 properly filled out; remind your CFII about it as you schedule your practical test.

B. If you are enrolled in a Part 141 flight school, the Air Agency Recommendation block of information on the back side may be completed by the chief instructor of your Part 141 flight school. (S)he, rather than a designated examiner or an FAA inspector, will administer the flight test if flight test examining authority has been granted to your flight school.

C. Your examiner or Part 141 flight school chief instructor will forward this and other required forms (listed on the bottom of the back side) to the nearest FSDO for review and approval.

1. Then they will be sent to Oklahoma City.
2. From there, your new pilot certificate will be issued and mailed to you.
   a. Adding an instrument rating does not require a new certificate; however, the certificate (private or commercial) must be reissued with the instrument rating reflected.

AIRMAN CERTIFICATE AND/OR RATING APPLICATION
INSTRUCTIONS FOR COMPLETING FAA FORM 8710-1

I. APPLICATION INFORMATION. *Check appropriate block(s).*

Block A. Name. Enter legal name. Use no more than one middle name for record purposes. Do not change the name on subsequent applications unless it is done in accordance with 14 CFR Section 61.25. If you do not have a middle name, enter "NMN." If you have a middle initial only, indicate "Initial only." If you are a Jr., or a II, or III, so indicate. If you have an FAA certificate, the name on the application should be the same as the name on the certificate unless you have had it changed in accordance with 14 CFR Section 61.25.

Block B. Social Security Number. Optional: See supplemental Information Privacy Act. Do not leave blank: Use only **US Social Security Number.** Enter either "SSN" or the words "Do not Use" or "None." SSN's are not shown on certificates.

Block C. Date of Birth. Check for accuracy. Enter eight digits; Use numeric characters, i.e., 07-09-1925 instead of July 9, 1925. Check to see that DOB is the same as it is on the medical certificate.

Block D. Place of Birth. If you were born in th USA, enter the city and state where you were born. If the city is unknown, enter the county and state. If you were born outside the USA, enter the name of the city and country where you were born.

Block E. Permanent Mailing Address. Enter residence number and street, P.O. Box or rural route number in the top part of the block above the line. The City, State, and ZIP code go in the bottom part of the block below the line. Check for accuracy. Make sure the numbers are not transposed. FAA policy requires that you use your permanent mailing address. **Justification must be provided on a separate sheet of paper signed and submitted with the application when a PO box or rural route number is used in place of your permanent physical address. A map or directions must be provided if a physical address is unavailable.**

Block F. Citizenship. Check USA if applicable. If not, enter the country where you are a citizen.

Block G. Do you read, speak, write and understand the English language? Check yes or no.

Block H. Height. Enter your height in inches. Example: 5'8" would be entered as 68 in. No fractions, use whole inches only.

Block I. Weight. Enter your weight in pounds. No fractions, use whole pounds only.

Block J. Hair. Spell out the color of your hair. If bald, enter "Bald." Color should be listed as black, red, brown, blond, or gray. If you wear a wig or toupee, enter the color of your hair under the wig or toupee.

Block K. Eyes. Spell out the color of your eyes. The color should be listed as blue, brown, black, hazel, green, or gray.

Block L. Sex. Check male or female.

Block M. Do You Now Hold or Have You Ever Held An FAA Pilot Certificate? Check yes or no. (NOTE: A student pilot certificate is a "Pilot Certificate.")

Block N. Grade of Pilot Certificate. Enter the grade of pilot certificate (i.e., Student, Recreational, Private, Commercial, or ATP). Do NOT enter flight instructor certificate information.

Block O. Certificate Number. Enter the number as it appears on your pilot certificate.

Block P. Date Issued. Enter the date your pilot certificate was issued.

Block Q. Do You Now Hold A Medical Certificate? Check yes or no. If yes, complete Blocks R, S, and T.

Block R. Class of Certificate. Enter the class as shown on the medical certificate, i.e., 1st, 2nd, or 3rd class.

Block S. Date Issued. Enter the date your medical certificate was issued.

Block T. Name of Examiner. Enter the name as shown on medical certificate.

Block U. Narcotics, Drugs. Check appropriate block. Only check "Yes" if you have actually been convicted. If you have been charged with a violation which has not been adjudicated, check "No."

Block V. Date of Final Conviction. If block "U" was checked "Yes" give the date of final conviction.

II. CERTIFICATE OR RATING APPLIED FOR ON BASIS OF:

Block A. Completion of Required Test.
1. AIRCRAFT TO BE USED. (If flight test required)—Enter the make and model of each aircraft used. If simulator or FTD, indicate.
2. TOTAL TIME IN THIS AIRCRAFT (Hrs.)—(a) Enter the total Flight Time in each make and model. (b) Pilot-in-Command Flight Time —In each make and model.

Block B. Military Competence Obtained In. Enter your branch of service, date rated as a military pilot, your rank, or grade and service number. In block 4a or 4b, enter the make and model of each military aircraft used to qualify (as appropriate).

Block C. Graduate of Approved Course.
1. NAME AND LOCATION OF TRAINING AGENCY/CENTER. As shown on the graduation certificate. Be sure the location is entered.
2. AGENCY SCHOOL/CENTER CERTIFICATION NUMBER. As shown on the graduation certificate. Indicate if 142 training center.
3. CURRICULUM FROM WHICH GRADUATED. As shown on the graduation certificate.
4. DATE. Date of graduation from indicated course. Approved course graduate must also complete Block "A" COMPLETION OF REQUIRED TEST.

Block D. Holder of Foreign License Issued By.
1. COUNTRY. Country which issued the license.
2. GRADE OF LICENSE. Grade of license issued, i.e., private, commercial, etc.
3. NUMBER. Number which appears on the license.
4. RATINGS. All ratings that appear on the license.

Block E. Completion of Air Carrier's Approved Training Program.
1. Name of Air Carrier.
2. Date program was completed.
3. Identify the Training Curriculum.

III. RECORD OF PILOT TIME. The minimum pilot experience required by the appropriate regulation must be entered. It is recommended, however, that ALL pilot time be entered. If decimal points are used, be sure they are legible. Night flying must be entered when required. You should fill in the blocks that apply and ignore the blocks that do not. Second In Command "SIC" time used may be entered in the appropriate blocks. Flight Simulator, Flight Training Device and PCATD time may be entered in the boxes provided. Total, Instruction received, and Instrument Time should be entered in the top, middle, or bottom of the boxes provided as appropriate.

IV. HAVE YOU FAILED A TEST FOR THIS CERTIFICATE OR RATING? Check appropriate block.

V. APPLICANT'S CERTIFICATION.
A. SIGNATURE. The way you normally sign your name.
B. DATE. The date you sign the application.

TYPE OR PRINT ALL ENTRIES IN INK

Form Approved OMB No: 2120-0021

DEPARTMENT OF TRANSPORTATION
FEDERAL AVIATION ADMINISTRATION

Airman Certificate and/or Rating Application

I Application Information ☐ Student ☐ Recreational ☐ Private ☐ Commercial ☐ Airline Transport ☐ Instrument
☐ Additional Rating ☐ Airplane Single-Engine ☐ Airplane Multiengine ☐ Rotorcraft ☐ Balloon ☐ Airship ☐ Glider ☐ Powered-Lift
☐ Flight Instructor ____ Initial ____ Renewal ____ Reinstatement ☐ Additional Instructor Rating ☐ Ground Instructor
☐ Medical Flight Test ☐ Reexamination ☐ Reissuance of ____________________ certificate ☐ Other ________

A. Name (Last, First, Middle) | B. SSN (US Only) | C. Date of Birth Month Day Year | D. Place of Birth

E. Address | F. Citizenship ☐ USA ☐ Other Specify | G. Do you read, speak, write, & understand the English language? ☐ Yes ☐ No

City, State, Zip Code | H. Height | I. Weight | J. Hair | K. Eyes | L. Sex ☐ Male ☐ Female

M. Do you now hold, or have you ever held an FAA Pilot Certificate? ☐ Yes ☐ No | N. Grade Pilot Certificate | O. Certificate Number | P. Date Issued

Q. Do you hold a Medical Certificate? ☐ Yes ☐ No | R. Class of Certificate | S. Date Issued | T. Name of Examiner

U. Have you ever been convicted for violation of any Federal or State statutes relating to narcotic drugs, marijuana, or depressant or stimulant drugs or substances? ☐ Yes ☐ No | V. Date of Final Conviction

II. Certificate or Rating Applied For on Basis of:

☐ A. Completion of Required Test — 1. Aircraft to be used (if flight test required) | 2a. Total time in this aircraft / SIM / FTD ____ hours | 2b. Pilot in command ____ hours

☐ B. Military Competence Obtained In — 1. Service | 2. Date Rated | 3. Rank or Grade and Service Number | 4a. Flown 10 hours PIC in last 12 months in the following Military Aircraft. | 4b. US Military PIC & Instrument check in last 12 months (List Aircraft)

☐ C. Graduate of Approved Course — 1. Name and Location of Training Agency or Training Center | 1a. Certification Number | 2. Curriculum From Which Graduated | 3. Date

☐ D. Holder of Foreign License Issued By — 1. Country | 2. Grade of License | 3. Number | 4. Ratings

☐ E. Completion of Air Carrier's Approved Training Program — 1. Name of Air Carrier | 2. Date | 3. Which Curriculum ☐ Initial ☐ Upgrade ☐ Transition

III RECORD OF PILOT TIME (Do not write in the shaded areas.)

	Total	Instruction Received	Solo	Pilot in Command (PIC)	Cross Country Instruction Received	Cross Country Solo	Cross Country PIC	Instrument	Night Instruction Received	Night Take-off Landings	Night PIC	Night Take-Off Landing PIC	Number of Flights	Number of Aero-Tows	Number of Ground Launches	Number of Powered Launches
Airplanes				PIC			PIC				PIC	PIC				
				SIC			SIC				SIC	SIC				
Rotorcraft				PIC			PIC				PIC	PIC				
				SIC			SIC				SIC	SIC				
Powered Lift				PIC			PIC				PIC	PIC				
				SIC			SIC				SIC	SIC				
Gliders																
Lighter Than Air																
Simulator																
Training Device																
PCATD																

IV. Have you failed a test for this certificate or rating? ☐ Yes ☐ No

V. Applicants's Certification -- I certify that all statements and answers provided by me on this application form are complete and true to the best of my knowledge and I agree that they are to be considered as part of the basis for issuance of any FAA certificate to me. I have also read and understand the Privacy Act statement that accompanies this form.

Signature of Applicant | Date

FAA Form 8710-1 (4-00) Supersedes Previous Edition

NSN: 0052-00-682-5007

Instructor's Recommendation

I have personally instructed the applicant and consider this person ready to take the test.

Date	Instructor's Signature (Print Name & Sign)	Certificate No:	Certificate Expires

Air Agency's Recommendation

The applicant has successfully completed our ____________________ course, and is recommended for certification or rating without further ____________________ test.

Date	Agency Name and Number	Officials Signature
		Title

Designated Examiner or Airman Certification Representative Report

- ☐ Student Pilot Certificate Issued (Copy attached)
- ☐ I have personally reviewed this applicant's pilot logbook and/or training record, and certify that the individual meets the pertinent requirements of 14 CFR Part 61 for the certificate or rating sought.
- ☐ I have personally reviewed this applicant's graduation certificate, and found it to be appropriate and in order, and have returned the certificate.
- ☐ I have personally tested and/or verified this applicant in accordance with pertinent procedures and standards with the result indicated below.
  - ☐ Approved -- Temporary Certificate Issued (Original Attached)
  - ☐ Disapproved -- Disapproval Notice Issued (Original Attached)

Location of Test (Facility, City, State)	Duration of Test		
	Ground	Simulator/FTD	Flight

Certificate or Rating for Which Tested	Type(s) of Aircraft Used	Registration No.(s)

Date	Examiner's Signature (Print Name & Sign)	Certificate No.	Designation No.	Designation Expires

Evaluator's Record (Use For ATP Certificate and/or Type Ratings)

	Inspector	Examiner	Signature and Certificate Number	Date
Oral	☐	☐	____________	____________
Approved Simulator/Training Device Check	☐	☐	____________	____________
Aircraft Flight Check	☐	☐	____________	____________
Advanced Qualification Program	☐	☐	____________	____________

Aviation Safety Inspector or Technician Report

I have personally tested this applicant in accordance with or have otherwise verified that this applicant complies with pertinent procedures, standards, policies, and or necessary requirements with the result indicated below.

☐ Approved -- Temporary Certificate Issued (Original Attached) ☐ Disapproved -- Disapproval Notice Issued (Original Attached)

Location of Test (Facility, City, State)	Duration of Test		
	Ground	Simulator/FTD	Flight

Certificate or Rating for Which Tested	Type(s) of Aircraft Used	Registration No.(s)

- ☐ Student Pilot Certificate Issued
- ☐ Examiner's Recommendation
  - ☐ Accepted ☐ Rejected
- ☐ Reissue or Exchange of Pilot Certificate
- ☐ Special Medical test conducted -- report forwarded to Aeromedical Certification Branch, AAM-330

- ☐ Certificate or Rating Based on
  - ☐ Military Competence
  - ☐ Foreign License
  - ☐ Approved Course Graduate
  - ☐ Other Approved FAA Qualification Criteria

- ☐ Flight Instructor ☐ Ground Instructor
  - ☐ Renewal
  - ☐ Reinstatement
- Instructor Renewal Based on
  - ☐ Activity ☐ Training Course
  - ☐ Test ☐ Duties and Responsibilities

Training Course (FIRC) Name	Graduation Certificate No.	Date

Date	Inspector's Signature (Print Name & Sign)	Certificate No.	FAA District Office

Attachments:

- ☐ Student Pilot Certificate (Copy)
- ☐ Knowledge Test Report
- ☐ Temporary Airman Certificate
- ☐ Notice of Disapproval
- ☐ Superseded Airman Certificate

☐ Airman's Identification (ID)

Form of ID ____________

Number ____________

Expiration Date ____________

Telephone Number ____________

ID:

Name: ____________

Date of Birth: ____________

Certificate Number: ____________

E-Mail Address ____________

FAA Form 8710-1 (4-00) Supersedes Previous Edition NSN: 0052-00-682-5007

4.6 AUTHORIZATION TO TAKE THE PRACTICAL TEST

A. Before you take your practical test, your CFII will endorse your logbook to certify that you are prepared for the practical test and that you have received and logged training in the following areas of operation (which are explained in Chapters I through VIII of this book).

1. *Preflight preparation*
2. *Preflight procedures*
3. *Air traffic control clearances and procedures*
4. *Flight by reference to instruments*
5. *Navigation systems*
6. *Instrument approach procedures*
7. *Emergency operations*
8. *Postflight procedures*

B. Your logbook must contain the following endorsement from your CFII certifying that (s)he has found that you are prepared for the practical test, given you at least 3 hr. of instrument training in preparation for the practical test within the preceding 60 days, and found that you have demonstrated knowledge of the subject areas in which you were shown to be deficient in your knowledge test report.

I certify that (First name, MI, Last name) has received the required training of Sec. 61.65(c) and (d). I have determined he/she is prepared for the Instrument-Airplane practical test. He/She has demonstrated satisfactory knowledge of the subject areas found deficient on his/her knowledge test.

Date	Signature	CFI No.	Expiration Date

4.7 ORAL PORTION OF THE PRACTICAL TEST

A. Your practical test will probably begin in your examiner's office.

1. You should have with you
   a. This book
   b. Your *Pilot's Operating Handbook (POH)* for your airplane (including weight and balance data)
   c. Your copy of Gleim's *FAR/AIM*
   d. All of the items listed on page 34 (This is the FAA's list, and they omitted the weight and balance data and the FARs.)
   e. A positive attitude
2. Your examiner will probably begin by reviewing your paperwork (FAA Form 8710-1, Airman Knowledge Test Report, logbook endorsement, etc.) and receiving payment for his/her services.
3. Typically, your examiner will begin with questions about your preplanned IFR cross-country flight with discussion of weather, charts, FARs, etc. When you schedule your practical test, your examiner will probably assign a cross-country flight for you to plan and bring to your practical test. You should also ask your examiner if (s)he wants you to file an IFR flight plan.
   a. When you file your flight plan, ask your examiner if (s)he would like you to note in the remarks section that this is an instrument practical test.
      1) This should alert ATC to be helpful in arranging the holding and instrument approaches required.
4. As your examiner asks you questions, follow the guidelines listed below:
   a. Attempt to position yourself in a discussion mode with him/her rather than being interrogated by the examiner.
   b. Be respectful but do not be intimidated. Both you and your examiner are professionals.
   c. Draw on your knowledge from this book and other books, your CFII, and your prior experience.
   d. Ask for amplification of any points you feel uncertain about.
   e. If you do not know an answer, try to explain how you would research the answer.
5. Be confident that you will do well. You are a good pilot. You have thoroughly prepared for this discussion by studying the subsequent pages and have worked diligently with your CFII.

B. After you discuss various aspects of the four "knowledge only" tasks, you will move out to your airplane to begin the flight portion of your practical test, which consists of 20 "knowledge and skill" tasks.

1. If possible and appropriate in the circumstances, thoroughly preflight your airplane just before you go to your examiner's office.
2. As you and your examiner approach your airplane, explain that you have already preflighted the airplane (explain any possible problems and how you resolved them).
3. Volunteer to answer any questions.
4. Make sure you walk around the airplane to observe any possible damage by ramp vehicles or other aircraft while you were in your examiner's office.
5. As you enter the airplane, make sure that your cockpit is organized and you feel in control of your charts, view-limiting device, clocks, paper to write down clearances, etc.

4.8 FLIGHT PORTION OF THE PRACTICAL TEST

A. As your begin the flight portion of your practical test, your examiner will have you depart on the IFR cross-country flight you previously planned.

1. You will taxi out, depart, and proceed on course to your destination.
2. Your departure procedures usually permit demonstration/testing of many of the tasks in Areas of Operation II through V. When VFR, your examiner will probably have you cancel your IFR flight plan so you can demonstrate additional flight maneuvers.

B. Note that you are required to perform all 24 tasks during your practical test. The first four tasks are usually completed in the examiner's office, and the remaining 20 tasks must be done in the airplane.

1. You will be required to perform three instrument approaches:
   a. One precision approach, which will be an ILS
   b. Two nonprecision approaches using two different approach systems
      1) Nonprecision approaches include VOR, LOC, NDB, LDA, SDF, or GPS.
      2) You may do a VOR and a LOC approach, but you may not do a VOR and a VOR/DME approach.

C. Remember that at all times you are the pilot in command of this flight. Take polite, but firm, charge of your airplane and instill in your examiner confidence in you as a safe and competent pilot.

4.9 USE OF FLIGHT SIMULATOR OR TRAINING DEVICE

You must demonstrate all of the instrument approach procedures (Chapter VI, beginning on page 215) required by FAR Part 61. At least one instrument approach procedure must be demonstrated in your airplane.

At the discretion of your examiner, the instrument approach(es) and missed approach(es) not selected for actual flight demonstration may be performed in a flight simulator or training device that meets the required level of simulation. If you are planning on using a simulator, you should check with your examiner to ensure that it meets all requirements.

4.10 YOUR TEMPORARY PILOT CERTIFICATE

A. When you successfully complete your practical test, your examiner will prepare a temporary pilot certificate similar to the one illustrated below.

1. The temporary certificate is valid for 120 days.

B. Your permanent certificate will be sent to you directly from the FAA Aeronautical Center in Oklahoma City in about 60 to 90 days.

1. If you do not receive your permanent certificate within 120 days, your examiner can arrange an extension of your temporary certificate.

I. UNITED STATES OF AMERICA
DEPARTMENT OF TRANSPORTATION—FEDERAL AVIATION ADMINISTRATION
II. TEMPORARY AIRMAN CERTIFICATE

III. CERTIFICATE NO.

THIS CERTIFIES THAT IV.
V.

DATE OF BIRTH	HEIGHT	WEIGHT	HAIR	EYES	SEX	NATIONALITY	VI.
	IN.						

IX. has been found to be properly qualified and is hereby authorized in accordance with the conditions of issuance on the reverse of this certificate to exercise the privileges of

RATINGS AND LIMITATIONS
XII.

XIII.

THIS IS ☐ AN ORIGINAL ISSUANCE ☐ A REISSUANCE OF THIS GRADE OF CERTIFICATE — DATE OF SUPERSEDED AIRMAN CERTIFICATE

BY DIRECTION OF THE ADMINISTRATOR — EXAMINER'S DESIGNATION NO. OR INSPECTOR'S REG. NO.

X. DATE OF ISSUANCE — X. SIGNATURE OF EXAMINER OR INSPECTOR — DATE DESIGNATION EXPIRES

VII. AIRMAN'S SIGNATURE

FAA Form 8060-4 (4-69) Supersedes Previous Edition

4.11 FAILURE ON THE PRACTICAL TEST

A. About 90% of applicants pass their instrument rating practical test the first time, and virtually all who experienced difficulty on their first attempt pass the second time.

B. If, in the judgment of your examiner, you do not meet the standards of performance of any task performed, the associated area of operation is failed, and thus the practical test is failed.

1. You or your examiner may discontinue the test after the failure of an area of operation.

   a. The test will be continued only with your consent. It is recommended that you discontinue the test after it is known that you have failed. Usually, confidence has been lost.

2. The following are typical areas of unsatisfactory performance and grounds for disqualification:

   a. Any action, or lack of action, by you that requires corrective intervention by your examiner to maintain safe flight

   b. Failure to use proper and effective visual scanning techniques, when applicable, to clear the area before and while performing maneuvers

   c. Consistently exceeding tolerances stated in the tasks

   d. Failure to take prompt corrective action when tolerances are exceeded

C. When on the ground, your examiner will complete the Notice of Disapproval of Application, FAA Form 8060-5, which appears below, and will indicate the areas necessary for re-examination.

1. Your examiner will give you credit for those tasks that you have successfully completed.
   a. However, during the retest and at the discretion of your examiner, any task may be reevaluated, including those previously passed.

D. You should do the following:

1. Indicate your intent to work with your instructor on your deficiencies.
2. Inquire about rescheduling the next practical test.
   a. Many examiners have a reduced fee for a retake (FAA inspectors do not charge for their services).
3. Inquire about having your flight instructor discuss your proficiencies and deficiencies with the examiner.

UNITED STATES OF AMERICA
DEPARTMENT OF TRANSPORTATION–FEDERAL AVIATION ADMINISTRATION

NOTICE OF DISAPPROVAL OF APPLICATION

NOTE
PRESENT THIS FORM UPON APPLICATION FOR REEXAMINTION

NAME AND ADDRESS OF APPLICANT

CERTIFICATE OR RATING SOUGHT

On the date shown, you failed the examination indicated below:

☐ FLIGHT ☐ ORAL ☐ PRACTICAL

AIRCRAFT USED *(Make and Model)*

FLT. TIME RECORDED IN LOGBOOK

PILOT-IN-COMM. OR SOLO	INSTRUMENT	DUAL

UPON REAPPLICATION YOU WILL BE REEXAMINED ON THE FOLLOWING:

I have personally tested this applicant and deem his performance unsatisfactory for the issuance of the certificate or rating sought.

DATE OF EXAMINATION	SIGNATURE OF EXAMINER OR INSPECTOR	DESIGNATION OR OFFICE NO.

FAA Form 8060–5 (5-80)

END OF CHAPTER

This is the end of Part I. Part II consists of Chapters I through VIII. Each chapter covers one Area of Operation in the Instrument Rating Practical Test Standards.

PART II
FLIGHT MANEUVERS AND FAA PRACTICAL TEST STANDARDS: DISCUSSED AND EXPLAINED

Part II of this book (Chapters I through VIII) provides an in-depth discussion of flight maneuvers and the Instrument Rating Practical Test Standards (PTS). Each of the eight areas of operation with its related task(s) is presented in a separate chapter.

	No. of Tasks	No. of Pages
I. Preflight Preparation	2	20
II. Preflight Procedures	3	34
III. Air Traffic Control Clearances and Procedures	3	36
IV. Flight by Reference to Instruments	7	56
V. Navigation Systems	1	22
VI. Instrument Approach Procedures	5	112
VII. Emergency Operations	2	9
VIII. Postflight Procedures	1	2
	24	291

Each task, reproduced verbatim from the PTS, appears in a shaded box within each chapter. General discussion is presented under "A. General Information." This is followed by "B. Task Objectives," which is a detailed discussion of each element of the FAA's task. Additionally, each "knowledge and skill" task (e.g., flight maneuver) common errors are listed and briefly discussed under "C. Common Errors . . . ."

Each objective of a task lists, in sequence, the important elements that must be satisfactorily performed. The objective includes

1. Specifically what you should be able to do
2. The conditions under which the task is to be performed
3. The acceptable standards of performance

Be confident. You have prepared diligently and are better prepared and more skilled than the average instrument rating applicant. Satisfactory performance to meet the requirements for certification is based on your ability to safely

1. Perform the approved areas of operation for the certificate or rating sought within the approved standards
2. Demonstrate mastery of the airplane with the successful outcome of each task performed never seriously in doubt
3. Demonstrate satisfactory proficiency and competency within the approved standards
4. Demonstrate sound judgment and aeronautical decision making (ADM)
5. Demonstrate single-pilot competence if the airplane is type certificated for single-pilot operations

Each task has an FAA reference list that identifies the publication(s) that describe(s) the task. Our discussion is based on the current issue of these references. The following FAA references are used in the Instrument Rating PTS:

14 CFR Part 61 -- Certification: Pilots, Flight Instructors, and Ground Instructors
14 CFR Part 91 -- General Operating and Flight Rules
AC 00-6 -- *Aviation Weather*
AC 00-45 -- *Aviation Weather Services*
AC 60-28 -- *English Language Skill Standards Required by 14 CFR Parts 61, 63, and 65*
AC 61-21 -- *Flight Training Handbook*
AC 61-23 -- *Pilot's Handbook of Aeronautical Knowledge*
AC 61-27 -- *Instrument Flying Handbook*
AC 61-84 -- *Role of Preflight Preparation*
AC 90-48 -- *Pilot's Role in Collision Avoidance*
AC 90-94 -- *Guidelines for Using Global Positioning Systems*
AIM -- *Aeronautical Information Manual*
DPs -- Instrument Departure Procedures
STARs -- Standard Terminal Arrivals
AFD -- *Airport/Facility Directory*
FDC NOTAMs -- National Flight Data Center Notices to Airmen
IAP -- Instrument Approach Procedures
En Route Low Altitude charts
Navigation Equipment Operation Manuals
Pilot's Operating Handbook (FAA-approved *Airplane Flight Manual*)

In each task, as appropriate, we will provide you with the chapter and/or module from Gleim's *Pilot Handbook* for additional discussion of an element (or concept) of the task, along with the approximate number of pages of discussion.

FARs ARE NOW REFERRED TO AS CFRs

The FAA has recently begun to abbreviate Federal Aviation Regulations as "14 CFR" rather than "FARs." The Office of Management and Budget uses FAR as an acronym for Federal Acquisition Regulations. CFR stands for Code of Federal Regulations and Federal Aviation Regulations are in Title 14. For example, FAR Part 1 and FAR 61.109 are now referred to as 14 CFR Part 1 and 14 CFR Sec. 61.109, respectively.

Due to CFIs' and pilots' widespread use of the acronym FAR, we continue to use FAR rather than CFR.

CHAPTER I
PREFLIGHT PREPARATION

This chapter explains the two tasks (A and B) of Preflight Preparation. These tasks are "knowledge only." Your examiner is required to test you on both of these tasks.

WEATHER INFORMATION

I.A. TASK: WEATHER INFORMATION

REFERENCES: 14 CFR Part 61; AC 00-6, AC 00-45; AIM.

NOTE: Where current weather reports, forecasts, or other pertinent information is not available, this information will be simulated by the examiner in a manner which will adequately measure the applicant's competence.

Objective. To determine that the applicant:

1. Exhibits adequate knowledge of the elements related to aviation weather information by obtaining, reading, and analyzing the applicable items such as --

a. Weather reports and forecasts.
b. Pilot and radar reports.
c. Surface analysis charts.
d. Radar summary charts.
e. Significant weather prognostics.
f. Winds and temperatures aloft.
g. Freezing level charts.
h. Stability charts.
i. Severe weather outlook charts.
j. Tables and conversion graphs.
k. SIGMETs and AIRMETs.
l. ATIS reports.

2. Correctly analyzes the assembled weather information pertaining to the proposed route of flight and destination airport, and determines whether an alternate airport is required, and, if required, whether the selected alternate airport meets the regulatory requirement.

A. General Information

1. The objective of this task is to determine your knowledge of aviation weather information by obtaining, reading, and correctly analyzing the applicable reports, forecasts, and charts.
2. Gleim's *Aviation Weather and Weather Services* is a 442-page book in outline format which combines the FAA's *Aviation Weather* (AC 00-6A) and *Aviation Weather Services* (AC 00-45E) and numerous FAA publications into one easy-to-understand book. It will help you to learn all aspects of weather, weather reports, and weather forecasts. It is a single easy-to-use reference that is more up-to-date than the FAA's weather books. The table of contents is on the next page.
3. Flight Service Stations (FSSs) are the primary source for obtaining preflight briefings and in-flight weather information.
    a. Prior to your flight, and before you meet with your examiner, you should visit or call the nearest FSS for a complete briefing.
    b. There are three basic types of preflight briefings to meet your needs:
        1) Standard briefing
            a) A standard briefing should be requested any time you are planning a flight and have not received a previous briefing.
        2) Abbreviated briefing
            a) Request an abbreviated briefing when you need information to supplement mass disseminated data (e.g., TIBS, DUATs) or to update a previous briefing, or when you need only one or two specific items.
        3) Outlook briefing
            a) Request an outlook briefing whenever your proposed time of departure is 6 hr. or more from the time of the briefing.

GLEIM'S
AVIATION WEATHER AND WEATHER SERVICES

Table of Contents

B. Task Objectives

1. **Exhibit your knowledge of the elements related to aviation weather information by obtaining, reading, and analyzing the applicable items such as:**

   a. **Weather reports and forecasts**

      1) Weather reports

         a) METARs provide surface weather information.

         b) Weather radar observations provide detailed information about precipitation, winds (by use of Doppler technology), and weather systems.

         c) Satellite imagery provides visible and infrared images of clouds.

         d) Upper-air observations are provided by radiosonde-equipped weather balloons and pilot weather reports (PIREPs).

      2) Weather forecasts

         a) A terminal aerodrome forecast (TAF) is a concise statement of the expected meteorological conditions at an airport.

         b) An aviation area forecast (FA) provides a forecast of visual meteorological conditions, clouds, and weather over an area of several states.

            i) The FA is used in conjunction with the aviation in-flight weather advisories to determine en route weather and to interpolate conditions at airports for which no TAF is issued.

         c) Transcribed weather broadcasts (TWEB) provide route and vicinity forecasts and synopses for more than 200 defined routes within the contiguous U.S.

         d) In-flight aviation weather advisories (AIRMETs, SIGMETs, and convective SIGMETs) are forecasts that advise en route aircraft of the development of potentially hazardous weather.

         e) A convective outlook (AC) is a forecast of severe and nonsevere thunderstorms across the U.S.

   b. **Pilot and radar reports**

      1) Pilot weather report (PIREP)

         a) No more timely or helpful weather observation fills the gaps between reporting stations than observations and reports made by fellow pilots during flight. Aircraft in flight are the **only** source of directly observed cloud tops, icing, and turbulence.

      2) Radar weather report (SD)

         a) A radar weather report is automatically generated from WSR-88D (Doppler) weather radar data and includes the type, intensity, and location of the echo top of the precipitation.

   c. **Surface analysis chart**

      1) The surface analysis chart provides a ready means of locating pressure systems and fronts.

         a) It also gives an overview of winds, temperatures, and dew point temperatures at chart time.

      2) The surface analysis chart is transmitted every 3 hr.

d. Radar summary chart

1) A radar summary chart is a computer-generated graphical display of a collection of automated radar weather reports (SDs).

2) The radar summary chart helps preflight planning by identifying areas of precipitation as well as information about type, intensity, configuration, coverage, echo top, and cell movement of precipitation.

   a) Severe weather watches are plotted if they are in effect when the chart is valid.

3) The radar summary chart is available hourly with a valid time of H+35 (35 min. past each hour).

e. Significant weather prognostics

1) The low-level significant weather prognostic chart (called a prog for brevity) is a four-panel chart.

   a) The two lower panels are 12- and 24-hr. surface progs (SFC PROG).

   b) The two upper panels are 12- and 24-hr. progs of significant weather (SIG WX) from the surface to 24,000 ft. MSL.

   c) The charts show conditions as they are forecast to be at the valid time (VT) of the chart.

2) The two surface prog panels use standard symbols for fronts, significant troughs, and pressure centers.

   a) The surface prog also outlines areas of forecast precipitation and/or thunderstorms.

      i) Solid lines enclose areas of forecast precipitation.

3) The two significant weather panels depict IFR, MVFR, turbulence, and freezing levels.

   a) Forecast areas of nonconvective turbulence of moderate or greater intensity are enclosed by long, dashed lines.

      i) A symbol entered within a general area of forecast turbulence denotes intensity.

      ii) Numbers below and above a short line show expected bases and tops of the turbulent layer in hundreds of feet MSL.

   b) Freezing-level height contours for the **highest** freezing level are drawn at 4,000-ft. intervals.

      i) Contours are labeled in hundreds of feet MSL.

      ii) The zig-zag line shows where the freezing level is forecast to be at the surface and is labeled "SFC."

4) The 36- and 48-hr. surface weather prog is an extension of the 12- and 24-hr. surface prog.

   a) The 36- and 48-hr. surface prog should be used only for outlook purposes, that is, just to get a very general picture of the weather conditions that are in the relatively distant future.

5) The high-level significant weather prog encompasses airspace from flight level (FL) 250 to FL 600.

f. Winds and temperatures aloft

1) Winds and temperatures aloft are available for both (1) forecasts and (2) observations.

2) Forecast winds and temperatures aloft are available in table or chart format.

   a) The table format lists specific locations in the U.S., including Alaska and Hawaii.

   b) The forecast winds and temperatures aloft chart is a graphic representation of the information presented in the table format.

3) Observed winds aloft chart is information collected from radiosonde balloons.

4) The winds and temperatures aloft forecast, in the table format, is the most commonly used to select the most favorable altitude for a proposed flight.

g. Freezing level charts

1) The freezing level panel of the composite moisture stability chart provides information about the observed freezing-level data from upper-air observations.

   a) The contour analysis shows an overall view of the lowest observed freezing level.

2) AIRMET ZULU (for icing and freezing level) will state the forecasted areas of expected icing.

3) The low-level significant weather prog will depict the forecasted highest freezing level.

h. Stability charts

1) The stability level of the composite moisture stability chart provides two stability indexes for each station.

   a) The lifted index (LI) is plotted above a short line and is a common measure of stability.

   b) The K index, plotted below the line, is used mainly by meteorologists.

2) A positive LI means stable air.

3) A negative LI means unstable air.

4) Stability is also very important when considering the type, extent, and intensity of aviation weather hazards. For example, a quick estimate of areas of probable convective turbulence can be made by associating the areas with unstable air.

   a) An area of extensive icing would be associated with stratiform clouds and steady precipitation, which are characterized by stable air.

i. Severe weather outlook charts

1) The convective outlook chart (formerly the severe weather outlook chart) is for severe and general thunderstorm activity.

   a) The chart consists of the Day 1 Convective Outlook Chart and the Day 2 Convective Outlook Chart.

2) This chart should be used only for planning purposes.

   a) Additional and more specific information is available from convective SIGMETs, the radar summary chart, and "live" radar images.

j. Tables and conversion graphs

1) Various tables and conversion graphs are available for you to further understand the weather. A partial list includes the following:

a) Density altitude chart
b) Standard conversion table (between metric and English units)
c) Turbulence-reporting criteria table
d) Icing-reporting criteria table

k. SIGMETs and AIRMETs

1) **Convective SIGMET (WST)**

a) Convective SIGMETs are issued in the contiguous 48 states (i.e., none for Alaska and Hawaii) for any of the following:

i) Severe thunderstorm due to

- Surface winds greater than or equal to 50 kt.
- Hail at the surface greater than or equal to ¾ in. in diameter
- Tornadoes

ii) Embedded thunderstorms

iii) A line of thunderstorms

iv) Thunderstorms producing precipitation greater than or equal to heavy precipitation affecting 40% or more of an area of at least 3,000 square mi.

b) Any convective SIGMET implies severe or greater turbulence, severe icing, and low-level wind shear.

i) A convective SIGMET may be issued for any convective situation that the forecaster feels is hazardous to all categories of aircraft.

c) Convective SIGMET bulletins are issued for the eastern (E), central (C), and western (W) United States.

i) Bulletins are issued hourly at H+55. Special bulletins are issued at any time as required and updated at H+55.

ii) If no criteria meeting a convective SIGMET are observed or forecast, the message "CONVECTIVE SIGMET...NONE" will be issued for each area at H+55.

iii) Individual convective SIGMETs for each area (E, C, W) are numbered sequentially from one each day, beginning at 0000Z.

- A continuing convective SIGMET phenomenon will be reissued every hour at H+55 with a new number.

iv) The text of the bulletin consists of either an observation and a forecast or just a forecast. The forecast is valid for up to 2 hr.

2) **SIGMET (WS)**

a) A SIGMET advises of nonconvective weather that is potentially hazardous to all aircraft.

i) In the conterminous U.S., SIGMETs are issued when the following phenomena occur or are expected to occur:
- Severe icing not associated with thunderstorms
- Severe or extreme turbulence or clear air turbulence (CAT) not associated with thunderstorms
- Duststorms, sandstorms, or volcanic ash lowering surface or in-flight visibilities to below 3 SM
- Volcanic eruption

ii) In Alaska and Hawaii, SIGMETs are also issued for
- Tornadoes
- Lines of thunderstorms
- Embedded thunderstorms
- Hail greater than or equal to ¾ in. in diameter

b) SIGMETs are unscheduled products that are valid for 4 hr. unless conditions are associated with a hurricane. Then the SIGMETs are valid for 6 hr.

i) Unscheduled updates and corrections are issued as necessary.

c) A SIGMET is identified by an alphabetic designator from NOVEMBER through YANKEE, excluding SIERRA and TANGO.

3) **AIRMET (WA).** AIRMETs are advisories of significant weather phenomena but describe conditions at intensities lower than those requiring SIGMETs to be issued. AIRMETs are intended for dissemination to all pilots in the preflight and en route phase of flight to enhance safety.

a) AIRMET bulletins are issued on a scheduled basis every 6 hr.

i) Unscheduled updates and corrections are issued as necessary.

b) Each AIRMET bulletin contains

i) Any current AIRMETs in effect
ii) An outlook for conditions expected after the AIRMET valid period

c) There are three AIRMETs:

i) AIRMET Sierra describes
- IFR weather conditions -- ceilings less than 1,000 ft. and/or visibility less than 3 SM affecting over 50% of the area at one time
- Extensive mountain obscuration

NOTE: AIRMET Sierra is referenced in the area forecast.

ii) AIRMET Tango describes
- Moderate turbulence
- Sustained surface winds of 30 kt. or greater
- Low-level wind shear

iii) AIRMET Zulu describes
- Moderate icing
- Freezing-level heights

d) After the first issuance each day, scheduled or unscheduled bulletins are numbered sequentially for easier identification.

I. ATIS reports

1) The automatic terminal information service (ATIS) provides a continuous transmission that provides information for arriving and departing aircraft, including

a) Time of the latest weather report
b) Sky conditions, visibility, and obstructions to visibility

i) The absence of a sky condition or ceiling and/or visibility and obstructions to visibility on ATIS indicates a sky condition of 5,000 ft. or above and visibility of 5 SM or more.

- A remark on the broadcast may state, "The weather is better than 5,000 and 5," or the existing weather may be broadcast.

c) Temperature and dew point (degrees Celsius)
d) Wind direction (magnetic) and velocity
e) Altimeter
f) Other pertinent remarks, instrument approach, and runway in use

i) The departure runway will be given only if it is different from the landing runway, except at locations having a separate ATIS for departure.

2) The ATIS broadcast is updated whenever any official weather is received, regardless of content or changes, or when a change is made in other pertinent data, such as a runway change. Each new broadcast is labeled with a letter of the alphabet at the beginning of the broadcast; e.g., "This is information alpha" or "information bravo."

2. Correctly analyze the assembled weather information pertaining to the proposed route of flight and destination airport, and determine whether an alternate airport is required, and, if required, whether the selected alternate airport meets the regulatory requirements.

a. An alternate airport is NOT required (FAR 91.169) when

1) Your destination airport has a published IAP, and

2) The 1-2-3 rule is met -- at least 1 hr. before and 1 hr. after your ETA, the weather reports or forecasts, or any combination of them, indicate

a) Ceiling of at least 2,000 ft. above the airport elevation, and
b) Visibility of at least 3 SM.

NOTE: Any remark in a forecast, such as a "chance of" or "occasionally" less than these minimums, causes an alternate to be required.

b. To list an airport as an alternate airport on your IFR flight plan, current weather forecasts must indicate that, at your ETA at the alternate airport, the ceiling and visibility will be at or above the following:

1) If the airport has a published IAP, the alternate airport minimums are those specified for that procedure or, if none are specified, the following standard alternate minimums:

a) For a precision approach procedure (e.g., ILS)

i) Ceiling 600 ft.
ii) Visibility 2 SM

b) For a nonprecision approach (e.g., VOR, NDB, GPS)

i) Ceiling 800 ft.
ii) Visibility 2 SM

2) If the airport does not have a published IAP, the ceiling and visibility minimums are those allowing descent from the MEA, approach, and landing under basic VFR weather minimums.

c. Finally, you must make a go/no-go decision.

1) In a well-equipped airplane with a proficient pilot flying, any ceiling and visibility within legal minimums should be flyable. In a poorly equipped airplane or with a new or rusty pilot, flying in LIFR (low IFR) should be avoided. This is not to say that you must be an ATP to fly LIFR, but if your last actual approach was 4 months ago and only to 1,500 ft., it is not a good idea to tackle LIFR.

2) Another factor to consider in your go/no-go decision is the weather. Low IFR in smooth air caused by a stalled front is considerably different from heavy turbulence ahead of a strong front or in a squall line. The following forecast conditions may lead to a no-go decision:

a) Thunderstorms

b) Embedded thunderstorms

c) Lines of thunderstorms

d) Fast-moving fronts or squall lines

e) Flights that require you to cross strong or fast-moving fronts

f) Extensive IFR that would require long periods of instrument flying (less of a problem with an autopilot)

g) Reported turbulence that is moderate or greater

i) Remember, moderate turbulence in a Boeing 727 is usually severe in a Cessna 172.

h) Icing

i) Fog

i) Unlike when in a ceiling, you usually cannot break out and land with ground fog. This condition is especially important if sufficient fuel may be a concern.

3) These factors must be considered in relation to the equipment to be flown. Thunderstorms are less of a problem in a radar-equipped airplane. The only way to fly safely is to be able to weigh each factor against your capabilities and those of your airplane. This is done only by using common sense and gaining experience.

4) Flying IFR is a continuing process of aeronautical decision making throughout the whole flight. You must use your rating to gain experience, but you must also temper your pursuit of experience so you do not get in beyond your capabilities or the capabilities of your airplane.

5) A final factor that must be considered in the go/no-go decision is your physical and mental condition. Are you sick, tired, upset, depressed, etc.? This list can be expanded. These factors greatly affect your ability to handle normal and abnormal problems.

   a) A good method to ensure safety is the "I'm Safe" checklist: I am not being impaired by

      I llness
      M edication

      S tress
      A lcohol
      F atigue
      E motion

END OF TASK

CROSS-COUNTRY FLIGHT PLANNING

I.B. TASK: CROSS-COUNTRY FLIGHT PLANNING

REFERENCES: 14 CFR Parts 61, 91; AC 61-27, AC 61-23, AC 90-94; AFD; AIM.

Objective. To determine that the applicant:

1. Exhibits adequate knowledge of the elements by presenting and explaining a preplanned cross-country flight, as previously assigned by the examiner (preplanning at examiner's discretion). It should be planned using real time weather and conform to the regulatory requirements for instrument flight rules within the airspace in which the flight will be conducted.
2. Exhibits adequate knowledge of the aircraft's performance capabilities by calculating the estimated time en route and total fuel requirement based upon factors, such as --
   - **a.** Power settings.
   - **b.** Operating altitude or flight level.
   - **c.** Wind.
   - **d.** Fuel reserve requirements.
3. Selects and correctly interprets the current and applicable en route charts, instrument departure procedures (DPs), Standard Terminal Arrival (STAR), and Standard Instrument Approach Procedure Charts (IAP).
4. Obtains and correctly interprets applicable NOTAM information.
5. Determines the calculated performance is within the aircraft's capability and operating limitations.
6. Completes and files a flight plan in a manner that accurately reflects the conditions of the proposed flight. (Does not have to be filed with ATC.)
7. Demonstrates adequate knowledge of Global Positioning Systems (GPS) and Receiver Autonomous Integrity Monitoring (RAIM) capability, when aircraft is so equipped.

A. General Information

1. The objective of this task is for you to demonstrate your knowledge by planning a cross-country IFR flight.
2. For additional reading, see *Pilot Handbook* for the following:
   a. Chapter 5, Airplane Performance and Weight and Balance for a 17-page discussion on airplane performance and a 12-page discussion on weight and balance.
   b. Chapter 9, Navigation: Charts, Publications, Flight Computers (Modules 9.4 through 9.24), for an eight-page discussion on various flight publications and a 22-page discussion on using a manual flight computer.

B. Task Objectives

1. **Exhibit knowledge of the elements by presenting and explaining your preplanned cross-country flight, as previously assigned by your examiner. It should be planned using real time weather and conform to the regulatory requirements for instrument flight rules within the airspace in which the flight will be conducted.**
   a. The flight planning procedures you learned as a student pilot will have prepared you thoroughly for your IFR cross-country flight planning.
      1) Some differences include route and altitude selection.
         a) Route selection may be a preferred IFR route, tower en route control routes, an airway route, or a direct route.
         b) Altitude selection will be affected by minimum IFR altitudes.
      2) Some pilots use a navigation log as shown on page 59.

FLIGHT LOG

DEPARTURE POINT	VOR IDENT. FREQ.	RADIAL TO FROM	DISTANCE LEG REMAINING	TIME POINT-POINT CUMULATIVE	TAKEOFF	GROUND SPEED
CHECK POINT					ETA	
					ATA	
DESTINATION						
			TOTAL			

PREFLIGHT CHECK LIST

DATE

EN ROUTE WEATHER/WEATHER ADVISORIES

DESTINATION WEATHER

WINDS ALOFT

ALTERNATE WEATHER

FORECASTS

NOTAMS/AIRSPACE RESTRICTIONS

b. Preferred IFR routes are established between busier airports to increase system efficiency and capacity.

1) Preferred IFR routes normally extend through one or more air route traffic control center (ARTCC) areas and are designed to achieve balanced traffic flows among high-density terminals.

2) IFR clearances are issued on the basis of these routes except when severe weather avoidance procedures or other factors dictate otherwise.

3) Preferred IFR routes are listed in the *Airport/Facility Directory* (*A/FD*).

a) If your flight is planned to or from an area having preferred IFR routes but the departure or arrival airport is not listed in the *A/FD*, you may use that part of a preferred IFR route that is appropriate for the departure or arrival airport listed.

4) Preferred IFR routes are correlated with instrument departure procedures (DPs) and standard terminal arrival routes (STARs) and may be defined by airways, jet routes, direct routes between navigational aids (NAVAIDS), waypoints, NAVAID radials/DME, or any combination.

c. Tower en route control (TEC) routes are designed to expedite traffic by linking designated approach control areas by a network of identified routes made up of the existing airway structure.

1) TEC routes are, generally, for nonturbojet aircraft operating at and below 10,000 ft. MSL.

2) TEC routes are entirely within the approach control airspace of multiple terminal facilities and are designed for relatively short flights of 2 hr. or less.

3) There are no unique requirements among pilots to use the TEC routes.

a) Pilots should include the acronym "TEC" in the remarks section of the IFR flight plan when requesting TEC.

4) All approach controls in the system may not operate to the maximum TEC altitude of 10,000 ft. MSL.

a) IFR flight may be planned to any satellite airport in proximity to the primary airport via the same routing.

d. Airway routes are depicted on your IFR en route chart.

e. Direct routing could be either flying from airport to airport using an area navigation (RNAV) system, such as GPS, or flying from VOR to VOR on a route other than an airway route.

1) Ensure that your equipment is capable of providing navigational guidance on the selected route, and identify the off-route obstruction clearance altitudes (OROCA) depicted on your IFR en route chart.

f. You must plan to maintain an altitude during your flight (except for takeoff or landing) that is above the published minimum IFR altitudes (FAR 91.177).

1) If no minimum altitudes are prescribed, you must remain at least 1,000 ft. (2,000 ft. in designated mountainous areas) above the highest obstacle within a horizontal distance of 4 NM from the course to be flown.

g. If both an MEA and an MOCA are prescribed for a particular route or route segment, you **may** operate below the MEA down to, but not below, the MOCA when within 22 NM of the VOR concerned (based on your reasonable estimate of that distance).

h. Altitudes (odd or even thousands) are normally selected the same as for Class G airspace as described on the following page.

i. In controlled airspace, you must maintain the altitude assigned by ATC, which may be different from the one you requested.

j. When operating under IFR in level cruising flight in Class G (uncontrolled) airspace, you must maintain the appropriate altitude, except while in a holding pattern of 2 min. or less, as shown below.

IFR ALTITUDES CLASS G AIRSPACE	
If your magnetic course (ground track) is:	And you are below 18,000 ft. MSL, fly:
0° to 179°	Odd thousands MSL, (3,000, 5,000, 7,000, etc.)
180° to 359°	Even thousands MSL (2,000, 4,000, 6,000, etc.)

k. A review of special-use airspace during IFR operation follows:

1) **Prohibited areas** contain airspace of defined dimensions within which flight is prohibited.

   a) These areas are depicted on en route charts.

2) **Restricted areas** contain airspace of defined dimension in which the flight of aircraft, while not wholly prohibited, is subject to restrictions.

   a) ATC applies the following procedures to aircraft operating on an IFR clearance (including VFR on top) via a route that lies within joint-use restricted airspace.

      i) If the restricted area is not active and has been released to the controlling agency (FAA), ATC will allow the aircraft to operate in the restricted area without issuing a specific clearance to do so.

      ii) If the restricted area (to include nonjoint-use airspace) is active, ATC will issue a clearance which will ensure that the aircraft avoids the restricted airspace unless it is on an approved altitude reservation mission or has obtained its own permission to operate in the airspace and informs ATC of that permission.

   b) Restricted airspace is depicted on en route charts.

      i) For joint-use restricted areas, the name of the controlling agency is shown.

      ii) For all prohibited or nonjoint-use restricted areas, unless otherwise requested by the using agency, the phrase "No A/G" is shown.

3) **MOAs** consist of airspace of defined dimensions established for the purpose of separating certain military training activities from IFR traffic.

   a) Whenever an MOA is being used, nonparticipating IFR traffic may be cleared through an MOA if IFR separation can be provided by ATC.

      i) Otherwise, ATC will reroute or restrict nonparticipating IFR traffic.

   b) MOAs are depicted on en route charts.

2. **Exhibit adequate knowledge of your airplane's performance capabilities by calculating the estimated time en route (ETE) and total fuel requirements based upon factors, such as the following:**
   a. **Power settings**
      1) Cruise power settings are obtained from your airplane's *POH*.
      2) The power setting, at your selected altitude, will determine your true airspeed (TAS) and fuel flow.
   b. **Operating altitude or flight level**
      1) When selecting an altitude, consider the following:
         a) Airplane performance and equipment
         b) Favorable winds
         c) Freezing level, icing, turbulence, and cloud tops
         d) Duration of the flight
      2) Your proposed altitude should be an IFR altitude that is above the minimum IFR altitude.
   c. **Wind.** Once you have selected your altitude, you can use the forecasted wind direction and speed to determine the ETE and your fuel requirements.
   d. **Fuel reserve requirements** are regulatory in nature (FAR 91.167), and you must determine whether your proposed flight can be conducted within the following requirements:
      1) If an alternate airport is not required
         a) Complete the flight to the first airport of intended landing, and
         b) Fly after that for 45 min. at normal cruise speed.
      2) If an alternate is required
         a) Complete the flight to the first airport of intended landing,
         b) Fly from that airport to the alternate airport, and
         c) Fly after that for 45 min. at normal cruise speed.
3. **Select and correctly interpret the current and applicable en route charts, instrument departure procedures (DPs), standard terminal arrival routes (STAR), and standard instrument approach procedure (IAP) charts.**
   a. Check to ensure that you have the correct en route and terminal procedures publication (which includes DPs, STARs, and IAPs) for your flight.
      1) If you are using NACO charts, check the effective dates to ensure that you have the current charts.
      2) If you are using Air Chart Systems or Jeppesen charts, ensure that you have checked the manufacturer's NOTAMs for changes.
   b. Review the en route chart to become familiar with your route, minimum IFR altitudes, and possible alternative.
      1) You must understand how to read the information and interpret the symbology used on the chart.

c. Review any DPs, STARs, and IAPs that you may use on your flight, at your departure, destination, and alternate airports.

1) The time to become familiar with the procedures is on the ground during your planning, not in the air.

2) Once again, you must be able to interpret all the information on the chart.

d. You should also check the *A/FD* for any preferred IFR routes.

e. Finally, you should have appropriate VFR charts with you for your trip.

1) VFR charts are useful in the event you cancel your IFR flight plan and continue the flight under VFR.

2) VFR charts are helpful in visualizing and locating unfamiliar airports.

3) VFR charts can provide you with information not available on IFR en route charts.

4. Obtain and correctly interpret applicable NOTAM information.

a. The National Notice to Airmen (NOTAM) System disseminates time-critical aeronautical information which either is of a temporary nature or is not sufficiently known in advance to permit publication on aeronautical charts or in other operational publications.

1) NOTAM information is that aeronautical information that could affect the decision to make a flight.

b. NOTAM information is classified into three categories.

1) **NOTAM (D)**, or distant, includes such information as airport or primary runway closures, changes in the status of navigational aids, ILSs, radar service availability, and other information essential to planned en route, terminal, or landing operations.

a) NOTAM (D) information is distributed automatically via the telecommunication system used to obtain a weather briefing.

i) These NOTAMs will remain available in this manner for the duration of their validity or until published.

2) **NOTAM (L)**, or local, includes such information as taxiway closures, personnel and equipment near or crossing runways, airport rotating beacon outages, and airport lighting that does not affect instrument approach procedure (IAP) criteria (e.g., VASI).

a) NOTAM (L) information is distributed locally only and is not attached to the hourly weather reports.

b) A separate file of local NOTAMs is maintained at each FSS for facilities in its area only.

i) NOTAM (L) information for other FSS areas must be specifically requested directly from the FSS that has responsibility for the airport concerned.

c) Direct User Access Terminal System (DUATS) vendors are not required to provide NOTAM (L) information.

3) A **Flight Data Center (FDC) NOTAM** is regulatory in nature and includes such information as amendments to published IAPs and other current aeronautical charts. It also advertises temporary flight restrictions caused by such things as natural disasters or large-scale public events that may generate a congestion of air traffic over a site.

a) FSSs are responsible for maintaining a file of current, unpublished FDC NOTAMs concerning conditions within 400 NM of their facilities.

b) FDC information that concerns conditions beyond 400 NM from the FSS or that is already published is given to you only when you request it.

c) DUATS vendors will provide FDC NOTAMs only upon site-specific requests using a location identifier.

c. The ***Notices to Airmen Publication (NTAP)*** is issued every 28 days and is an integral part of the NOTAM System. Once a NOTAM is published in the *NTAP*, the NOTAM is not provided during pilot weather briefings unless specifically requested by you.

1) The *NTAP* consists of two sections.

a) The first section contains NOTAMs (D) that are expected to remain in effect for an extended period and FDC NOTAMs that are current at the time of publication.

b) The second section contains special notices that either are too long or concern a wide or unspecified geographic area.

2) All information contained in the *NTAP* will be carried until the information expires, is canceled, or, in the case of permanent conditions, is published in other publications (e.g., *A/FD*, aeronautical charts, etc.).

3) The NTAP is also available on the Internet at:

http://www.faa.gov/NTAP/default.htm

d. If your airplane has an IFR-approved GPS receiver, you need to obtain any GPS NOTAMs and GPS RAIM aeronautical information during your preflight briefing with the FSS specialist.

5. Determine that the calculated performance is within your airplane's capability and operating limitations.

a. Throughout your cross-country flight planning process, you must be constantly checking to ensure that you remain within your airplane's capabilities and limitations as per your *POH*.

b. This should include weight and balance computations (for takeoff and landing); takeoff and landing distances required; and climb, cruise, and descent performance.

c. Avoid planning a flight into known or forecast conditions that may put undue stress on your airplane (e.g., turbulence) or violate your airplane's certification (e.g., flying into known icing conditions).

6. Complete and file a flight plan in a manner that accurately reflects the conditions of your proposed flight.

a. Prior to departure from within or prior to entering controlled airspace, you must submit a complete IFR flight plan and receive ATC clearance if weather conditions are below VFR minimums (FAR 91.173).

b. You should file your flight plan at least 30 min. prior to estimated time of departure.

c. Ask your examiner if you should actually file your flight plan for the practical test.

d. Your examiner may have you file an IFR flight plan for this cross-country flight to evaluate your ability to obtain and comply with ATC clearances.

7. Demonstrate adequate knowledge of Global Positioning Systems (GPS) and receiver autonomous integrity monitoring (RAIM) capability, when the aircraft is so equipped.

a. GPS is a U.S. satellite-based radio navigational, positioning, and time transfer system operated by the Department of Defense (DOD).

b. Authorization to conduct any GPS operation under IFR requires that

   1) Only specified GPS receivers (with a specific installation) are approved for IFR use.

   2) Airplanes navigating by IFR-approved GPS are considered to be RNAV airplanes, and you should file the appropriate equipment suffix on your flight plan.

      a) /Y for GPS with no transponder
      b) /C for GPS with transponder with no Mode C
      c) /I for GPS with transponder with Mode C
      d) /G for GPS with en route and terminal capability

         i) /G implies a transponder with Mode C.

   3) Airplanes using GPS under IFR must be equipped with an approved and operational alternate means of navigation appropriate to the flight.

      a) Active monitoring of alternate navigation is not required unless the receiver autonomous integrity monitoring (RAIM) capability is lost.

c. VFR and hand-held GPS units are not authorized for IFR navigation, for instrument approaches, or as a primary instrument flight reference.

   1) During IFR operations, VFR and hand-held GPS units may be considered only as aids to situational awareness.

d. The GPS receiver verifies the integrity (usability) of the signals received from the GPS satellites through RAIM to determine if a satellite is providing bad information.

   1) RAIM needs a minimum of five satellites in view, or four satellites and a barometric altimeter setting (baro-aiding), to detect a problem.

      a) To ensure baro-aiding, the current altimeter setting must be entered into the receiver as described in the operating manual.

   2) For receivers capable of doing so, RAIM needs six satellites in view (or five satellites with baro-aiding) to isolate the corrupt satellite signal and remove it from the navigation solution.

e. Without RAIM capability, you have no assurance of the accuracy of the GPS position.

END OF TASK -- END OF CHAPTER

CHAPTER II
PREFLIGHT PROCEDURES

This chapter explains the three tasks (A-C) of Preflight Procedures. Tasks A and B are knowledge only, while Task C includes both knowledge and skill. Your examiner is required to test you on all three tasks.

AIRCRAFT SYSTEMS RELATED TO IFR OPERATIONS

II.A. TASK: AIRCRAFT SYSTEMS RELATED TO IFR OPERATIONS

REFERENCES: 14 CFR Parts 61, 91; AC 61-27, AC 61-84.

Objective. To determine that the applicant exhibits adequate knowledge of the elements related to applicable aircraft anti-icing/deicing system(s) and their operating methods to include:

1. Airframe.
2. Propeller/intake.
3. Fuel.
4. Pitot-static.

A. General Information

1. The objective of this task is to determine your knowledge of airplane anti-icing/deicing system(s) and their operating methods.
2. For additional reading, see Part I/Chapter 10, Icing, of *Aviation Weather and Weather Services* for a 10-page discussion on types and location of icing and its effect on airplane performance and instruments.
3. You must be knowledgeable about the anti-icing/deicing system(s) on your airplane. This task covers Section 2 (to determine if flight into known icing conditions is permitted) and Section 7 of your airplane's *POH*.
   a. Section 2: Limitations
   b. Section 7: Systems Description
4. Two types of ice-protection systems are used (many airplanes use a combination of both systems as a total ice-protection system).
   a. Anti-icing equipment prevents ice formation, e.g., pitot heat.
   b. Deicing equipment removes ice after it has started accumulating, e.g., wing boots.
5. Your airplane may not have any anti-icing or deicing equipment; thus, you must be emphatic with your examiner that icing is to be avoided both in flight planning and in the air!
6. Here are a few specific points to remember about icing:
   a. Before takeoff, check weather for possible icing areas along your planned route. Check for pilot reports, and, if possible, talk to other pilots who have flown along your proposed route.
   b. If your aircraft is not equipped with deicing or anti-icing equipment, avoid areas of icing. Water (clouds or precipitation) must be visible, and outside air temperature must be near 0°C or colder for structural ice to form.
   c. Always remove ice or frost from airfoils before attempting takeoff.
   d. In cold weather, avoid, when possible, taxiing or taking off through mud, water, or slush. If you have taxied through any of these, make a preflight check to ensure freedom of controls.
   e. When climbing out through clouds where icing is possible, climb at an airspeed slightly faster than normal to avoid a stall in case icing does occur.

f. Use anti-icing equipment whenever the possibility for icing exists, before ice has begun to accumulate.

1) Deicing equipment must be continuously monitored for effectiveness. When such equipment becomes less than totally effective, change course or altitude to get out of the icing as rapidly as possible.

g. If your aircraft is not equipped with a pitot-static anti-icing system, be alert for erroneous readings from your ASI, VSI, and ALT.

h. In stratiform clouds, you can likely alleviate icing by changing to an altitude that has above-freezing temperatures or to one colder than −10°C. An altitude change also may take you out of clouds. Rime icing in stratiform clouds can be very extensive horizontally.

i. In frontal freezing rain, you may be able to climb or descend to a layer warmer than freezing. Temperature is always warmer than freezing at some higher altitude. If you are going to climb, move quickly; procrastination may leave you with too much ice. If you are going to descend, you must know the temperature and terrain below.

j. Avoid cumuliform clouds if at all possible. Clear ice may be encountered anywhere above the freezing level. Most rapid accumulations are usually at temperatures from 0°C to −15°C.

k. Avoid abrupt maneuvers when your aircraft is heavily coated with ice since the aircraft has lost some of its aerodynamic efficiency.

l. When "iced up," fly your landing approach with additional power.

m. It is ill-advised to change configuration, e.g., add flaps, when icing is severe.

7. Follow the recommended operating procedures in your *POH* for any anti-icing or deicing equipment in your airplane.

B. Task Objectives

1. **Airframe ice-protection systems**

a. Deicer boots are fabric-reinforced rubber sheets containing inflation tubes. They are normally cemented to the leading edges of the wings, vertical stabilizer, and horizontal stabilizer.

1) During normal operation, vacuum pressure holds the boots in the deflated position.

2) After ice accumulation of ¼ to ½ in. is present, the system is activated. A pneumatic pump controlled by a timer inflates segments of the boots. This inflation breaks off the accumulated ice.

b. The weeping wing system uses a special leading edge on the airplane's wings and stabilizers which is laser drilled with very small holes.

1) A deicing fluid is pumped out of these holes, causing any built-up ice to fall off. The remaining fluid on the flight surfaces deters further ice buildup.

c. Thermal anti-icing uses heated air flowing through passages in the leading edge of wings, stabilizers, and engine cowlings to prevent the formation of ice.

1) The heat source normally comes from combustion heaters in reciprocating-engine-powered airplanes and from engine bleed air in turbine-powered airplanes.

2. **Propeller/intake ice-protection systems**
   a. Two primary methods used in propeller ice protection
      1) Electric deicing
         a) The heating elements are enclosed in a rubber pad, which is normally cemented to the leading edge of the propeller blades, near the hub.
      2) Fluid (normally isopropyl alcohol) anti-icing
         a) Normally the fluid is released from a slinger ring assembly to the leading edge of the blades by centrifugal (i.e., outward) force.
   b. The most common intake ice protection is the use of heated air flowing around the engine cowling.
      1) You should be concerned that ice/water may be sucked into the induction system.
      2) Prevent induction system icing by use of carburetor heat or an alternate air source.
3. **Fuel system ice protection**
   a. Fuel system icing results from the presence of water in the fuel system. This condition may cause freezing of screens, strainers, and filters. When fuel enters the carburetor, the additional cooling may freeze the water.
   b. Normally, proper use of carburetor heat can warm the air sufficiently in the carburetor to prevent ice formation.
   c. Some airplanes are approved to use anti-icing fuel additives.
      1) Remember that an anti-icing additive is not a substitute for carburetor heat.
4. **Pitot-static ice-protection system**
   a. Pitot heat is an electrical system and may put a severe drain on the electrical system on some airplanes.
      1) When flying through visible moisture in temperatures near freezing, you should use the pitot heat, if your airplane is so equipped.
      2) Blockage of the pitot tube will affect only the airspeed indicator.
         a) If the pitot tube is blocked and its drain hole remains open, the air in the system will vent through the drain hole, and the airspeed indication will drop to zero.
            i) Thus, the instrument will sense no difference between the pitot and static pressure.
         b) If both the pitot tube and drain hole are blocked, the pitot pressure will not change, and the ASI will react as an altimeter.
            i) As altitude increases above the level where the blockage occurred, static pressure will decrease and cause an apparent increase in the pitot pressure. Thus, the airspeed indication will increase.
            ii) As altitude decreases below the level where the blockage occurred, static pressure will increase, and the airspeed indication will decrease.

b. Blockage of the static port by ice could affect the ASI, the ALT, and the VSI.

1) If the pitot tube remains open, the ASI will operate, but it will be inaccurate.

a) The airspeed indication decreases as the airplane climbs above the level where the blockage occurred because the trapped static pressure is higher than normal.

b) The airspeed indication increases as the airplane descends below the level where the blockage occurred because the trapped static pressure is lower than normal.

2) Since both the ALT and VSI operate only on the static system, the ALT will indicate the altitude at which the blockage occurred, and the VSI will indicate zero.

3) An alternate static source may be available in case of a blockage. In an unpressurized airplane, this alternate source normally measures the pressure in the cockpit, which is less than the external static pressure.

a) This will cause slight errors as the ALT and ASI will indicate slightly high, and the VSI will momentarily show a climb before returning to a level indication.

b) The glass on the vertical speed indicator can be cracked to supply cabin pressure to the static instruments.

i) The VSI is the most expendable static pressure instrument and the only one not required by the FARs.

END OF TASK

AIRCRAFT FLIGHT INSTRUMENTS AND NAVIGATION EQUIPMENT

II.B. TASK: AIRCRAFT FLIGHT INSTRUMENTS AND NAVIGATION EQUIPMENT

REFERENCES: 14 CFR Parts 61, 91; AC 61-27, AC 61-84, AC 90-48.

Objective. To determine that the applicant:

1. Exhibits adequate knowledge of the elements related to applicable aircraft flight instrument system(s) and their operating characteristics to include --

a. Pitot-static.
b. Altimeter.
c. Airspeed indicator.
d. Vertical speed indicator.
e. Attitude indicator.
f. Horizontal situation indicator.
g. Magnetic compass.
h. Turn-and-slip indicator/turn coordinator.
i. Heading indicator.
j. Electrical systems.
k. Vacuum systems.

2. Exhibits adequate knowledge of the applicable aircraft navigation system(s) and their operating characteristics to include --

a. VHF omnirange (VOR).
b. Distance measuring equipment (DME).
c. Instrument landing system (ILS).
d. Marker beacon receiver/indicators.
e. Transponder/altitude encoding.
f. Automatic direction finder (ADF).
g. Global positioning system (GPS).

A. General Information

1. The objective of this task is for you to demonstrate your knowledge of your airplane's flight instrument and navigation systems and their operating characteristics.

2. For additional reading, see *Pilot Handbook* for the following:

   a. Chapter 2, Airplane Instruments, Engines, and Systems, for a 22-page discussion on flight instrument systems and a three-page discussion on electrical systems.

   b. Chapter 10, Navigation Systems, for a 26-page discussion on various navigation systems and their operating characteristics.

3. To prepare for this task, systematically study, not just read, Sections 1, 7, and 9 of your *POH*.

   a. Section 1: General
   b. Section 7: Airplane and Systems Descriptions
   c. Section 9: Supplement (Optional Systems Description and Operating Procedures)

B. Task Objectives

1. **Exhibit adequate knowledge of the elements related to applicable aircraft flight instrument system(s) and their operating characteristics.**

   a. **Pitot-static system**

      1) The pitot-static system provides the source of atmospheric air pressure for operation of the

         a) Altimeter
         b) Vertical speed indicator
         c) Airspeed indicator

      2) The two major parts of the pitot-static system are

         a) The pitot pressure chamber and lines
         b) The static pressure chamber and lines

3) The pitot pressure (also called the ram, impact, or total pressure) is taken from a pitot tube, which is normally mounted on or beneath the leading edge of the left wing (so it can be seen easily by the pilot, especially in icing conditions) and aligned with the relative wind.
   a) The pitot line is connected only to the airspeed indicator.
   b) Some pitot tubes are equipped with an electric heating element to prevent ice from blocking the pitot tube.
   c) Pitot tubes also have a drain opening to remove water.
4) The static pressure (pressure of the still air) is usually taken from the static line attached to a vent or vents mounted flush with the side of the fuselage.
   a) The static pressure lines provide static air pressure to the altimeter, vertical speed indicator, and airspeed indicator.
5) An alternate source for static pressure is provided in most airplanes in the event the static ports become clogged. This source is selected manually with a valve and usually is vented to the inside of the cockpit.

b. Altimeter

1) The **pressure altimeter** in your airplane operates through the response of trapped air within the instrument to changes in atmospheric pressure.
   a) The pressure altimeter is a barometer that senses changes in atmospheric pressure and, through a gearing mechanism, converts the pressure to an altitude indication in number of feet.
      i) The conversion is based upon a fixed set of values known as the International Standard Atmosphere (ISA).
   b) Two essential facts apply:
      i) Conditions are rarely standard.
      ii) The altimeter presents you with standard information even when it senses nonstandard conditions.
   c) The basic component of the pressure altimeter is a stack of hollow, elastic metal aneroid wafers, which expand or contract as the atmospheric pressure changes, and through a shaft and gearing linkage, rotate the pointers (needles) on the dial of the instrument.
      i) For each pressure level, the aneroid wafers assume a definite size and cause the needles to indicate height above whatever pressure level is set into the altimeter setting window.
   d) The altimeter setting dial provides a means of adjusting the altimeter for nonstandard pressure.
   e) The altimeter setting dial is calibrated from 28.00 in. to 31.00 in. of Hg.
      i) Rotating the setting knob simultaneously rotates the scale and the altimeter pointers at a rate of 1 inch per 1,000 ft. of indicated change in altitude.
      ii) Setting the current altimeter setting in the window adjusts the altimeter to a desired indication for the size of the aneroid wafers at existing pressure.

f) The altimeter setting system provides you with the means that must be used to correct your altimeter for pressure variations.

i) The system is necessary to ensure safe terrain clearance for instrument approaches and landings and to maintain vertical separation between aircraft during instrument weather conditions.

2) The **encoding altimeter** operates in conjunction with the transponder in your airplane.

a) When the transponder is operated in the ALT (or Mode C) position, the encoding altimeter supplies the transponder reply code.

i) This code contains information on the airplane's altitude.

b) Encoding altimeters are available as indicating instruments or blind instruments.

i) An indicating instrument is basically a pressure altimeter that also contains the electronic circuitry necessary to produce the altitude code for the transponder.

ii) A blind instrument may be located anywhere in the airplane where electrical power and a static pressure line are available.

- The blind encoder has no external display. It functions only as an electronic sensor of static air pressure and provides the altitude code for the transponder.

c) The computer at the ground radar site as well as the encoding altimeter electronics are referenced to 29.92 in. of Hg.

i) The ground computer automatically corrects the altitude for the local barometric pressure before displaying the "MSL altitude" on the radar screen.

ii) Since the altimeter electronics are referenced to 29.92 in. of Hg, changing the altimeter setting does not change the altitude read-out viewed by ATC.

3) The **radar altimeter**, also known as a radio altimeter, provides a continuous indication of the airplane's height above ground.

a) The radar altimeter is a "down-looking" device that accurately measures the distance between the airplane and the highest object on the terrain.

i) The time interval between a transmitted radio signal and a received radio signal is automatically converted into an indication.

b) The radar altimeter may have a dial-type readout or a digital readout.

i) Some models may have a warning light and/or an aural tone to alert the pilot when the airplane reaches a preselected altitude.

c) The radar altimeter provides you rate information on the progress of the final approach and an accurate indication and warning when reaching the minimum descent altitude (MDA) or decision height (DH) during an instrument approach.

i) The radar altimeter also increases your situational awareness of ground proximity.

c. Airspeed indicator

1) The airspeed indicator (ASI) indicates the speed at which the airplane is moving through the air.

2) The ASI is a differential air pressure instrument that measures the difference between the total pressure (measured from the pitot line) and static pressure. This difference is called **dynamic pressure**.

   a) To measure the dynamic pressure, the ASI is constructed as a sealed case in which a diaphragm is mounted.

      i) The pitot line (total pressure) is connected to one side of the diaphragm.

      ii) The static line is connected to the other side of the diaphragm.

   b) As the airplane moves, total (or impact) pressure becomes greater than static pressure, causing the diaphragm to expand.

      i) Expansion or contraction of the diaphragm moves the indicator needle by means of gears and levers.

   c) The airspeed dial may be calibrated to convert dynamic pressure into units of knots (kt.), miles per hour (mph), or both.

   d) The ASI is calibrated to display an airspeed representative of a given dynamic pressure only at ISA sea-level values; thus it does not reflect changes in density altitude.

3) Indicated airspeed (IAS) is the direct instrument reading obtained from the ASI, uncorrected for variations in air density or installation and instrument errors.

   a) Your airplane's *Pilot's Operating Handbook (POH)* will list airspeed limitations and performance airspeeds based on IAS.

   b) The FARs and ATC will also use IAS for speed limitations.

d. Vertical speed indicator

1) The vertical speed indicator (VSI) indicates whether the airplane is climbing, descending, or flying level. The rate of climb or descent is indicated in feet per minute (fpm). If properly calibrated, the indicator will register zero in level flight.

2) Although the VSI operates solely from static pressure, it is a differential pressure instrument.

   a) The case of the instrument is airtight except for a restricted passage (also known as a calibrated leak) to the static line of the pitot-static system. The sealed case contains a diaphragm with connecting linkage and gearing to the indicator pointer. The diaphragm also receives air from the static line, but this is not a restricted passage.

   b) When the airplane is on the ground or in level flight, the pressures inside the diaphragm and the instrument case remain the same, and the pointer indicates zero.

   c) When the airplane climbs or descends, the pressure inside the diaphragm changes immediately. But the restricted passage causes the pressure of the rest of the case to remain higher or lower for a short time. This differential pressure causes the diaphragm to contract or expand. The movement of the diaphragm is indicated on the instrument needle as a climb or a descent.

3) Limitations in the use of the VSI are due to the calibrated leak.

a) Sudden or abrupt changes in the airplane's altitude causes erroneous instrument readings as the air flow fluctuates over the static ports.

b) Both rough control technique and turbulence result in unreliable indications.

4) When used properly, the VSI provides reliable information to establish and maintain level flight and rate climbs and descents.

e. **Attitude indicator**

1) The attitude indicator (AI), with its miniature aircraft and horizon bar, depicts the attitude of the airplane.

a) The relationship of the miniature airplane to the horizon bar is the same as the relationship of the real airplane to the actual horizon.

b) The instrument gives an instantaneous indication of even the smallest changes in attitude.

c) In most light airplanes, the AI is powered by the vacuum system.

2) The gyro in the attitude indicator is mounted on a horizontal plane and depends upon rigidity in space for its operation. The horizon bar is fixed to the gyro. It remains in a horizontal plane as the airplane is pitched or banked about its lateral or longitudinal axis. The dial (banking scale) indicates the bank angle.

3) The AI is highly reliable and the most realistic flight instrument on the instrument panel. Its indications are very close approximations of the actual attitude of the airplane.

f. **Horizontal situation indicator** (HSI) -- a combination of the heading indicator (HI) and the VOR/ILS indicator

1) An HSI is illustrated below and discussed on pages 77 and 78.

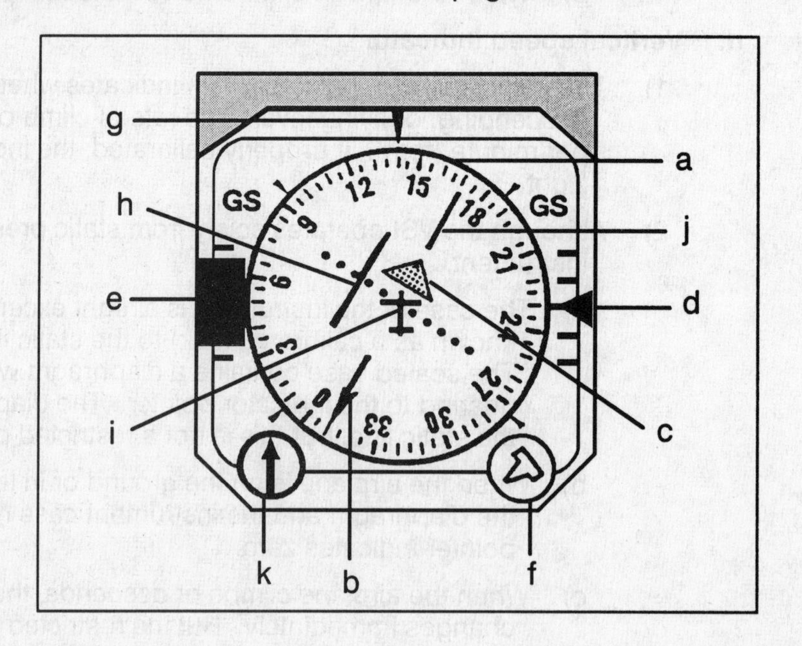

a) The azimuth card -- a circular dial that rotates to show the heading under the index at the top of the instrument

i) The azimuth card may be part of a remote indicating compass (RIC), or

ii) The azimuth card must be checked against the magnetic compass and reset with a heading set knob.

b) The course indicating arrow -- the VOR (OBS) indicator

c) The TO/FROM indicator for the VOR -- an indicator that shows whether the selected course is TO or FROM the station

d) Glide slope deviation pointer -- an index that indicates above or below the glide slope, which is the longer center line.

e) Glide slope warning flag -- an indicator that comes out when reliable signals are not received by the glide slope receiver

f) Heading set knob -- an adjustment knob used to coordinate the heading indication (g) with the actual compass

i) If the azimuth card is part of an RIC, this is normally a heading bug (pointer) set knob that moves a bug around the periphery of the azimuth card for use with an autopilot.

g) Lubber line -- an index at the top of the HSI that shows the current heading

h) Course deviation bar -- an indicator that shows the direction you would have to turn to intercept the desired radial if you were on the approximate heading of the OBS selection

i) The airplane symbol -- a fixed symbol that shows the airplane relative to the selected course as though you were above the airplane looking down

j) The tail of the course indicating arrow -- the reciprocal of the OBS heading (b)

k) The course setting knob -- a knob used to adjust the OBS

2) By combining the HI and the VOR indicator, the HSI provides the pilot with a concise navigational picture in one instrument. It reduces pilot workload by decreasing the required scan and instrument interpretation.

3) The HSI azimuth card is often part of a remote indicating compass (RIC) system. The system consists of a magnetic slaving transmitter and a directional gyro unit.

a) Once the system is on and fully stabilized, the azimuth card should present correct heading throughout the flight without the need to periodically reset the HI, as on normal HI systems.

b) If the slaved HI does precess from the correct heading, the system (depending on models) provides two ways of resetting through the use of a slaving control and compensator unit.

i) To reset many systems, the pilot selects a free gyro mode, which renders the card just like a normal HI, and the heading is adjusted accordingly. Once reset, the unit is placed back in the slaved gyro position.

ii) Other systems provide a slaving meter and two manual heading drive buttons. The slaving meter indicates the difference between the displayed heading and the magnetic heading.

- To reset the card, place the system in the free gyro mode, and press the appropriate heading drive button. If the needle is to the left, press the left drive button; if to the right, press the right drive button. Once reset, the system is placed back in the slaved gyro position.

4) On advanced HSI systems (i.e., those found in high-performance aircraft), additional information may be displayed.

a) The course selected is shown in digital format.

b) DME distance information is also shown in digital format.

c) Speed variations (in reference to a set bug speed) are shown visually through a fast/slow format.

d) Changeable CDI input allows the display to relay NDB, GPS, INS (inertial navigation system), or other navigational information in addition to the VOR/ILS.

e) For approaches beyond CAT I, the course deviation bar has tighter tolerance indications, i.e., is more sensitive.

5) Flying with an HSI is very easy.

a) The course-indicating arrow is set to the desired OBS setting just as on a VOR. The difference is that the HSI will pictorially show you where the course is and where you are in relation to that course. The course deviation bar will show off-course indications in dot format just as a VOR does.

b) Your current heading is always displayed under the top index (lubber line). You can thus always have a pictorial representation of your airplane and its relation to the desired radial.

c) Flying with the HSI is simply a matter of selecting the desired course or radial, intercepting it just as for any OBS, and keeping the needle centered just as for any OBS. The benefit is that there is no need to interpret two instruments. The VOR, HI, TO/FROM indicator, and localizer/glide-slope functions are all combined into one instrument.

g. Magnetic compass

1) The magnetic compass (the only direction-seeking instrument in the airplane) is used primarily to set the heading indicator prior to flight and to verify its continued accuracy during flight. It contains two steel magnetized needles fastened to a float around the edge of which is mounted a compass card.

a) The needles are parallel, with their north-seeking ends pointed in the same direction.

2) The float assembly, consisting of the magnetized needles, compass card, and float, is mounted on a pedestal and sealed in a chamber filled with white kerosene.

a) This fluid decreases oscillations and lubricates the pivot point on the pedestal, and, due to buoyancy, part of the weight of the card is taken off the pivot that supports the card.

b) The pedestal is the mount for the float assembly. The float assembly is balanced on the pivot, which allows free rotation of the card and allows it to tilt at an angle of up to 18°.

3) The magnetic compass is also important as a standby, or an emergency directional indicator, in the event the heading indicator fails.

h. Turn-and-slip indicator/turn coordinator

1) The turn coordinator (TC) is a type of turn indicator commonly used in airplanes to indicate rate and quality of turn and to serve as an emergency source of bank information if the attitude indicator fails.

a) The TC is actually a combination of two instruments: a miniature airplane and an inclinometer (or ball).

i) The miniature airplane is gyro-operated to show the rate of turn.

ii) The inclinometer reacts to gravity and/or centrifugal force to indicate the need for rudder to maintain coordinated flight.

b) The miniature airplane is connected to a gyro (usually driven by electricity). Its design tilts the gimbal axis of the gyro up about 30° so that the gyro precesses in reaction to movement about both the vertical (yaw) and the longitudinal (roll) axes.

i) This precession allows the TC to show rate of roll as well as rate of turn.

ii) The TC indicates direction of roll or yaw and rate of turn.

- The TC does not give a direct indication of the banked attitude of the airplane.
- The miniature airplane will show a turn in a wings-level yaw or during a turn while taxiing.

c) When the turn needle points to one of the small side marks, it indicates that the airplane is turning at a standard rate, i.e., 3° per sec.

d) The inclinometer of the turn-and-slip indicator consists of a sealed, curved glass tube containing kerosene and a ball that is free to move inside the tube. The fluid provides a dampening action, which ensures smooth and easy movement of the ball.

i) The ball then is a visual aid to determine coordinated use of the aileron and rudder control. During a turn, it indicates the quality of the turn, i.e., whether the airplane has the correct rate of turn for the angle of bank.

2) The turn-and-slip indicator (T&SI) is another type of turn indicator used in some older airplanes. The T&SI has a needle instead of a miniature airplane and indicates movement only around the vertical (yaw) axis, not the longitudinal (roll) axis.

a) The turn needle indicates only the rate at which the airplane is rotating about its vertical axis.

b) The needle will deflect in a wings-level yaw or during a turn while taxiing.

c) The inclinometer in the T&SI works in the same manner as in the TC.

i. Heading indicator

1) The heading indicator (HI) commonly used in light aircraft is a gyroscopic instrument that has no direction-seeking properties and must be set to headings shown on the magnetic compass.
   a) Thus, the HI is fundamentally a mechanical instrument designed to facilitate the use of the magnetic compass.
   b) Errors in the magnetic compass are numerous, making straight flight and turns to headings difficult to accomplish, especially in turbulent air.
      i) A HI is not affected by the forces that make the magnetic compass difficult to interpret.
2) The HI is normally powered by the vacuum system.
3) Operation of the HI depends upon the gyroscopic principle of rigidity in space.
   a) The rotor turns in a vertical plane. Fixed to the rotor is a compass card.
   b) Since the rotor remains rigid in space, the points on the card hold the same position in space relative to the vertical plane.
   c) Once the HI is operating (i.e., minutes after the airplane engine is started), the compass card must be set to the heading shown on the magnetic compass.
4) Because of precession, caused chiefly by bearing friction or improper vacuum pressure, the HI may creep or drift from a heading to which it is set.
   a) Among other factors, the amount of drift depends largely upon the condition of the instrument. If the bearings are worn, dirty, or improperly lubricated, drift may be excessive.
   b) For accuracy, the HI should be compared to the magnetic compass at 15-min. intervals or after prolonged turns.
      i) This comparison can be done accurately only when the airplane is in straight, level, and unaccelerated flight.

j. Electrical systems (source of power for gyro operation)

1) The principal value of the electric gyro in light airplanes is a safety factor.
2) In single-engine airplanes equipped with vacuum-driven attitude and heading indicators, the turn coordinator (or turn-and-slip indicator) is commonly operated by an electric gyro.
   a) In the event of a vacuum system failure and loss of two gyro instruments, you still have a reliable instrument for emergency operation.
3) The electric gyro is operated on current directly from the battery. The turn coordinator is reliable as long as current is available, regardless of alternator or vacuum system malfunction.

k. Vacuum systems

1) The vacuum system spins the gyros by sucking a stream of air against the rotor vanes to turn the rotor at high speed, essentially as a water wheel or turbine operates.
   a) The attitude and heading indicators are normally operated by the vacuum system.
2) Air at atmospheric pressure drawn into the instruments through a filter(s), drives the rotor vanes and is sucked from the instrument case through a line to the vacuum pump and vented into the atmosphere.

NOTE: While not specifically mentioned in this task, the following three systems warrant your attention.

l. **Radio magnetic indicator (RMI)** (not specifically mentioned in this task, but considered valuable enough by your author to warrant inclusion here)

1) The RMI is a single instrument display that combines three navigational instruments to provide you with an easy-to-read orientation system.

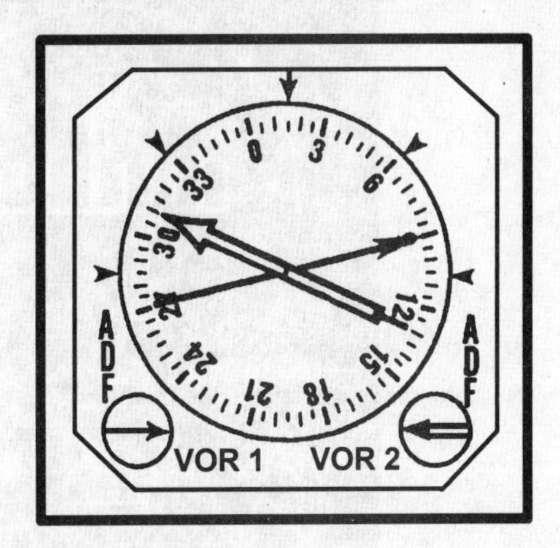

2) The instrument is a slaved compass card that has two bearing pointers, each of which can be set to an ADF or a VOR superimposed over it.

a) The head of each needle will always point to an NDB or a VOR.

b) The slaving system rotates the compass card, maintaining the current heading under the top index (without precession).

3) Unlike a CDI, however, the VOR needle has no TO/FROM indicator and there are no deviation dots. The head of the needle simply shows the relationship of the VOR to your airplane's heading. The tail shows the radial your airplane is on.

4) The best way to understand and use an RMI is to compare it to an ADF. It presents the same type of information (i.e., the needle will always point to the station). The difference is that you can select whether the station is a VOR or an NDB. Since the RMI is also tied into a compass slaving system that keeps the current heading under the top index, a pictorial display is provided of the aircraft's position in relation to the station or desired radial or course.

m. **Remote indicating compass (RIC)**

1) The RIC combines the functions of the magnetic compass and the gyro-powered heading indicator.

2) The two panel-mounted components of a RIC are the heading indicator and the slaving control and compensator unit.

a) The heading indicator is normally an HSI.

b) The slaving control and compensator unit (shown below) has a push button that provides the pilot a means of selecting either the "slaved gyro" or "free gyro" mode.

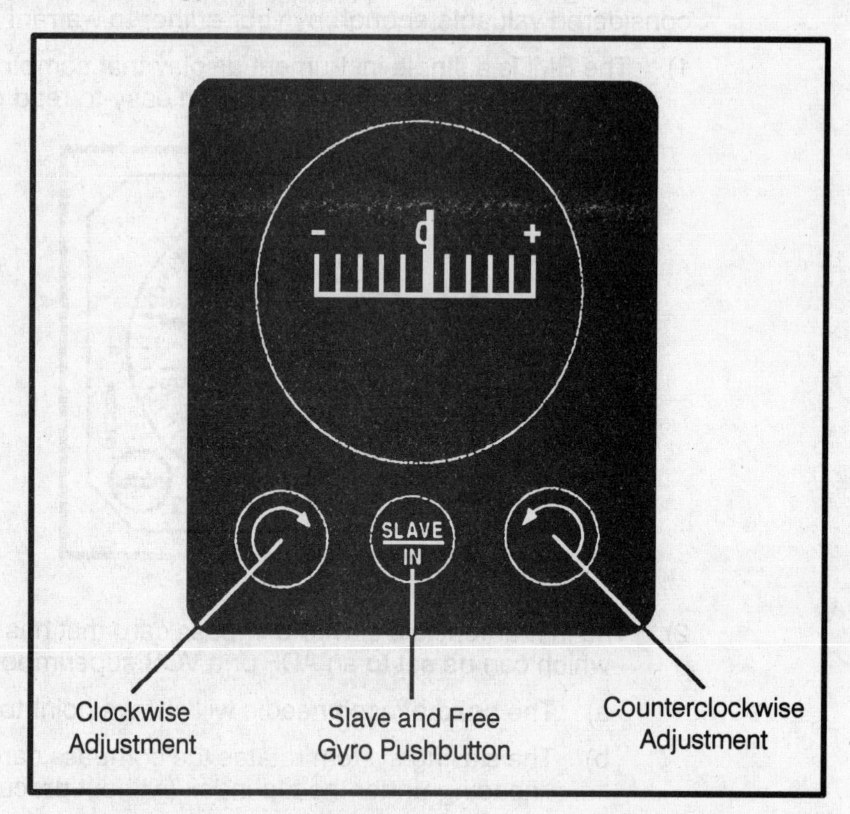

i) The RIC is normally operated in the "slaved gyro" mode.

ii) When in the "free gyro" mode, the compass card on the HSI may be manually adjusted by depressing the appropriate heading drive button on the HSI.

3) The slaving control and compensator unit also has a slaving meter and two manual heading drive buttons.

a) This slaving meter contains a slaving meter needle, which indicates the difference between the displayed heading and the actual magnetic heading.

i) A right deflection (+) indicates a clockwise (right) error in the heading indicator compass card (i.e., the correct magnetic heading is to the right of the indicated heading).

- Depressing the right (counterclockwise) heading drive button will move the heading indicator compass card to the left, thus increasing (+) the indicated heading toward the correct value (i.e., from 180° to 190°).

ii) A left deflection (–) indicates a counterclockwise (left) error in the heading indicator compass card (i.e., the correct magnetic heading is to the left of the indicated heading).

- Depressing the left (clockwise) heading drive button will move the heading indicator compass card to the right, thus decreasing (–) the indicated heading toward the correct value (i.e., from 190° to 180°).

b) To make corrections to the RIC, the system must be placed in the free gyro mode.

i) After corrections are made, the system is returned to the slaved mode, which is the normal mode of operation.

c) Whenever the aircraft is in a turn and the card rotates, the slaving meter will show a full deflection to one side or the other.

4) The magnetic slaving transmitter and directional gyro units are remotely mounted, thus giving the instrument the name of remote indicating compass.

a) The magnetic slaving transmitter is usually mounted in a wingtip to eliminate the possibility of magnetic interference, and this unit contains the flux valve, which is the direction-sensing device of the system.

i) The flux valve detects the lines of magnetic force (direction) in the unit.

ii) The concentration of lines of magnetic force, after being amplified, is relayed electronically to the directional gyro unit.

b) The signals received by the directional gyro unit operate a torque motor.

i) The torque motor precesses the gyro unit until it is aligned with the transmitter signal.

ii) The torque motor is connected electronically to the HSI.

n. Flight director system (integrated flight system)

1) A flight director system consists of electronic components that compute and indicate the airplane's attitude (pitch and bank) to attain and maintain a preselected flight condition.

a) Command indicators on the attitude indicator tell the pilot in which direction and how much to change the airplane's attitude to achieve the desired result.

b) The computed command indications relieve the pilot of many of the mental calculations required for instrument flight.

2) A flight director/autopilot system will normally consist of a mode controller, a flight command indicator (attitude indicator), an HSI, and an annunciator panel.

2. Exhibit adequate knowledge of the applicable aircraft navigation system(s) and their operating characteristics.

a. VHF omnirange (VOR)

1) The VOR equipment in your airplane includes an antenna, a receiver with a tuning device, and a VOR navigation instrument.

a) VOR signals from the ground station are received through the antenna to the receiver, which interprets and separates the navigation information. Then this information is displayed on the navigation instrument.

2) The VOR navigation instrument consists of

a) An **omnibearing selector (OBS)**, sometimes referred to as the course selector

i) By turning the OBS knob (lower left of the diagram below), the desired course is selected. In the diagram below, the course is under the index at the top of the instrument, i.e., 360°.

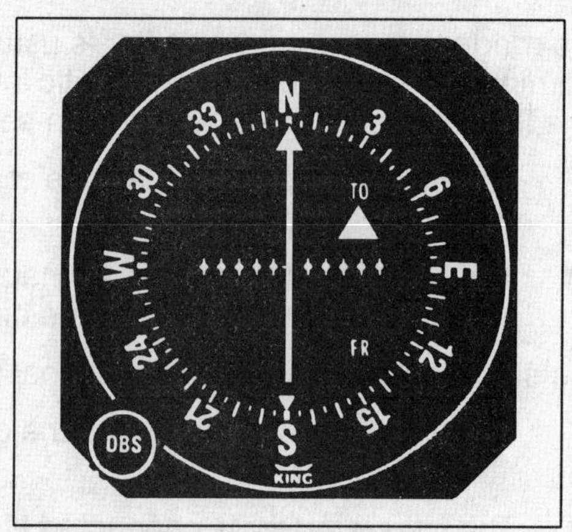

Source: AlliedSignal General Aviation Avionics

b) A **course deviation indicator (CDI)**, referred to as the needle

i) The CDI needle is hinged to move laterally across the face of the instrument.

ii) It indicates the position of the selected course relative to your airplane.

iii) The CDI needle centers when your airplane is on the selected radial, as shown in the diagram above.

c) A **TO/FROM indicator**, also called a sense indicator or ambiguity indicator

i) The TO/FROM indicator shows whether the course selected in the OBS will take your airplane TO or FROM the station.

- It does not indicate whether the airplane is heading to or from the station.

3) Most VOR ground stations have a designated standard service volume (SSV).

a) The SSV defines the reception limits of VORs, which are usable for random/unpublished route navigation.

b) SSV limitations do not apply to published IFR routes or procedures.

c) The SSV classification of a VOR is found in the *Airport/Facility Directory*.

i) Restrictions to service volumes are first published as a NOTAM and then with the alphabetical listing of the NAVAID in the *Airport/Facility Directory*.

4) The SSV altitude and range boundaries are described in the chart below and graphically shown on page 87.

SSV Class Designator	Altitudes (AGL)	Distance (NM)
T (Terminal) -- See top figure on page 87.	1,000 ft. to 12,000 ft.	25
L (Low Altitude) -- See middle figure on page 87.	1,000 ft. to 18,000 ft.	40
H (High Altitude) -- See bottom figure on page 87.	1,000 ft. to 14,500 ft. 14,500 ft. to 18,000 ft. 18,000 ft. to 45,000 ft. 45,000 ft. to 60,000 ft.	40 100 130 100

b. Distance measuring equipment (DME)

1) The DME in the airplane includes a transceiver and a small shark fin-type antenna. The DME display is on the face of the transceiver and may be part of the VOR receiver or a separate unit.

2) In the operation of DME, your airplane first transmits a signal (interrogation) to the ground station. The ground station (transponder) then transmits a signal back to your airplane.

a) The DME in your airplane records the round-trip time of this signal exchange. From this it can compute

i) Distance (NM) to the station
ii) Groundspeed (kt.) relative to the station
iii) Time (min.) to the station at the current groundspeed

3) The mileage readout is the direct distance from the airplane to the DME ground facility. This is commonly referred to as **slant-range** distance.

a) The difference between a measured distance on the surface and the DME slant-range distance is known as slant-range error.

i) Slant-range error is smallest at low altitude and long range.

ii) This error is greatest when the airplane is at a high altitude close to or over the ground station, at which time the DME receiver will display altitude in NM above the station.

iii) Slant-range error is negligible if the airplane is 1 NM or more from the ground facility for each 1,000 ft. of altitude above the elevation of the facility.

4) To use the groundspeed and/or time-to-station function of the DME, you must be flying directly to or from the station.

a) Flying in any other direction will provide you with false groundspeed and time-to-station information.

c. **Instrument landing system (ILS)**

1) The ILS is designed to provide an approach path for exact alignment and descent of an aircraft on final approach to a runway.

2) The ground equipment consists of two highly directional transmitting systems (the localizer and glide slope) and, along the approach, three (or fewer) marker beacons.

3) The system may be divided functionally into three parts:

a) Guidance information -- localizer, glide slope

b) Range information -- marker beacon, DME, compass locator (NDB)

c) Visual information -- approach lights, touchdown and centerline lights, runway lights

4) The localizer provides course guidance to the runway centerline.

a) The approach course of the localizer is called the **front course** and is used with other functional parts.

i) The localizer antenna is located at the far end of the runway and is adjusted for a course width of 3° to 6° (i.e., full scale left to a full scale right of 700 ft. at the runway threshold).

b) The course line along the extended centerline of a runway, in the opposite direction from the front course, is called the **back course.**

c) The localizer provides course guidance throughout the descent path to the runway threshold from a distance of 18 NM from the antenna and from an altitude between 1,000 ft. above the highest terrain along the course line and 4,500 ft. above the elevation of the antenna site.

i) Proper off-course indications are provided throughout the angular areas of the operational service volume, as shown below. Unreliable signals may be received outside of these areas.

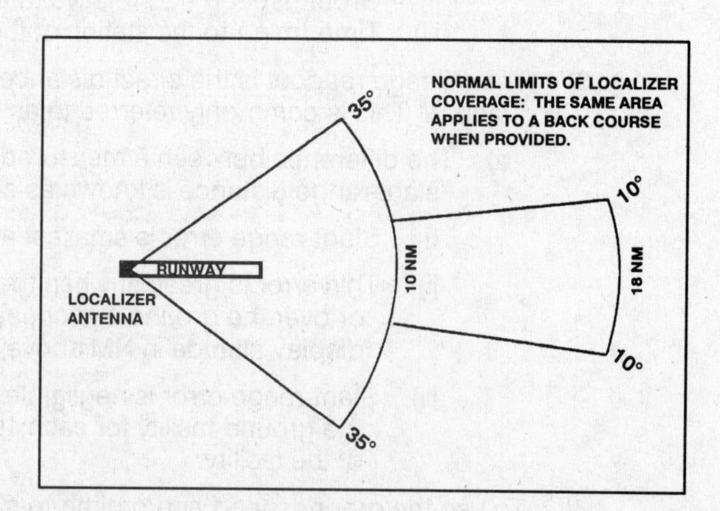

d) The localizer frequency is tuned into your VOR receiver, and the CDI provides course correction information.

i) Unless your airplane is equipped with an HSI, you will need to correct away from the CDI (i.e., reverse sensing) when flying inbound on the back course and outbound on the front course.

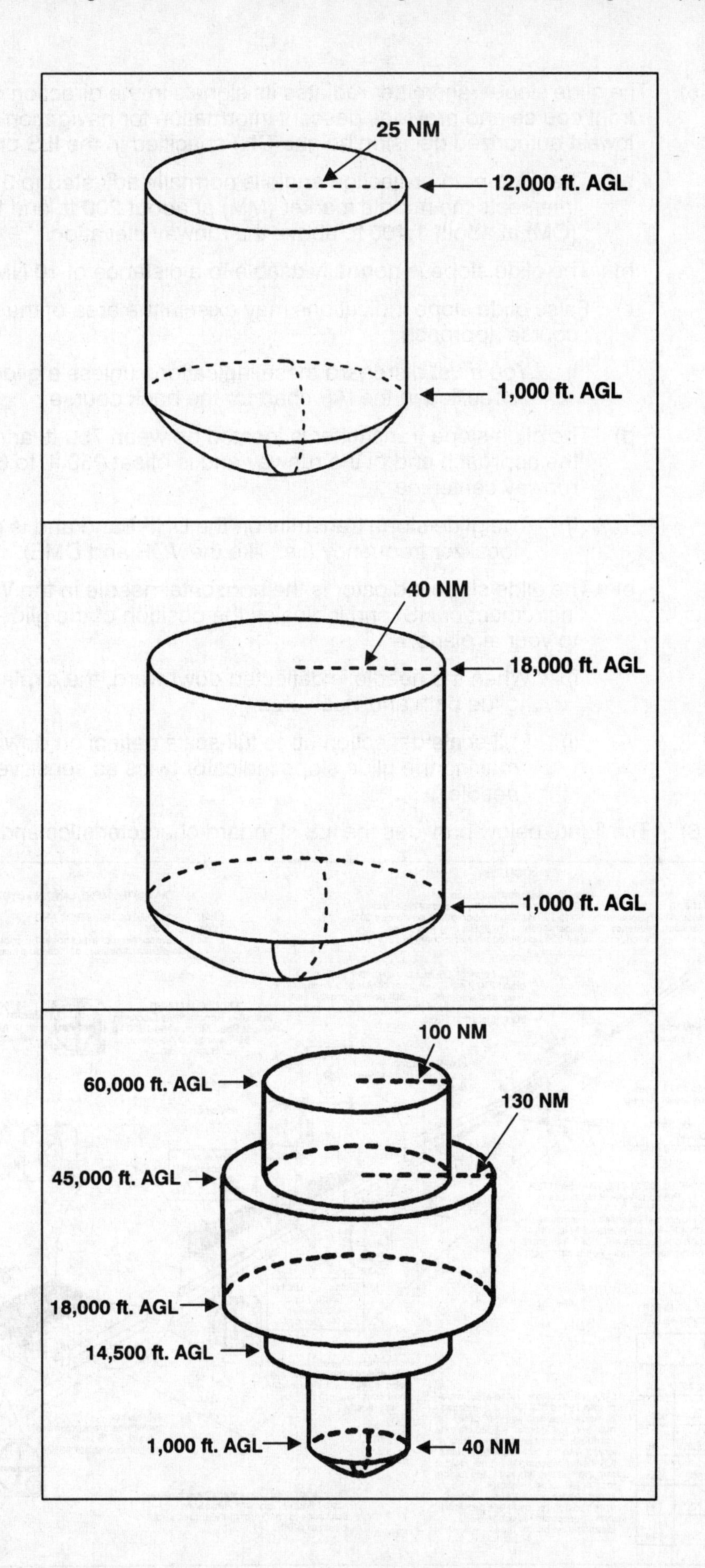
25 NM
12,000 ft. AGL
1,000 ft. AGL
40 NM
18,000 ft. AGL
1,000 ft. AGL
100 NM
60,000 ft. AGL
130 NM
45,000 ft. AGL
18,000 ft. AGL
14,500 ft. AGL
1,000 ft. AGL
40 NM

5) The glide slope transmitter radiates its signals in the direction of the localizer front course and provides descent information for navigation down to the lowest authorized decision height (DH) specified in the ILS procedure.

   a) The glide path projection angle is normally adjusted to 3° so that it intersects the middle marker (MM) at about 200 ft. and the outer marker (OM) at about 1,400 ft. above the runway elevation.

   b) The glide slope is normally usable to a distance of 10 NM.

   c) False glide slope indications may exist in the area of the localizer back course approach.

      i) You must disregard these indications unless a glide slope is specified in the IAP chart for the back course.

   d) The glide slope transmitter is located between 750 ft. and 1,250 ft. from the approach end of the runway and is offset 250 ft. to 650 ft. from the runway centerline.

      i) The glide slope transmits on the UHF band and is paired to the localizer frequency (i.e., like the VOR and DME).

   e) The glide slope indicator is the horizontal needle in the VOR navigation instrument or HSI and indicates the position of the glide slope in relation to your airplane.

      i) When the needle is deflected downward, the airplane is above the glide path and vice versa.

      ii) Full-scale deflection up to full-scale deflection down is 1.4°, thus making the glide slope indicator twice as sensitive as the localizer needle.

6) The figure below provides the ILS standard characteristics and terminology.

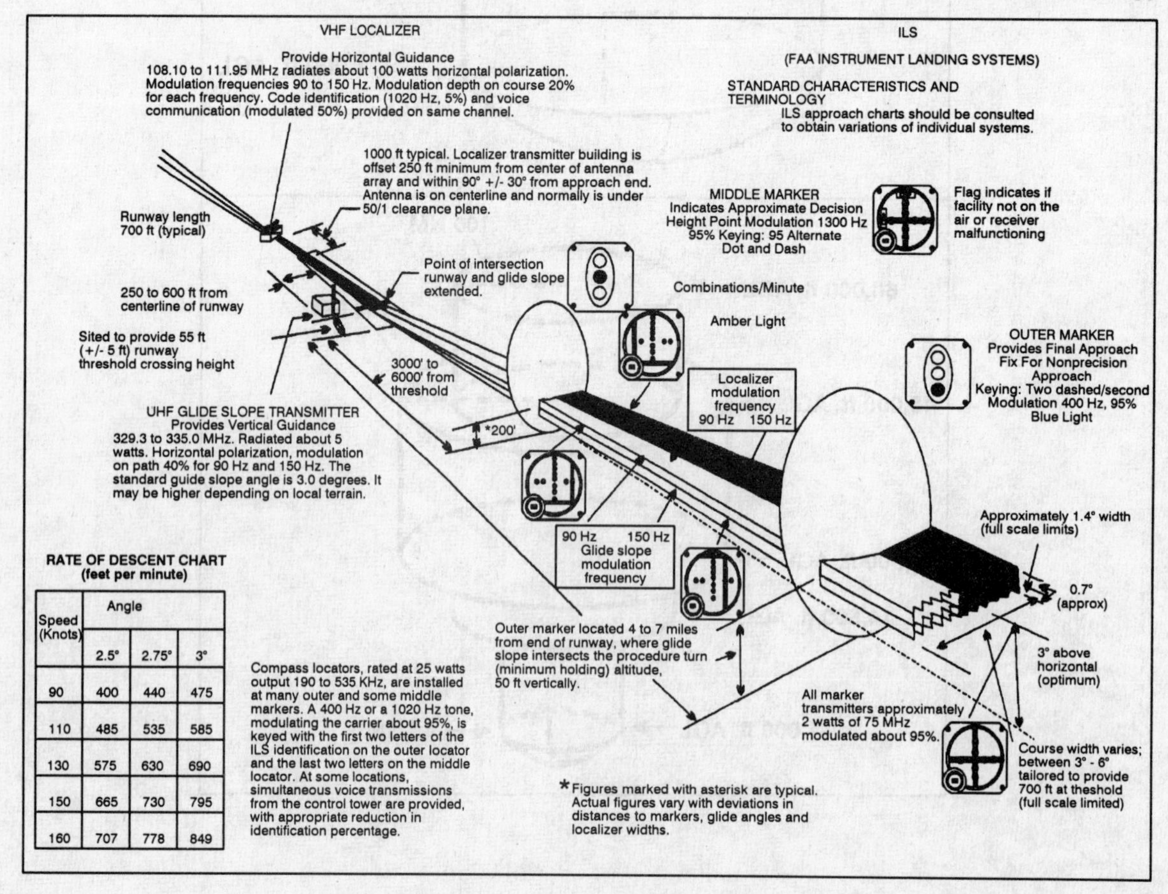

Speed (Knots)	Angle 2.5°	2.75°	3°
90	400	440	475
110	485	535	585
130	575	630	690
150	665	730	795
160	707	778	849

d. **Marker beacon receiver/indicators**

1) The marker beacon receiver in your airplane is permanently tuned to 75 MHz. All marker beacons used in instrument approaches transmit on this frequency.

a) Passage over a marker beacon is indicated by an audio tone and a light, which are activated by the frequency modulation.

b) If your marker beacon receiver has a selective sensitivity feature, it should be operated in the "low" sensitivity position for proper reception of ILS marker beacons.

i) Low sensitivity provides a more accurate indication of when you pass over the marker beacon.

ii) Some pilots prefer to use the high sensitivity position, which provides a better chance of receiving a weak signal or a strong signal with a weak receiver.

2) Normally two marker beacons are associated with an ILS, the OM and the MM. A category (CAT) II or III ILS will also have an inner marker (IM).

a) The OM normally indicates the approximate position at which the airplane at the appropriate altitude on the localizer course will intercept the ILS glide path.

i) Normally intercepted at 4 to 7 NM from the threshold

ii) Identified by continuous dashes at the rate of 2/sec. and by a blue light

b) The MM indicates a position approximately 3,500 ft. from the landing threshold.

i) When on the glide path, your airplane will be about 200 ft. above the touchdown zone elevation (TDZE).

ii) MM can be identified by alternate dots and dashes at a rate of 95 dot/dash combinations per min. and by an amber light.

c) The IM will indicate a point at which an aircraft is at a designated DH on the glide path between the MM and the landing threshold during a CAT II ILS approach.

i) Identified by continuous dots at a rate of 6/sec. and by a white light

3) Some localizer back courses have a back course (BC) marker that indicates the final approach fix (FAF) where approach descent is commenced.

a) Identified by two dots at a rate of 72 to 75 two-dot combinations per minute and by a white light

4) Summary of marker beacon indications

Marker	Code	Light
OM	– – –	Blue
MM	. – . –	Amber
IM	. . . .	White
BC	. . . .	White

e. Transponder/altitude encoding

1) The transponder in your airplane is the airborne radar beacon receiver/transmitter portion of the Air Traffic Control Radar Beacon System (ATCRBS).

a) The transponder automatically receives a signal (interrogation) from ATC's ground-based radar and replies with a specific pulse or pulse group only to those interrogations being received to which it is set to respond.

i) When the transponder is in the "ON" position, the transponder replies to Mode A (no altitude) interrogations.

ii) When the transponder is in the "ALT" position, the transponder replies to Mode C (altitude-reporting) interrogations.

b) When operating under IFR, ATC will provide you with the transponder code to be set in the transponder.

2) The Mode C (automatic altitude-reporting) system converts your airplane's altitude in 100-ft. increments to coded digital information, which is transmitted in the reply to the interrogating radar facility.

a) If your airplane is Mode C-equipped, you must set your transponder to reply Mode C (i.e., set function switch to ALT) unless ATC requests otherwise.

i) If ATC requests that you "STOP ALTITUDE SQUAWK," you should set the function switch from ALT to ON.

b) An instruction by ATC: "STOP ALTITUDE SQUAWK, ALTITUDE DIFFERS (number of feet) FEET," may be an indication that your transponder is transmitting incorrect altitude information or that you have an incorrect altimeter setting. Plus or minus 300 feet is the normal limit of error that ATC will accept.

i) The encoding altimeter equipment of the Mode C function is preset to a setting of 29.92. Computers at the radar facility correct for current altimeter settings and display indicated altitudes on the radarscope.

ii) Although an incorrect altimeter setting has no effect on the Mode C altitude information transmitted by the transponder, it will cause you to fly at a different altitude.

iii) When a controller indicates that an altitude readout is invalid, you should check to verify that your airplane's altimeter is set correctly.

c) For IFR operations in controlled airspace, the automatic pressure altitude-reporting system (Mode C) must be tested and inspected every 24 calendar months.

f. Automatic direction finder (ADF)

1) Some airplanes are equipped with an ADF radio, which receives radio signals in the low-to-medium frequency bands of 190 kHz to 1750 kHz.

a) Two types of ground stations may be used with the ADF.

i) Nondirectional radio beacons (NDB), which operate in the frequency band of 190 to 535 kHz

ii) Commercial broadcast (AM) radio stations, which operate in the frequency band of 540 to 1620 kHz

2) The equipment in the airplane includes two antennas, a receiver with a tuning device, and a navigational display.

a) The two antennas are the loop antenna and the sense antenna.

i) A loop antenna is used as the directional antenna.

- The loop antenna determines the direction in which the signal is the strongest, but it cannot determine whether the station is in front of or behind the airplane (known as loop ambiguity).

ii) The sense antenna is nondirectional and allows the ADF to solve the problem of loop ambiguity, thus enabling the ADF to determine the direction of the signal.

b) The receiver allows you to tune the correct frequency and function selectors.

c) The navigational display consists of a dial upon which the azimuth (0° to 360°) is printed and a needle that rotates around the dial and points to the station to which the receiver is tuned.

i) Some ADF dials can be rotated to align the azimuth with the airplane heading. This is called a **movable card indicator**.

ii) Other ADF dials are fixed, with the 0° - 180° points on the azimuth aligned with the longitudinal axis of the airplane. On these dials, the 0° position on the azimuth represents the nose of the airplane. This is called a **fixed-card indicator**.

g. Global Positioning System (GPS)

1) GPS operation is based on the concept of ranging and triangulation from a group of satellites in space that act as precise reference points.

a) A GPS receiver measures distance from a satellite using the travel time of a radio signal.

i) Each satellite transmits a specific code, called a **course/acquisition (CA) code**, which contains information on the satellite's position, the GPS system time, and the health and accuracy of the transmitted data.

ii) If the receiver knows the speed at which the signal traveled (approximately 186,000 miles per second) and the exact broadcast time, it can compute the distance traveled by the signal from the arrival time.

b) The GPS receiver matches each satellite's CA code with an identical copy of the code contained in the receiver's database.

i) By shifting its copy of the satellite's code in a matching process, and by comparing this shift with its internal clock, the receiver can calculate how long it took the signal to travel from the satellite to the receiver.

ii) The distance derived from this method of computing distance is called a **pseudo-range** because it is not a direct measurement of distance but a measurement based on time.

- Pseudo-range is subject to several error sources, e.g., ionospheric and tropospheric delays and multipath.

c) In addition to knowing the distance to a satellite, a receiver needs to know the satellite's exact position in space; this is known as its **ephemeris**.

i) Each satellite transmits information about its exact orbital location.

ii) The GPS receiver uses this information to precisely establish the position of the satellite.

d) Using the calculated pseudo-range and position information supplied by the satellite, the GPS receiver mathematically determines its position by triangulation.

i) The GPS receiver needs at least four satellites to yield a three-dimensional position (latitude, longitude, and altitude) and time solution.

ii) The GPS receiver computes navigational values, such as distance and bearing to a waypoint, ground speed, etc., by using the aircraft's known latitude/longitude and referencing these to a database built into the receiver.

2) The GPS constellation of 24 satellites is designed so that a minimum of five are always observable by a user anywhere on earth.

a) The receiver uses data from a minimum of four satellites above the mask angle (the lowest angle above the horizon at which it can use a satellite).

3) The GPS receiver verifies the integrity (usability) of the signals received from the GPS constellation through **receiver autonomous integrity monitoring (RAIM)** to determine if a satellite is providing corrupted information.

a) At least one satellite, in addition to those required for navigation, must be in view for the receiver to perform the RAIM function; thus, RAIM needs a minimum of five satellites in view, or four satellites and a barometric altimeter (baro-aiding), to detect an integrity anomaly.

b) For receivers capable of doing so, RAIM needs six satellites in view (or five satellites with baro-aiding) to isolate the corrupt satellite signal and remove it from the navigation solution.

i) **Baro-aiding** is a method of augmenting the GPS integrity solution by using a nonsatellite input source.

- GPS-derived altitude should not be relied upon to determine aircraft altitude since the vertical error can be quite large.

ii) To ensure that baro-aiding is available, the current altimeter setting must be entered into the receiver as described in the operating manual.

4) RAIM messages vary somewhat between receivers; however, generally there are two types.

a) One type indicates that not enough satellites are available to provide RAIM integrity monitoring.

b) The other type indicates that the RAIM integrity monitor has detected a potential error that exceeds the limit for the current phase of flight.

c) Without RAIM capability, the pilot has no assurance of the accuracy of the GPS position.

5) Authorization to conduct any GPS operation under IFR requires that

   a) GPS navigation equipment used must be FAA-approved and the installation must be done in accordance with published FAA requirements.

      i) Approval for the use of the GPS for IFR operations, and any limitations, will be found in the airplane's *POH* (also called the FAA-approved Airplane Flight Manual) and the airplane's logbook.

      ii) VFR and hand-held GPS systems are not authorized for IFR navigation, for instrument approaches, or as a principle instrument flight reference. During IFR operations, they may be considered only an aid to situational awareness.

6) Aircraft using GPS navigation equipment under IFR must be equipped with an approved and operational alternate means of navigation appropriate to the flight.

   a) Active monitoring of alternative navigation equipment is not required if the GPS receiver uses RAIM for integrity monitoring.

   b) Active monitoring of an alternate means of navigation is required when the RAIM capability of the GPS equipment is lost.

7) The GPS operation must be conducted in accordance with the FAA-approved aircraft flight manual (AFM) or flight manual supplement.

   a) You must be thoroughly familiar with the particular GPS equipment installed in the aircraft, the receiver operation manual, and the AFM or flight manual supplement.

   b) Unlike ILS and VOR, the basic operation, receiver presentation to the pilot, and some capabilities of the equipment can vary greatly.

   c) Due to these differences, operation of different brands, or even models of the same brand, of GPS receiver under IFR should not be attempted without thorough study of the operation of that particular receiver and installation.

      i) Most receivers have a built-in simulator mode that will allow you to become familiar with operation prior to attempting operation in the aircraft.

   d) Using the equipment in flight under VFR conditions prior to attempting IFR operation will allow further familiarization.

END OF TASK

INSTRUMENT COCKPIT CHECK

II.C. TASK: INSTRUMENT COCKPIT CHECK

REFERENCES: 14 CFR Parts 61, 91; AC 61-27.

Objective. To determine that the applicant:

1. Exhibits adequate knowledge of the elements related to preflighting instruments, avionics, and navigation equipment cockpit check by explaining the reasons for the check and how to detect possible defects.

2. Performs the preflight on instruments, avionics, and navigation equipment cockpit check by following the checklist appropriate to the aircraft flown.

3. Determines that the aircraft is in condition for safe instrument flight including --

a. Radio communications equipment.

b. Radio navigation equipment including the following, as appropriate to the aircraft flown:

1) VOR/VORTAC.
2) ADF.
3) ILS.
4) GPS.
5) LORAN.

c. Magnetic compass.
d. Heading indicator.
e. Attitude indicator.
f. Altimeter.
g. Turn-and slip indicator/turn coordinator.
h. Vertical speed indicator.
i. Airspeed indicator.
j. Clock.
k. Power source for gyro-instruments.
l. Pitot heat.

4. Notes any discrepancies and determines whether the aircraft is safe for instrument flight or requires maintenance.

A. General Information

1. The objective of this task is to determine your knowledge of the preflight instrument, avionics, and navigation equipment cockpit check.
2. For additional reading, see your airplane's *POH* for the following:
   a. Section 4: Normal Procedures
   b. Section 9: Supplement (Optional Systems Description and Operating Procedures)
3. Remember that you must ensure that all required documents are aboard the aircraft before beginning your flight.
   a. An easy way to remember the required documents is by using the memory aid **ARROW**.

A irworthiness certificate
R egistration
R adio station license
O perating limitations
W eight and balance

NOTE: A radio station license is required only if the airplane is flown outside of U.S. airspace (i.e., to another country). Additionally, on these flights you are required to have a restricted radiotelephone operator permit. These are requirements of the Federal Communications Commission (FCC), not FAA requirements.

B. Task Objectives

1. **Exhibit adequate knowledge of the elements related to preflighting instruments, avionics, and navigation equipment cockpit check by explaining the reasons for the check and how to detect possible defects.**
   a. The reason for the instrument cockpit check is to ensure that all of your airplane's instruments (flight and engine), avionics, and navigation equipment are functioning properly and in good working order.
      1) You will be depending on these to make a safe flight in IFR conditions.
   b. You must have a knowledge of how your instruments, avionics, and navigation equipment operate so when you perform the check you will be able to detect possible defects.

2. Perform the preflight on instruments, avionics, and navigation equipment cockpit check by following the checklist appropriate for your airplane.

a. Always use the checklist from your *POH* and follow it item by item. If you are interrupted, mark your place so you can return to that item.

1) The checklist will ensure that you check every single item at the appropriate time.

b. If your *POH* does not have an instrument cockpit checklist, you must develop your own. Use and follow it exactly the same every time.

1) You may start at the upper left and systematically check and touch each instrument one by one, moving to the right, then back to the left on the second row.

a) Follow this procedure until you have checked every item on the instrument panel.

2) Your instructor will assist you in learning the proper instrument cockpit check procedures.

c. Check your instruments in a logical manner, not in a haphazard fashion. Do not skip over instruments.

1) During your flight training, as you perform the instrument cockpit check, explain each item you are checking, why you are checking it, and what indication you are expecting.

a) Follow the same procedure with your examiner during your practical test.

2) These checks will be performed throughout the time from before engine start until the takeoff roll.

3. Determine that your airplane is in condition for safe instrument flight by checking the following instruments, avionics, and navigation equipment.

a. Radio communications equipment

1) Check to ensure that all of your radio(s) are operating properly.

a) If you have more than one communication radio, you should use both to ensure that they are working; i.e., use one to call ground control and one to call the tower.

2) The transponder should be on standby.

a) While the transponder is warming up, the reply light should be on. Once the unit is ready to use, the reply light will go out.

b) Use the transponder's self-test function to determine that the transponder is operating properly.

i) This test does not provide you with any indication that the Mode C is operating properly.

b. Radio navigation equipment including the following, as appropriate, to your airplane

1) **VOR/VORTAC receiving equipment** is required to be operationally checked every 30 days for IFR flight (FAR 91.171). There are various methods of accomplishing this check.

   a) FAA VOR test facilities (VOTs) are available on a specified frequency at certain airports. The facility permits you to check the accuracy of your VOR receiver on the ground and, in some cases, while airborne.

      i) The receiver is tested by tuning the VOR to the specified frequency and then turning the OBS until the CDI needle centers.
         - The indicated course should be either 0° or 180°.
         - If 0°, the TO/FROM indicator should indicate FROM.
         - If 180°, the TO/FROM indicator should indicate TO.
         - Accuracy of the VOR receiver should be ±4°.

      ii) Should the VOR receiver operate an RMI, the needle should point to 180°.

      iii) The VOT can be identified by either a series of dots or a continuous tone.

      iv) Information concerning an individual VOT can be obtained from the local FSS.

   b) Certified airborne and ground checkpoints consist of certified radials that should be received (i.e., CDI centered and FROM indicated) over specific points or landmarks while airborne in the immediate vicinity of the airport or at a specific point on the airport surface. Accuracy of the VOR receiver should be

      i) ±6° for airborne checks
      ii) ±4° for ground checks

   c) If you have dual VORs (everything separate except antenna), you can check one VOR against the other. Tune in the same VOR and center the CDI on each VOR; the OBS headings should not differ by more than 4°.

      i) This method is the least reliable (e.g., if one VOR has an error of 8° and the second VOR has an error of 10°, they would pass the dual VOR check).

      ii) It is recommended that you periodically use a VOT or a certified airborne or ground checkpoint so you can check each VOR instrument independently.

   d) A list of VOTs, airborne checkpoints, and ground checkpoints is published in the *Airport/Facility Directory*.

   e) Each person making this check should enter the date, place, and bearing error, and sign the airplane log or other record.

      i) Check the log or record for the last VOR check.

   f) During the before-takeoff cockpit check, tune the proper VOR/VORTAC frequencies in your navigation equipment and, if possible, identify the station.

      i) Check for CDI and TO/FROM flag movement; then set OBS to initial en route course.

g) Should an error in excess of ±4° be indicated by a ground or dual VOR check, or ±6° by an airborne check, IFR flight should not be attempted without first correcting the source of the error.

2) **ADF** (if installed)

a) If possible, tune and identify an NDB on your ADF and verify that the needle is pointing toward the NDB.

b) Some ADFs will have a test button that you depress, which swings the needle from its position.

i) When you release the button, the needle should return to its original position toward the station.

c) If the ADF has a movable compass card that is slaved to the heading indicator, ensure that the compass card moves freely with the heading indication and check against known headings.

3) **ILS**

a) If the airport has a localizer (LOC), you can test the LOC indication on your VOR instrument.

i) Tune and identify the LOC.

ii) Determine that the indication is consistent with your airplane's position relative to the LOC.

b) Most airplanes have a test button that will light up the marker beacon indicators to ensure that they are working properly.

4) **GPS** (if installed and IFR-approved)

a) Prior to an IFR flight using GPS, you should ensure that the GPS equipment and installation are approved and certified for the intended IFR operation.

i) The GPS equipment should be operating in accordance with the airplane's *POH*, *AFM* (FAA-approved *Airplane Flight Manual*), or flight manual supplement.

b) Follow the specific start-up and self-test procedures for the GPS receiver, as stated in the manufacturer's operating manual.

5) **LORAN** (if installed and IFR-approved)

a) The checks for a LORAN receiver are the same as described for a GPS receiver.

c. **Magnetic compass**. Check the card for freedom of movement and be sure the bowl is full of fluid.

1) Determine compass accuracy by comparing the indicated heading against a known heading while the airplane is stopped or taxiing straight.

2) Remote indicating compasses (RIC) should also be checked against known headings.

d. **Heading indicator**. Allow 5 min. after engine start for the gyro of the vacuum HI (or 3 min. for an electric gyro) to attain normal operating speed.

1) Ensure the HI is uncaged if it has a caging feature.

2) Before taxiing, or while taxiing straight, set the HI to correspond with the magnetic compass heading.

   a) Normally, this is done prior to taxiing.

3) While taxiing, ensure that the HI is working properly.

   a) During a right turn, the heading increases.
   b) During a left turn, the heading decreases.

4) Check against known headings.

5) RIC headings should be compared with those of the magnetic compass while taxiing.

6) Any slaved compass, such as an RMI, should be checked for slaving action and its indications compared with those of the magnetic compass.

7) Before takeoff (and while not moving), recheck the HI against the magnetic compass heading and the runway's published heading.

e. **Attitude indicator**. Allow the same amount of time for the gyros to attain normal speed as for the HI.

1) If the horizon bar erects to the horizontal position and remains at the correct position for the attitude of your airplane, or if it begins to vibrate after this attitude is reached and then slowly stops vibrating altogether, the instrument is working properly.

2) If the horizon bar fails to remain in the horizontal position during straight taxiing or indicates a bank in excess of 5° during taxi turns, the instrument is to be considered unreliable.

3) Adjust the miniature airplane with reference to the horizon bar while on the ground.

f. **Altimeter**. Set the altimeter to the current reported altimeter setting, and note any variation between the known field elevation and the altimeter indication.

1) If the variation is greater than 75 ft., the accuracy of the altimeter is questionable, and it should be referred to an appropriate repair facility for evaluation and/or correction.

2) If an altimeter setting is not available, adjust the altimeter to indicate the airport elevation.

   a) When airborne, request an official altimeter setting from ATC.

g. **Turn coordinator/turn-and-slip indicator**. During taxi turns, check the miniature airplane/turn needle for proper turn indications.

1) The ball should move freely to the outside of the turn.

2) When taxiing straight, the miniature airplane/turn needle should be level/straight up.

h. **Vertical speed indicator**. The instrument should indicate zero. If it does not, you may want to tap the panel gently.

1) If the reading remains off the zero reading and is not adjustable, the ground indication will have to be interpreted as the zero position in flight.

i. **Airspeed indicator**. Before taxiing, ensure that the ASI indicates zero.

1) While you accelerate during your initial takeoff roll, check for an increasing rate on the ASI.

2) As your have done throughout your flying career, if the ASI is not functioning properly at that time, discontinue the takeoff roll immediately.

j. **Clock**. Set the clock to the appropriate time, and check to ensure that it is operating.

k. **Power source for gyro-instruments**. Check the suction gauge for the correct operating levels in accordance with your *POH*.

1) If the gyros are electrically driven, check the ammeter gauge for correct operating levels in accordance with your *POH*, and ensure that no low voltage or other warning lights are on.

l. **Pitot heat**. Check by turning on the pitot heat (if equipped), and confirm its operational status by correct ammeter indications of a load on the system.

1) Any additional deice/anti-ice equipment should also be checked for proper operation.

4. **Note any discrepancies and determine whether your airplane is safe for instrument flight or requires maintenance.**

a. You should stop at each discrepancy and note its effect(s). How is any problem covered by another instrument, piece of equipment, pilot workload, etc.? Relate problems to FARs.

b. You, as pilot in command, are the final authority as to the airworthiness of your airplane. Do not take chances; if in doubt, turn back and go to the ramp to get the discrepancy checked out.

1) Explain your position to your instructor or examiner, and give the reason(s) that you feel the airplane is not safe for instrument flight.

END OF TASK -- END OF CHAPTER

CHAPTER III
AIR TRAFFIC CONTROL CLEARANCES AND PROCEDURES

This chapter explains the three tasks (A-C) of Air Traffic Control Clearances and Procedures. These tasks include both knowledge and skill. Your examiner is required to test you on all three tasks.

The ATC clearance to depart on the IFR cross-country flight you previously planned may be an actual or simulated ATC clearance based on whether or not you filed an IFR flight plan.

AIR TRAFFIC CONTROL CLEARANCES

III.A. TASK: AIR TRAFFIC CONTROL CLEARANCES

REFERENCES: 14 CFR Parts 61, 91; AC 61-27; AIM.

Objective. To determine that the applicant:

1. Exhibits adequate knowledge of the elements related to ATC clearances and pilot/controller responsibilities to include tower en route control and clearance void times.
2. Copies correctly, in a timely manner, the ATC clearance as issued.
3. Determines that it is possible to comply with ATC clearance.
4. Interprets correctly the ATC clearance received and, when necessary, requests clarification, verification, or change.
5. Reads back correctly, in a timely manner, the ATC clearance in the sequence received.
6. Uses standard phraseology when reading back clearance.
7. Sets the appropriate communication and navigation frequencies and transponder codes in compliance with the ATC clearance.

A. General Information

1. The objective of this task is to determine your knowledge of the elements of ATC clearances.
2. For additional reading, see Chapter 3, Airports, ATC, and Airspace, Module 3.8, Radio Communications and Phraseology, in *Pilot Handbook* for a three-page discussion on standard phraseology.
3. **ATC clearance** means an authorization by ATC, for the purpose of preventing a collision between known aircraft, for an aircraft to proceed under specified conditions within controlled airspace.
   a. An ATC clearance is NOT an authorization for you to deviate from any FAR or minimum altitude nor to conduct unsafe operations in your airplane.

4. Before you can operate your airplane in controlled airspace under IFR, you must have filed an IFR flight plan and received an appropriate ATC clearance.
   a. An ATC clearance issued to aircraft operating under IFR is commonly known as an IFR clearance.
5. During your flight training, your instructor will provide you with simulated ATC clearances so you will become comfortable with copying the clearance.
   a. When appropriate, your instructor will have you obtain actual ATC clearances.
6. Normally, you will obtain your IFR departure clearance in one of the following ways:
   a. At a controlled airport:
      1) In class B or C airspace, on the clearance delivery frequency.
      2) In class D airspace, or in class E or G airspace at an airport with an operating control tower, on the ground control or (frequency use permitting) tower frequency.
   b. At an uncontrolled airport:
      1) From the nearest FSS by telephone before boarding the aircraft. This clearance will have a void time.
         a) If an FSS is on the field, directly from the FSS by radio while on the ground. This clearance will have a void time.
      2) From the appropriate ATC facility by telephone. This clearance will have a void time.
         a) If you are in range of a Remote Communications Outlet (RCO) for that facility, by radio while on the ground. This clearance will have a void time.
      3) Obtain the frequency for the appropriate ATC facility from the FSS when filing your flight plan; contact that facility by radio while airborne, after departing VFR. This clearance will not have a void time.
7. The ATC clearance during your practical test may be an actual or simulated ATC clearance based upon the flight plan and your examiner's discretion.

B. Task Objectives

1. **Exhibit adequate knowledge of the elements related to ATC clearances and pilot/controller responsibilities to include tower en route control and clearance void times.**
   a. An initial IFR (ATC) clearance contains the following items, as appropriate, in the order shown:
      1) **Airplane identification.** This is your airplane's call sign, e.g., Skyhawk 2479H.

2) **Clearance limit.** The ATC clearance issued prior to departure will normally authorize flight to the airport of intended landing; e.g., "Cleared to the Gainesville airport."
   a) At some locations and under certain conditions, a short-range clearance procedure is utilized whereby a clearance is issued to a fix within or just outside of the terminal area.
      i) You will be advised of the frequency on which you will receive your long-range clearance direct from the center controller; e.g., "After departure, turn left to 090, cleared direct Chanute VOR; obtain further clearance Kansas City Center 126.6."
3) **Departure procedure.** Headings to fly and altitude restrictions may be issued to separate a departure from other air traffic in the terminal area; e.g., "After departure, turn right to 300°."
   a) Where volume of traffic warrants, instrument departure procedures (DPs) have been developed.
4) **Route of flight.** Clearances are normally issued for the route and altitude you filed; e.g., "Cleared as filed."
   a) However, due to traffic conditions, it is frequently necessary for ATC to specify an altitude or a route different from that requested. For instance, instead of "as filed," you might receive a full route clearance; e.g., ". . . Victor 97 to La Belle (VOR), Victor 157 to Ocala (VOR), direct."
   b) In addition, flow patterns have been established in certain congested areas, or between congested areas, whereby traffic capacity is increased by routing all traffic on preferred routes. Preferred routes are listed in the *A/FD*.
   c) You must notify ATC immediately if your navigation equipment cannot receive the type of signals needed to comply with the clearance.
5) **Altitude data.** The altitude or flight level instructions in an ATC clearance normally require that you "MAINTAIN" the altitude or flight level at which you will operate when in controlled airspace; e.g., "Climb and maintain 6,000."
   a) You may receive a clearance of ". . . climb and maintain 2,000, expect 6,000 one-zero minutes after departure. . . ."
      i) ATC issues this type of initial altitude clearance when concerned with a communications failure.
         - If you do lose two-way radio communications shortly after takeoff, you can climb to the expected altitude 10 min. after your takeoff time.
   b) The term "CRUISE" may be used instead of "MAINTAIN" to assign a block of airspace from the minimum IFR altitude (e.g., MEA) up to and including the altitude specified in the cruise clearance.
      i) You may level off at any intermediate altitude within this block of airspace. Climb/descent within the block is at your discretion.
      ii) However, once you begin a descent from an altitude and report leaving that altitude, you may **not** return to that altitude without additional ATC clearance.

6) **Additional information.** This special information may be anything the ATC feels is necessary to assist you in understanding your IFR clearance.

7) **Departure frequency.** This frequency is used to contact your departure controller after takeoff.

   a) If departing from a tower controlled airport, switch to departure frequency only after receiving instructions to do so from the tower; e.g., "Skyhawk 2479H, contact departure."

   b) If departing from an uncontrolled airport, contact departure control as soon as possible after takeoff and after you are clear of any local traffic.

8) **Transponder code.** This is your assigned transponder code for your flight.

b. The following are abbreviated IFR departure clearance (cleared...as filed) procedures:

1) ATC will issue an abbreviated IFR clearance based on the route of flight filed in your IFR flight plan, provided the filed route can be approved with little or no revision. These procedures are based on the following conditions:

   a) You are on the ground or you have departed VFR and are requesting your IFR clearance while airborne.

   b) You will not accept an abbreviated clearance if you have changed the route or destination of a flight plan filed with ATC before departure.

   c) You are responsible to inform ATC in the initial call-up (for clearance) when the filed flight plan has been either amended or canceled and replaced with a new filed flight plan.

2) The controller is required to state the instrument departure procedure (DP) name, the current number, and the DP transition name after the phrase "Cleared to (destination) airport" and prior to the phrase "then as filed," for all departure clearances when the DP or DP transition is to be flown.

   a) The procedure applies whether or not the DP is filed in the flight plan.

3) "Cleared to (destination) airport as filed" does NOT include the en route altitude filed in your flight plan.

   a) An en route altitude will be stated in the clearance or you will be advised to expect an assigned or filed altitude within a given time frame or at a certain point after departure.

4) To ensure success of the program, you should

   a) Avoid making changes to a filed flight plan just prior to departure.

   b) State the following information in the initial call-up to ATC when no change has been made to the filed flight plan: airplane call sign, location, type of operation, and name of the airport (or fix) to which you expect clearance.

      i) EXAMPLE: "Tampa clearance delivery (or ground control if appropriate) Warrior 2479H at Raytheon, IFR Birmingham, information Bravo."

c) If the flight plan has been changed, state the change and request a full route clearance.

i) EXAMPLE: "Tampa clearance delivery, Warrior 2479H at Raytheon, IFR Jacksonville, information Bravo. My destination changed (or plan route amended). Request full route clearance."

d) Request verification or clarification from ATC if ANY portion of your IFR clearance is not clearly understood.

e) When requesting clearance for the IFR portion of a VFR/IFR flight, request your clearance prior to the fix where IFR operation is proposed to commence in sufficient time to avoid delay.

c. Amended clearances

1) Amendments to a prior clearance will be issued at any time ATC deems such action necessary to avoid possible conflict between aircraft.

2) You may request an amended clearance if

a) The clearance issued would cause you to violate an FAR or, in your opinion, would place your airplane in jeopardy.

b) You feel that another course of action is more practicable.

c) Airplane equipment limitations or company procedures forbid compliance with the clearance issued.

d. IFR clearance VFR-on-top

1) While operating on an IFR flight plan in visual meteorological conditions (VMC), you may request VFR-on-top in lieu of an assigned altitude.

a) This would permit your choice of altitude or flight level (subject to any ATC restrictions).

2) You may desire to climb through clouds, haze, smoke, or other meteorological formations then either cancel your IFR flight plan or operate VFR-on-top.

a) ATC clearance will contain both of the following:

i) A cloud top report or a statement that no top report is available

ii) A request to report reaching VFR-on-top

b) ATC clearance may contain a clearance limit, routing, and an alternate clearance if VFR-on-top is not reached at a specified altitude.

3) When cleared to "MAINTAIN VFR-ON-TOP/MAINTAIN VFR CONDITIONS" while on an IFR flight plan, you must

a) Fly at appropriate VFR cruising altitudes.

b) Comply with VFR visibility and distance from clouds criteria.

c) Comply with the IFRs that are applicable, i.e., minimum IFR altitudes, position reporting, adherence to ATC clearance, etc.

4) VFR-on-top permits operation above, below, or between layers, or in areas where there is no obscuration.

5) VFR-on-top does not imply that your IFR flight plan is canceled.

e. Tower en route control (TEC) is an ATC program to provide a service to aircraft proceeding to and from metropolitan areas.

1) It links designated approach control areas by a network of identified routes made up of the existing airway structure of the National Airspace System.

a) TEC routes are published for certain portions of the U.S. in the *A/FD*.

2) It is generally for nonturbojet aircraft flying a relatively short distance (usually less than 2 hr.) that can transition through approach control airspace (not requiring center control) for altitudes at or below 10,000 ft. MSL.

a) A flight is transferred from departure control at the departure airport to one or more successive approach control facilities.

3) Pilots requesting TEC are subject to the same delay factor at the destination airport as other aircraft in the ATC system, and departure and en route delays may occur depending upon individual facility workload.

4) No unique requirements are placed upon pilots to use the TEC program. Normal flight plan filing procedures will ensure proper processing.

a) Pilots should include the abbreviation "TEC" in the remarks section of the flight plan when requesting tower en route control.

b) IFR flight may be planned to any satellite airport in proximity to the major primary airport via the same routing.

f. ATC may assign departure restrictions, clearance void times, hold for release, and release times when necessary to separate departures from other traffic or to restrict or regulate the departure flow.

1) **Clearance void times.** When operating from an airport without a control tower, you may receive a clearance that contains a provision for your clearance to be void if you are not airborne by a specific time.

a) If you do not depart prior to the clearance void time, you must advise ATC as soon as possible as to your intentions.

b) ATC will normally advise you of the time allotted (not to exceed 30 min.) to notify ATC that you did not depart prior to the clearance void time.

i) Failure to do so will result in your being considered overdue, and search and rescue procedures will be initiated.

c) If you depart at or after your clearance void time, you will not be afforded IFR separation and may be in violation of FAR 91.173, which requires an ATC clearance prior to operating IFR in controlled airspace.

2) **Hold for release.** ATC may issue hold-for-release instructions in your clearance to delay your departure for traffic management reasons (i.e., weather, traffic volume, etc.).

a) You may not depart utilizing your clearance until a release time or additional instructions are issued by ATC.

b) You may cancel your IFR clearance with ATC and depart VFR (if appropriate), but an IFR clearance may not be available after departure.

3) **Release times.** A release time is a departure restriction issued by ATC, specifying the earliest time that you may depart.

a) ATC will use release times in conjunction with traffic management procedures and/or the separation of departing aircraft from other traffic.

4) If practical when you are departing from an airport without an operating control tower, you should obtain your IFR clearance prior to becoming airborne when two-way communications with the controlling ATC facility is available.

g. Pilot responsibilities are covered by FARs 91.3 and 91.123.

1) Responsibility and authority of the pilot in command

a) As the pilot in command of your airplane, you are directly responsible for, and are the final authority as to, the operation of that airplane.

b) Thus, in emergencies, you may deviate from the FARs to the extent needed to maintain the safety of the airplane and passengers.

c) If you do deviate from the FARs in such an emergency, you may be required to file a written report with the FAA.

2) Compliance with ATC clearances and instructions

a) Once you have been given ATC instructions or a clearance, you may not deviate from it unless you obtain amended instructions or clearance, an emergency exists, or the deviation is in response to a traffic alert and collision avoidance system (TCAS) resolution advisory.

i) If you deviate from a clearance in an emergency or in response to a TCAS resolution advisory, you must notify ATC as soon as possible.

ii) If you are given priority by ATC in an emergency, you must submit a detailed report of the emergency within 48 hr. to the manager of that ATC facility, if requested.

- The report may be requested even if you do not deviate from any rule of Part 91.

b) If you are uncertain about the meaning of an ATC clearance, you should immediately ask for clarification from ATC.

h. ATC responsibilities include

1) Issuing appropriate clearances for the operation to be conducted, in accordance with established criteria

2) Assigning altitudes in IFR clearances that are at or above the minimum IFR altitudes in controlled airspace

3) Ensuring acknowledgment by the pilot for issued information, clearances, or instructions

4) Ensuring that readbacks by the pilot of altitude, heading, or other items are correct, and making corrections as appropriate if the readbacks are incorrect, distorted, or incomplete

a) Editorial note: While there has been discussion on whether it is ATC's responsibility to ensure that the readback of information is correct or not, we feel that the majority of pilots and controllers are professionals and work together as a team to ensure the safety of air traffic.

2. Copy correctly, in a timely manner, the ATC clearance as issued.

a. The first step in obtaining your IFR clearance is simply to copy the clearance on your pad of paper.

1) Make sure you are ready to copy your clearance (i.e., pencil and paper in hand) before telling ATC that you are ready to copy.

2) Receive your clearance when you are ready, not when ATC informs you that they have your clearance.

a) If you are taxiing or are in the middle of a checklist, ask ATC to stand by.

3) In order to copy the IFR clearance in a timely manner, use a shorthand that only you need to be able to read. Your instructor will assist you at the beginning to get you started.

b. You may want to write the word "CRAFT" vertically on your pad of paper to assist you in copying all of the elements of the IFR clearance, as shown below.

C(leared to)
R(oute)
A(ltitude)
F(requency)
T(ransponder)

1) This format follows the order of the clearance as it is given to you and will ensure that you have all the information.

2) Additionally, you can make changes throughout your flight, such as altitude and frequency changes.

c. The specified conditions in your clearance may be different from those in your filed flight plan. The fact that ATC specifies different or additional conditions means that other aircraft are involved in the traffic situation.

d. The following is an example of an IFR clearance:

1) Skyhawk 2479H is cleared to the Tucson International Airport as filed; after departure, fly heading 120; climb and maintain 3,000; expect 7,000 in 10 min. Departure frequency is 124.3, squawk 4371.

2) The following is an example of how the IFR clearance in d.1) above may be written:

C - TUS
R - AF
A - 3 7/10
F - 124.3
T - 4371

e. Remember, the objective is to copy what the controller is saying. Do not attempt to interpret the clearance.

NOTE: We have placed items 3. through 5. of this task in the order that represents the proper sequence of events.

3. Read back correctly, in a timely manner, the ATC clearance in the sequence received.

a. After you have copied your IFR clearance, you will read back to the controller what you have written.

1) Simply read back what you have written, and inform the controller if you missed an item.

b. Reading back the clearance is a check between ATC and you to ensure that you have the correct information.

1) Reading back an initial clearance does not mean that you accept the clearance, only that you have recorded it correctly.

2) It is only on takeoff that you are committed to the clearance you have received.

c. ATC will ensure that your readback is correct. If items are incorrect, distorted, or incomplete, the controller will say those items again.

   1) Correct those items on your written copy and read them back.

d. When your readback is correct, the controller will state, "Readback correct."

4. Determine that it is possible to comply with the ATC clearance.

a. In this step, you will study your clearance, especially if it is different from what you filed on your flight plan.

   1) If the route is "as filed," you do not need to check the routing.

   2) If you are given different routing (called a full route clearance), you will need to look at the appropriate charts and check the route from start to finish. Check for the following:

      a) The route makes sense.
      b) Any airway routing leads to the specified fixes.
      c) The route takes you to your destination.

b. Ensure that your clearance does not exceed any limitations of your airplane, such as a minimum rate of climb required on a DP.

c. Check that the assigned altitude is appropriate for your airplane and that the altitude, if different from the one on your flight plan, is acceptable given the forecast winds.

d. Your clearance must be appropriate for the navigation system(s) in your airplane; e.g., you will be unable to identify a DME fix if your airplane does not have DME or an approved GPS.

5. Interpret correctly the ATC clearance received and, when necessary, request clarification, verification, or change.

a. After you have reviewed your clearance, you should request any needed clarification, verification, or change.

   1) Any items you found that would not allow you to comply with the clearance should be corrected at this time.

b. If you have any doubt about what you have written on your paper, ask ATC. It is better to ask immediately than to assume.

6. Use standard phraseology when reading back clearance.

a. Use standard phraseology when talking on the radio, just as you did during your private pilot training. Your procedures will become more natural during your training.

7. Set the appropriate communication and navigation frequencies and transponder codes in compliance with the ATC clearance.

a. Before takeoff, your navigation frequencies should be set in all available navigation equipment.

   1) Your primary navigation system (e.g., VOR, GPS) should be tuned to your first navigational facility or waypoint, and the appropriate OBS setting should be selected (if a VOR station).

   2) If your HI is equipped with a heading bug, it should be set to the runway heading or your assigned departure heading.

   3) Some pilots will use the second navigation radio (if available) to the approach facility to be used at the departure airport in the event of a takeoff emergency in IMC.

b. If you have dual communication equipment or one with the capability of selecting a standby frequency, set the departure control frequency in the second radio or on standby.

c. Set transponder to the assigned code. Transponder should be set to STANDBY until ready for takeoff, then switched to ON or ALT, as appropriate.

d. This element relates to the management of your resources in the cockpit. Exploit your assets to the maximum extent to help eliminate unnecessary work.

C. Common Errors with Air Traffic Control Clearances

1. **Not having paper and pencil ready to copy your clearance.**

   a. At busy airports, the controllers are busy and will usually rapid-fire your clearance to you.

   b. Be ready to copy before you call for your clearance (by stating your call sign and IFR destination).

2. **Not copying (writing) a clearance down, but instead trying to memorize it.**

   a. Write down each clearance, every time. Do not take a chance. With the clearance on paper, far fewer mistakes are likely to occur because you can always refer to it.

3. **Not reading a clearance back to ATC.**

   a. Readback is a measure of safety, as it confirms your understanding of ATC's expectation of your altitude, heading, etc.

4. **Accepting a clearance you are unclear about.**

   a. If you are not sure, tell the controller. Clarify uncertainties while still on the ground, not in the air, or before you violate the clearance assigned to you by ATC.

5. **Accepting a clearance that you cannot comply with.**

   a. A DP that requires DME fixes is unacceptable if your airplane has no DME or GPS.

   b. A clearance may be unacceptable for many reasons, such as climb rate, altitude, weather, speed, instrumentation, etc. If weather is a problem, say so.

      1) Forewarn ATC before you get yourself into trouble so that ATC can be more able to assist and accommodate you.

   c. Do not accept clearances just because ATC assigns them; they must be within your and your airplane's capabilities.

END OF TASK

COMPLIANCE WITH DEPARTURE, EN ROUTE, AND ARRIVAL PROCEDURES AND CLEARANCES

III.B. TASK: COMPLIANCE WITH DEPARTURE, EN ROUTE, AND ARRIVAL PROCEDURES AND CLEARANCES

REFERENCES: 14 CFR Parts 61, 91; AC 61-27; DPs; En Route Low Altitude Charts; STARs.

Objective. To determine that the applicant:

1. Exhibits adequate knowledge of the elements related to DPs, En Route Low Altitude Charts, STARs, and related pilot/controller responsibilities.
2. Uses the current and appropriate navigation publications for the proposed flight.
3. Selects and uses the appropriate communications frequencies; selects and identifies the navigation aids associated with the proposed flight.
4. Performs the appropriate aircraft checklist items relative to the phase of flight.
5. Establishes two-way communications with the proper controlling agency, using proper phraseology.
6. Complies, in a timely manner, with all ATC instructions and airspace restrictions.
7. Exhibits adequate knowledge of two-way radio communications failure procedures.
8. Intercepts, in a timely manner, all courses, radials, and bearings appropriate to the procedure, route, or clearance.
9. Maintains the applicable airspeed within 10 kt.; headings within 10°; altitude within 100 ft. (30 meters); and tracks a course, radial, or bearing.

A. General Information

1. The objective of this task is to determine your knowledge of and ability to comply with departure, en route, and arrival procedures and clearances.

B. Task Objectives

1. **Exhibit adequate knowledge of the elements related to DPs, En Route Low Altitude Charts, STARs, and related pilot/controller responsibilities.**
    a. An instrument departure procedure (DP) is a preplanned ATC departure procedure printed for pilot use in graphic and/or textual form. A DP provides a way to depart the airport and transition to the en route structure safely.
        1) A DP is established primarily to provide obstacle clearance protection to aircraft in instrument meteorological conditions (IMC).
            a) A secondary reason a DP is established is to increase efficiency and reduce communications and departure delays at busier airports.
        2) If an aircraft may turn in any direction from a runway and be clear of obstacles, no DP will be published unless needed for air traffic purposes.
        3) Obstacle clearance for all departures, including those that have no DP, is based on your crossing the end of the runway at least 35 ft. AGL, climbing to 400 ft. above the elevation at the departure end of runway before turning, and climbing at least 200 feet per nautical mile (FPNM), unless the DP specifies a higher climb gradient or unless a crossing restriction requires the aircraft to level off.
            a) Climb gradients greater than 200 FPNM are specified when required for obstacle clearance and/or ATC-required crossing restrictions.
            b) Climb gradients may be specified only to an altitude/fix, above which the normal gradient (200 FPNM) applies.

c) Some DPs established solely for obstacle avoidance require a climb in visual conditions to cross the airport or an on-airport NAVAID in a specified direction, at or above a specified altitude.

4) DPs are designed so that adherence to the procedure by the pilot will ensure obstacle protection.

a) Obstacle clearance responsibility also rests with you when you choose to climb in visual conditions rather than fly a DP and/or depart under increased takeoff minima rather than fly a DP.

i) Specified ceiling and visibility minima will allow visual avoidance of obstacles until you enter the standard obstacle protection area.

- Obstacle avoidance is not guaranteed if you maneuver farther from the airport than the specified visibility minima.

ii) That segment of a DP that requires you to see and avoid obstacles ends when your airplane crosses the specified point at the required altitude.

- At that time, standard obstacle protection is provided, and the standard climb gradient (200 FPNM) is required unless specified otherwise.

b) ATC may assume responsibility for obstacle clearance by providing you with vectors.

i) ATC may also vector you off a previously assigned DP.

ii) In all cases, the 200 FPNM climb gradient is assumed and obstacle clearance is not provided by ATC until the controller begins to provide navigational guidance in the form of radar vectors.

NOTE: When the controller uses the term “radar contact,” that term should not be interpreted as relieving you of your responsibility to maintain appropriate terrain and obstacle clearance. ATC responsibility begins when the controller gives you a heading to fly.

5) DPs will be listed by airport in Section C, IFR Take-off Minimums and Departure Procedures, of the NACO *Terminal Procedures Publication* (TPP).

a) If the DP is textual, it will be described in Section C.

b) Complex non-RNAV and all RNAV DPs will be published graphically and named.

i) The name will be listed by airport name and runway in Section C.

c) Graphic DPs developed solely for obstacle clearance will also have the term “(OBSTACLE)” printed on the chart.

6) A DP that has been developed solely for obstacle avoidance will be indicated with the symbol “T” on appropriate NACO IAP charts and DP charts for that airport.

a) The “T” symbol is to refer you to Section C of the TPP.

7) If ATC does not assign a DP, any published DP may be filed and flown.

a) As a general rule, ATC will assign a DP from an airport without an operating control tower only when compliance with the DP is necessary for separation of aircraft.

8) Responsibilities

a) Prior to departing an airport on an IFR flight, you should consider the type of terrain and other obstacles on or in the vicinity of the departure airport. Also determine

i) Whether a DP is available, and

ii) Whether obstacle avoidance can be maintained or whether the DP should be flown.

b) After you are established on a DP and then vectored or cleared off the DP or DP transition, you shall consider the DP canceled, unless the controller adds "expect to resume DP."

c) If you are instructed to resume a procedure that contains restrictions, such as a DP, ATC will issue/reissue all applicable restrictions or you will be advised to comply with those restrictions.

d) If an altitude to "maintain" is restated, whether prior to departure or while airborne, previously issued altitude restrictions are canceled, including any DP altitude restrictions, if any.

e) When operating from airports where DPs are effective, you should expect ATC clearances containing a DP.

i) Use of a DP requires that you possess the textual description or graphic depiction of the approved current DP, as appropriate.

ii) You must advise ATC if you do not possess a charted DP or a DP description, or if your airplane is not capable of flying the DP.

- If you cannot accept a DP, you should add "NO DP" in the remarks section of your IFR flight plan.

f) Adherence to all restrictions on the DP is required unless ATC clearance to deviate is received.

g) ATC may omit the departure control frequency if a DP clearance is issued and the departure control frequency is published on the graphic DP.

b. En Route Low Altitude Charts provide aeronautical information for navigation under IFR below 18,000 ft. MSL. These four-color charts include

1) Airways

2) Limits of controlled airspace

3) VHF radio aids to navigation (frequency, identification, and geographic coordinates)

4) Airports that have an IAP or a minimum 3,000-ft. hard surface runway

5) Off-route obstruction clearance altitudes (OROCA)

6) Reporting points

7) Special-use airspace areas

8) Military training routes (MTRs)

c. A STAR is an ATC arrival route that has been established for IFR aircraft destined for certain airports to simplify clearance delivery procedures and to facilitate the transition between en route and instrument approach procedures.

1) Some STARs may have mandatory speeds and/or crossing altitudes published.

2) Some STARs may have planning information depicted to inform pilots what clearances or restrictions to "expect."

a) "Expect" altitudes/speeds are not considered crossing restrictions until verbally issued by ATC.

b) "Expect" altitudes/speeds in a STAR should not be used in the event of lost communications unless ATC has specifically advised you to expect these altitudes/speeds as part of a further clearance.

3) You shall maintain the last assigned altitude until receiving clearance to change altitude.

a) At that time, you are expected to comply with all published/issued restrictions.

b) The clearance may be given as a normal descent clearance or the phraseology "DESCEND VIA."

4) A "descent via" clearance authorizes navigation vertically and laterally, in accordance with the STAR, to meet published restrictions.

a) Vertical navigation is at the pilot's discretion; however, adherence to published altitude crossing restrictions and speeds is mandatory unless otherwise cleared.

b) EXAMPLE: "Descend via the Civit One arrival."

c) Pilots cleared for vertical navigation using the phraseology "descend via" shall inform ATC upon initial contact with a new frequency.

i) EXAMPLE: "Cessna 1234 descending via the Civit One arrival."

5) Pilots of IFR aircraft destined to locations for which STARs have been published may be issued a clearance containing a STAR whenever ATC determines it is appropriate.

6) Use of a STAR requires that you possess at least the approved textual description.

a) As with any ATC clearance (or portion of a clearance), it is your responsibility to accept or refuse an issued STAR.

i) Notification may be accomplished by stating "NO STAR" in the remarks section of the flight plan or by the less desirable method of verbally advising ATC.

7) STAR charts are published in the NACO *Terminal Procedure Publication*..

2. Use the current and appropriate navigation publications for your proposed flight.

a. Aeronautical information changes rapidly, and it is vitally important that you check the effective dates on each NACO aeronautical chart and publication to be used.

1) Obsolete charts and publications should be discarded and replaced by current editions.

b. You should also consult the Aeronautical Chart Bulletin section contained in the *Airport/Facility Directory* and NOTAMs for changes essential to the safety of flight that may occur during the effective dates of a chart or publication.

c. NACO charts and publications are continually updated to reflect current terrain and cultural information. The following is an average number of changes per chart per cycle length (time between publication of new charts):

Product	Avg Changes per Chart per Cycle	Cycle Length
Airport/Facility Directory	775	56 days
En Route Low Altitude Chart	35	56 days
Terminal Procedure Publication	75	56 days

d. Chart revisions include changes in radio/navigation frequencies, new obstructions, changes in departure/arrival routes, changes in IAPs to include revised MDAs and DHs, and other temporary or permanent hazards to flight.

3. Select and use the appropriate communication frequencies; select and identify the navigation aids associated with your proposed flight.

a. During your preflight preparation, you should have the frequencies for ground operations (e.g., clearance delivery, ground control, and tower) and navigation facilities.

b. When instructed by ATC to change frequencies, acknowledge the instruction by reading it back to the controller. You should also write it down for your record.

1) If you select the new frequency without an acknowledgment, the controller's workload is increased because (s)he has no way of knowing whether you received the instruction or have had radio communication failure.

2) At times a controller may be working a sector with multiple frequencies and may request, "(Identification), change to my frequency, 123.4."

a) This request means that you are changing frequencies only, not controllers, and the initial callup phraseology may be abbreviated; e.g., "Skyhawk 2479H on 123.4."

3) Select the new frequency as soon as possible unless you were instructed to make the change at a specified time, fix, or altitude.

a) If you are to make a frequency change at a specific time, fix, or altitude, monitor the frequency you are on until reaching the specified time, fix, or altitude.

b) A delay in making a frequency change could result in an untimely receipt of important information.

c. Throughout your flight, you will be required to select the proper navigation facilities.

1) Tune the correct frequency and then positively identify that facility.

2) The Morse code identifier is the means to positively identify a facility.

a) Some VOR/VORTAC will have voice identification.
b) The name of the facility and the word V-O-R will be transmitted.

3) Proper preflight planning will assure you of having the correct navigational aid frequencies available.

4) Ensure that you have the correct GPS waypoints entered into the GPS receiver.

d. A **changeover point (COP)** is a point along the route or airway segment between two adjacent navigation facilities or waypoints where changeover in navigation guidance should occur.

   1) At this point, you should change navigation receiver frequency from the station behind your airplane to the station ahead.

   2) The COP is normally located midway between the navigation facilities, for straight route segments, or at the intersection of radials or courses forming a dogleg, in the case of dogleg route segments.

      a) When the COP is NOT located at the midway point, aeronautical charts will depict the COP location and give the mileage (NM) to the radio aids, such as 20/40.

   3) COPs are prescribed for federal airways, jet routes, area navigation routes, or other direct routes for which an MEA is designated.

   4) COPs are established to prevent loss of navigation guidance, to prevent frequency interference from other facilities, and to prevent use of different facilities by different aircraft in the same airspace.

   5) You are urged to observe COPs to the fullest extent.

4. Perform the appropriate airplane checklist items relative to the phase of flight.

a. Throughout your flight, you must continuously manage your resources. One means of doing this is by properly completing the appropriate checklist for your airplane. We emphasize the appropriate use of checklists throughout this book.

b. Checklists are a means of flying safely. Generally, they are to be used as specified in the *POH* to accomplish a safe flight.

5. Establish two-way communications with the proper controlling agency, using proper phraseology.

a. When establishing communication with a new controller in a radar environment, use the following:

   1) Name of controlling agency, e.g., "Miami Center"

   2) Airplane type and identification, e.g., "Bonanza 66421"

   3) Assigned altitude preceded by "level at," "climbing to," or "descending to," the present vacating altitude, if applicable, e.g., "leaving 7,000 ft. descending to 5,000 ft."

   4) Assigned heading, if applicable

b. When operating in a nonradar environment, on initial contact you should inform the controller of the aircraft's present position, altitude, and ETA to next reporting point.

   1) After initial contact, when a position report is made, you should give the controller a complete position report.

c. Position reporting provides ATC with a means to provide proper aircraft separation, to expedite aircraft movements, and to make accurate estimates of the progress of every aircraft operating on an IFR flight plan.

   1) **Position identification**

      a) When passing a VOR facility, the time reported should be the time at which the first complete reversal of the "to/from" indicator is accomplished.

b) When passing an NDB station, the time reported should be the time at which the ADF indicator in your airplane makes a complete reversal.

c) When an aural or light panel indication is used (e.g., marker beacon), the time should be noted when the signal is first received and again when it ceases. The mean of these two times should be used as the actual time over the fix.

d) If position is given with respect to distance and direction from a reporting point, the distance and direction should be computed as accurately as possible.

e) Except for terminal area transition purposes, position reports or navigation with reference to aids not established for use in the structure in which flight is being conducted will not normally be required by ATC.

2) **Position reporting points.** FARs require you to maintain a listening watch on the appropriate frequency and, unless operating under the provisions in item 3) below, to furnish position reports passing certain reporting points. Reporting points are indicated on en route charts.

a) Compulsory reporting point is a solid triangle (▲).

b) "On request" reporting point is an open triangle (△) and is used only when requested by ATC.

3) Position reporting requirements

a) **Flights along airways or routes.** A position report is required by all flights regardless of altitude (including VFR-on-top operations) over each designated compulsory reporting point along the route being flown.

b) **Flight along a direct route.** Regardless of altitude (including VFR-on-top operations), you need to report over each reporting point used in your flight plan to define the route of flight.

c) **Flight in a radar environment.** When informed by ATC that your airplane is in "RADAR CONTACT," you should discontinue position reports over designated reporting points.

i) You should resume normal position reporting when ATC advises "RADAR CONTACT LOST" or "RADAR SERVICE TERMINATED."

ii) ATC will inform you that you are in radar contact when

- Your airplane is initially identified in the ATC system.
- Radar identification is reestablished after radar service has been terminated or radar contact lost.

4) **Position report items.** Position reports should include the following:

a) Identification

b) Position

c) Time (minutes past the hour)

d) Altitude or flight level (including actual altitude or flight level when operating VFR-on-top)

e) Type of flight plan

i) Not required in reports made directly to ARTCCs or approach control

f) ETA and name of next reporting point

g) Name only of the next succeeding reporting point along the route of flight

h) Pertinent remarks (i.e., unforecast weather conditions)

d. **Additional reports.** Certain reports should be made to ATC or FSS facilities without a specific ATC request.

1) The following reports should be made at all times:

a) When vacating any previously assigned altitude or flight level to a newly assigned altitude or flight level

b) When making an altitude change while operating VFR-on-top

c) When unable to climb/descend at a rate of at least 500 feet per minute (fpm)

d) When approach has been missed (request clearance for specific action, i.e., to alternative airport, another approach, etc.)

e) When the average TAS (at cruising altitude) varies by 5% or 10 kt. (whichever is greater) from that filed in the flight plan

f) The time and altitude or flight level when reaching a holding fix or point to which cleared

g) When leaving any assigned holding fix or point

h) When any loss occurs, in controlled airspace, of VOR, DME, ADF, LORAN, or GPS navigation receiver capability; when complete or partial loss of ILS receiver capability occurs; or when air/ground communications capability is impaired

i) Include your airplane identification, the equipment affected, the degree to which the capability to operate IFR in the ATC system is impaired, and the nature and extent of assistance desired from ATC.

ii) Any other equipment installed in your airplane (e.g., weather radar) that may impair safety and/or ability to operate under IFR should also be reported.

i) When any other information relates to the safety of flight

2) The following reports should be made when not in radar contact:

a) When leaving the final approach fix (FAF) inbound on final approach (nonprecision approach) or when leaving the outer marker or fix used in lieu of the outer marker inbound on final approach (precision approach)

b) When an estimate should be corrected because it is apparent that the estimate as previously submitted is in error in excess of 3 min.

3) Pilots encountering weather conditions which have not been forecast, or hazardous conditions which have been forecast, are expected to forward a report to ATC.

6. **Comply, in a timely manner, with all ATC instructions and airspace restrictions.**

a. When an ATC clearance has been obtained, you must not deviate from the provisions of that clearance unless an amended clearance is obtained or an emergency arises. (FAR 91.123)

b. When ATC issues a clearance or instructions, you are expected to execute its provisions upon receipt.

   1) You are responsible for requesting clarification or amendment, as appropriate, anytime a clearance is not fully understood or is considered unacceptable from a safety standpoint.

   2) ATC, in certain situations, will include the word "IMMEDIATELY" in a clearance or an instruction to impress upon you the urgency of an imminent situation, and your immediate compliance is expected and necessary for safety.

      a) The term "EXPEDITE" is used by ATC when prompt compliance is required to avoid the development of an imminent situation.

c. When a heading is assigned or a turn is requested by ATC, you are expected to initiate the turn promptly, to complete the turn, and to maintain your new heading unless additional instructions are issued.

d. The term "AT PILOT'S DISCRETION" included in the altitude information of an ATC clearance means that ATC has offered you the option to start a climb or descent when you wish.

   1) You are authorized to conduct the climb or descent at any rate you wish and to temporarily level off at any intermediate altitude desired.

   2) However, once you have vacated an altitude, you may not return to that altitude.

e. When ATC has not used the term "AT PILOT'S DISCRETION" nor imposed any climb or descent restrictions, you should initiate the climb or descent on acknowledgment of the clearance.

   1) Descend or climb at an optimum rate consistent with the operating characteristics of the aircraft to 1,000 ft. above or below the assigned altitude, and then attempt to descend or climb at a rate of between 500 and 1,500 fpm until the assigned altitude is reached.

      a) If at any time you are unable to climb or descend at a rate of at least 500 fpm, you must advise ATC.

   2) If it is necessary to level off at an intermediate altitude during the climb or descent, you must advise ATC.

f. If the altitude information of an ATC descent clearance includes a provision to "CROSS (fix) AT" or "AT OR ABOVE/BELOW (altitude)," the manner in which the descent is made to comply with the crossing altitude is at your discretion.

   1) This authorization to descend at your discretion is applicable only to that portion of the flight to which the crossing altitude restriction applies.

g. The guiding principle is that the last ATC clearance has precedence over any previous clearance.

   1) When the route or altitude in a previously issued clearance is amended, the controller will restate applicable altitude restrictions.

h. In case emergency authority is used to deviate from provisions of an ATC clearance, you must notify ATC as soon as possible and obtain an amended clearance.

   1) In an emergency situation that requires ATC to give priority to your airplane, you must, when requested by ATC, make a report within 48 hr. of such emergency situation to the chief of that ATC facility (FAR 91.123).

7. **Exhibit adequate knowledge of two-way radio communication failure procedures.**
   a. This element is covered in detail in Task VII. A., Loss of Communications, beginning on page 328.

8. **Intercepts, in a timely manner, all courses, radials, and bearings appropriate to the procedure, route, or clearance.**
   a. On your IFR departure, you will normally receive navigational guidance from departure control by radar vector.
   b. When your departure is to be vectored immediately following takeoff, you will be advised before takeoff of the initial heading to be flown but may not be advised of the purpose of the heading.
      1) When operating in a radar environment, you should expect to associate the departure heading with vectors to your route given in your IFR clearance.
      2) When given a vector taking you off a previously assigned nonradar route, you will be advised briefly about what the vector is to achieve.
         a) Radar service will be provided until you have been reestablished "on course" using an appropriate navigation aid and you have been advised of your airplane's positions or until a handoff is made to another controller with further surveillance capabilities.
   c. The radar departure is normally simple. Following takeoff, you will contact departure control when advised to do so from the tower controller, or when able after takeoff at an airport without an operating control tower.
      1) At this time, departure control will verify radar contact; tell you briefly the purpose of the vector (airway, point, or route to which you will be vectored); and give headings, altitude and climb instructions, and other information to move you quickly and safely out of the terminal environment.
      2) You should listen to instructions and fly basic instrument maneuvers (climbs, level-offs, turns to predetermined headings, and straight-and-level flight) until the controller tells you your position with respect to the route given in your IFR clearance and the one to contact next and directs you to "resume own navigation."
   d. Departure control will vector you to either a navigation facility or an en route position appropriate to your departure clearance, or you will be transferred to another controller.
      1) The radar departure procedure is like having your instructor along to tell you what to do and when to do it.
      2) The procedure is so easy, in fact, that some pilots are often inclined to depend entirely on radar for navigational guidance, unconcerned about the consequences of loss of radar contact and indifferent to commonsense precautions associated with flight planning.
         a) A radar-controlled departure does NOT relieve you of your responsibilities as PIC. You must be prepared before takeoff to conduct your own navigation according to your IFR clearance.
   e. While en route, you are required to adhere to airways or routes being flown, especially during course changes. Remember that the width of the airway or route is 4 NM on each side of the centerline.
      1) Some variables that must be considered during a course change are turn radius, wind effect, airspeed/groundspeed, and navigation systems in the airplane.

2) Turning before arriving at the fix is one method of adhering to airways or routes.

   a) Turns that begin at or after fix passage may exceed airway or route boundaries.

3) You may use any available approved navigation system in your airplane (DME, GPS) to determine when to begin the turn for a course change.

   a) This is consistent with the intent of FAR 91.181, which requires pilots to operate along the centerline of an airway and along the direct course between navigation aids or fixes/waypoints.

4) If you undershoot or overshoot your new course, you must intercept it in a timely manner.

9. ***Maintain the applicable airspeed within 10 kt.; headings within 10°; and altitude within 100 ft.; and track a course, radial, or bearing.***

   a. If you are unable to maintain these limits because of turbulence or other flight situations beyond your control, notify ATC.

C. Common Errors with IFR Procedures and Clearances

1. **Not using current editions of the appropriate navigation publications for your proposed flight.**

   a. Always check the effective dates of your charts and ensure that you have the latest FDC NOTAMs.

2. **Incorrectly selecting communication frequencies.**

   a. Write down the frequency you are instructed to use to contact a new controller.
   b. Check the frequency you have tuned in on the radio before you transmit.

3. **Not positively identifying the selected navigational aids.**

   a. If the signal is not reliable, the identifier will be removed. Without identifying the station, you will not know if the signal is reliable.

   b. Ensure that you have the correct waypoint if you are using a GPS.

4. **Failure to intercept the desired courses, radials, or bearings in a timely manner.**

   a. Intercepting courses, radials, and bearings is a part of your clearance and needs to be done in a timely manner so you are where you should be in the big picture of air traffic.

END OF TASK

HOLDING PROCEDURES

III.C. TASK: HOLDING PROCEDURES

REFERENCES: 14 CFR Parts 61, 91; AC 61-27; AIM.

NOTE: Any reference to DME will be disregarded if the aircraft is not so equipped.

Objective. To determine that the applicant:

1. Exhibits adequate knowledge of the elements related to holding procedures.
2. Changes to the holding airspeed appropriate for the altitude or aircraft when 3 min. or less from, but prior to arriving at, the holding fix.
3. Explains and uses an entry procedure that ensures the aircraft remains within the holding pattern airspace for a standard, nonstandard, published, or nonpublished holding pattern.
4. Recognizes arrival at the holding fix and initiates prompt entry into the holding pattern.
5. Complies with ATC reporting requirements.
6. Uses the proper timing criteria, where applicable, as required by altitude or ATC instructions.
7. Complies with pattern leg lengths when a DME distance is specified.
8. Uses proper wind correction procedures to maintain the desired pattern and to arrive over the fix as close as possible to a specified time.
9. Maintains the airspeed within 10 kt.; altitude within 100 ft. (30 meters); headings within 10°; and tracks a selected course, radial, or bearing.

A. General Information

1. The objective of this task is to determine your knowledge of and ability to perform holding procedures.

NOTE: If your airplane is not equipped with DME, you may disregard any reference to that instrument.

2. A holding pattern involves a combination of simple basic maneuvers -- two turns and two legs in straight-and-level flight or descending when cleared by ATC.
    a. Although these maneuvers are far less difficult than some, holding procedures are a common source of confusion and apprehension among instrument pilot trainees, and even some instrument-rated pilots.
3. There are many reasons for this apprehension, among them the idea that holding implies uncertainty, delay, procedural complications, and generally an increased workload at a time when you are already busy reviewing details of your instrument approach.
    a. Another reason involves the normal psychological pressure attending approach to your destination, when you become increasingly conscious of the fact that your margin of error is narrowing.
        1) The closer you get to landing, the more decisions you must make, and your aeronautical decision making (ADM) needs to be quick, positive, and accurate as you may have fewer chances to correct the inaccuracies.
4. Like any other flight problem (remember learning to land as a student pilot), the "complicated" holding pattern becomes routine after sufficient study of the procedures in their normal sequence.

B. Task Objectives

1. Exhibit adequate knowledge of the elements related to holding procedures.

a. Holding instructions

1) Whenever you have been cleared to a fix other than the destination airport and delay is expected, it is the responsibility of ATC to issue complete holding instructions (unless the pattern is charted), an EFC (expect further clearance) time, and the best estimate of any additional en route/terminal delay.

2) If the holding pattern is charted and ATC does not issue complete holding instructions, you are expected to hold as depicted on the appropriate chart.

a) ATC may omit all holding instructions except the charted holding direction and the statement "as published"; e.g., "Hold east as published."

b) ATC will issue complete holding instructions if you request them.

3) If no holding pattern is charted and holding instructions have not been received, you should ask ATC for holding instructions prior to reaching the fix.

a) This procedure will eliminate the possibility of an aircraft entering a holding pattern other than the one desired by ATC.

b) If you are unable to obtain holding instructions prior to reaching the fix (due to frequency congestion, stuck microphone, etc.), you should hold in a standard pattern on the course on which you approached the fix and request further clearance as soon as possible.

i) In this event, the altitude of your airplane will be protected so that separation will be provided as required.

4) When no delay is expected, ATC should issue you a clearance beyond the fix as soon as possible and, whenever possible, at least 5 min. before you reach the clearance limit.

5) An ATC clearance requiring an airplane to hold at a fix where the pattern is not charted will include the following information:

a) Direction of holding from the fix in terms of the eight cardinal compass points (i.e., N, NE, E, SE, etc.)

b) Holding fix. The fix may be omitted if included at the beginning of the transmission as the clearance limit.

c) Radial, course, bearing, airway, or route on which the airplane is to hold

d) Leg length in miles if DME or RNAV is to be used. Leg length will be specified in minutes if the pilot requests it or the controller considers it necessary.

e) Direction of turn if left turns are to be made, the pilot requests this information, or the controller considers it necessary

f) Time to expect further clearance (EFC) and any pertinent additional delay information

b. The holding procedure is a predetermined maneuver, with respect to a holding fix, that keeps your airplane within prescribed airspace while awaiting further clearance from ATC.

1) The holding fix is identifiable by use of NAVAIDs and is used as a reference point in establishing and maintaining the position of your airplane while holding.

a) The holding fix can be at a station (VOR or NDB), an intersection, a DME fix, or a waypoint (GPS or other type of RNAV).

2) Examples of holding are presented below.

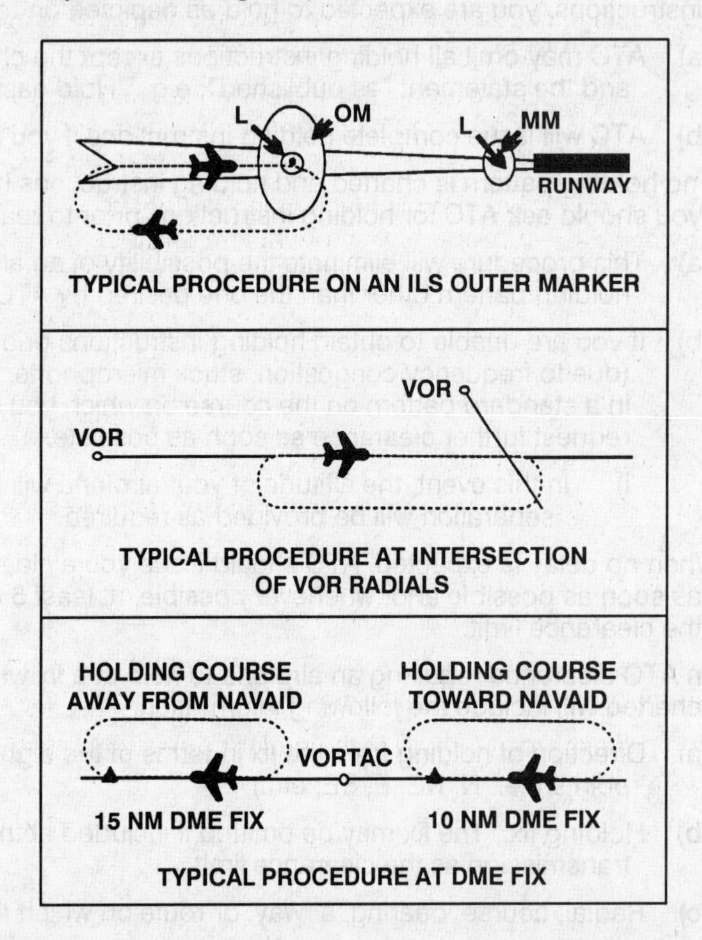

c. The holding pattern is a racetrack pattern (with no wind) in which you follow a specified course inbound to the holding fix, turn 180° and fly a parallel course outbound, and then turn 180° back to the inbound course to the holding fix. The holding pattern and descriptive terms are shown in the figure below.

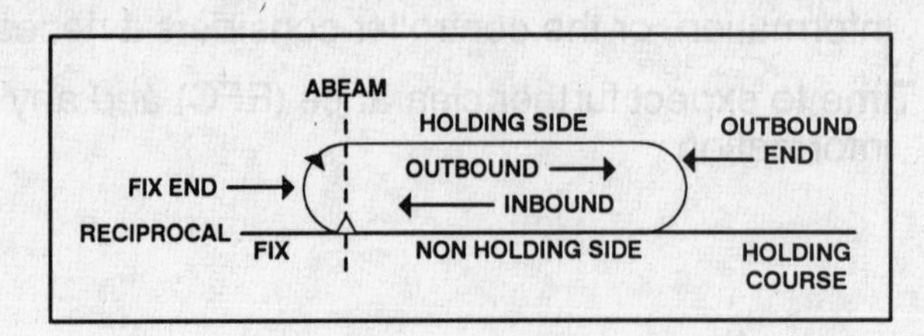

d. Whenever aircraft are holding at an outer fix, ATC will usually provide radar surveillance of the outer fix holding pattern airspace area, or any portion of it, if it is shown on the controller's radar display.

 1) The controller will attempt to detect any holding aircraft that stray outside the holding pattern airspace area and will assist any detected aircraft to return to the assigned airspace area.

 2) Many factors could prevent ATC from providing this additional service, such as workload, number of targets, precipitation, ground clutter, and radar system capability.

  a) These circumstances may make it unfeasible to maintain surveillance of the holding pattern airspace area.

  b) The provision of this service depends entirely upon whether controllers believe they are in a position to provide it and does not relieve you of your responsibility to adhere to an accepted ATC clearance.

e. If you are established in a published holding pattern at an outer fix at an assigned altitude above the published minimum holding altitude (MHA) and are subsequently cleared for the approach, you may descend to the published MHA.

 1) The holding pattern would be only a segment of the IAP if it is published on the IAP chart and is used in lieu of a procedure turn.

f. For those holding patterns in which there is no MHA, when you receive your approach clearance, you must maintain your last assigned altitude until you leave the holding pattern and are established on the inbound course.

 1) Once established, the published minimum altitude of the route segment being flown will apply.

 2) It is ATC's responsibility to assign you a holding altitude that will permit a normal descent on the inbound course.

g. Holding patterns at the most commonly used holding fixes are depicted (charted) on low/high altitude en route, area, and STAR charts.

 1) You are expected to hold in the pattern depicted unless specifically advised otherwise by ATC.

2. Change to the holding airspeed appropriate for the altitude or your airplane when 3 min. or less from, but prior to arriving at, the holding fix.

a. You are required to start a speed reduction when 3 min. or less from the holding fix.

1) You should cross the holding fix, initially, at or below the following maximum holding airspeeds (indicated) for the appropriate altitude:

a) MHA to 6,000 ft. MSL -- 200 kt.
b) 6,001 ft. MSL to 14,000 ft. MSL -- 230 kt.
c) 14,001 ft. MSL and above -- 265 kt.

2) Nonstandard maximum holding airspeeds will be depicted by an icon on the IFR chart.

a) The icon is a standard holding pattern symbol (racetrack) with the airspeed restriction shown either in the center or next to the holding pattern symbol.

b. In your airplane, you will most likely use the approach airspeed for your holding airspeed.

1) Your instructor will provide you with airspeed, configuration, and power setting for your airplane.

2) In your airplane, holding airspeed is __________.

c. While not required, the further reduced airspeed/power setting will provide you with greater fuel savings and endurance.

3. Explain and use an entry procedure that ensures that your airplane will remain within the holding pattern airspace for a standard, nonstandard, published, or nonpublished holding pattern.

a. While other entry procedures may enable you to enter the holding pattern and remain within protected airspace, the parallel, teardrop, and direct entries are the procedures for entry and holding recommended by the FAA.

b. The entry procedure sectors are based upon the direction of the inbound holding course and a line drawn 70° to it from the holding fix, as shown in the FAA figure below.

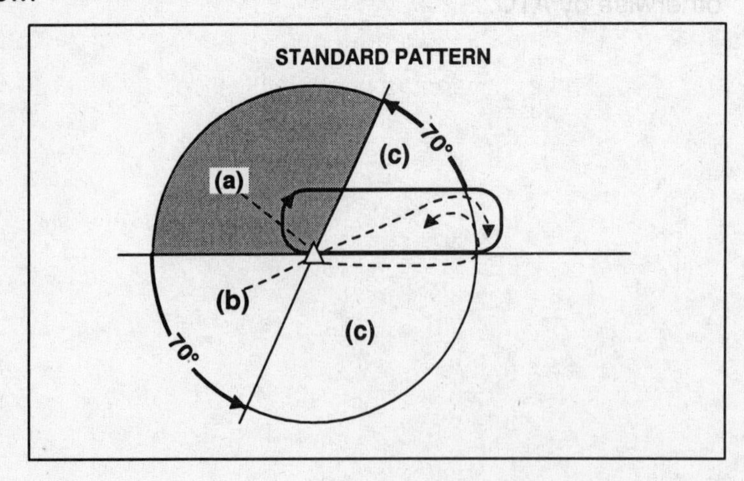

1) **Standard pattern** -- All turns are to the right (as shown on the previous page).
2) **Nonstandard pattern** -- All turns are to the left.
   a) A nonstandard pattern is drawn in the same manner as the standard pattern, except at the fix the turn is left (not right).
   b) From the inbound course, a line is drawn 70° to it from the fix, cutting the outbound leg at about one-third of its length.
3) All turns are made at standard rate, i.e., 3° per second.

c. The FAA holding pattern entry diagram on the previous page and FAA entry procedures described in d., e., and f. below confuse many pilots because they are based on the position from which you are approaching the holding fix, rather than on your heading to the holding fix and the desired outbound course in the holding pattern.

d. **Parallel entry procedure.** When approaching the holding fix from anywhere in sector (a), as illustrated on the previous page , you should use the parallel entry procedure. Once you have crossed the holding fix, turn to a heading to parallel the holding course outbound on the non-holding side for 1 min.; then turn in the direction of the holding pattern through more than 180°, and return to the holding fix or intercept the holding course inbound.

e. **Teardrop entry procedure.** When approaching the holding fix from anywhere in sector (b), as illustrated on the previous page, you should use the teardrop entry procedure. Once you have crossed the holding fix, turn outbound to a heading (outbound heading minus 30° for a standard pattern or outbound heading plus 30° for a nonstandard pattern) for a 30° teardrop entry within the pattern (on the holding side) for a period of 1 min.; then turn in the direction of the holding pattern to intercept the inbound holding course.

f. **Direct entry procedure.** When approaching the holding fix from anywhere in sector (c), as illustrated on the previous page, you should use the direct entry procedure. Once you have crossed the holding fix, you would turn to follow the holding pattern.

g. The better you visualize your airplane's ground track to the holding fix and the ground track of the holding pattern, the easier it is to select and execute the proper entry.

h. You may allow ±5° in heading when determining the appropriate entry procedure.
   1) Thus, if you are approaching the holding fix on an entry sector boundary, you may use either of the two entries.

i. One popular method with many CFIIs and IFR pilots is to use the HI to help visualize the entry procedure.

1) When heading toward the holding fix, note your heading.
2) Next locate the holding pattern outbound course on your HI.
3) For a **nonstandard pattern** (left turns), illustrated on the left below
    a) If the outbound course is located between your heading and 70° to the left, use a teardrop entry.
    b) If the outbound course is located between your heading and 110° to the right, use a parallel entry.
    c) If the outbound course is located anyplace else, use a direct entry.
4) For a **standard pattern** (right turns), illustrated on the right below
    a) If the outbound course is located between your heading and 70° to the right, use a teardrop entry.
    b) If the outbound course is located between your heading and 110° to the left, use a parallel entry.
    c) If the outbound course is located anyplace else, use a direct entry.

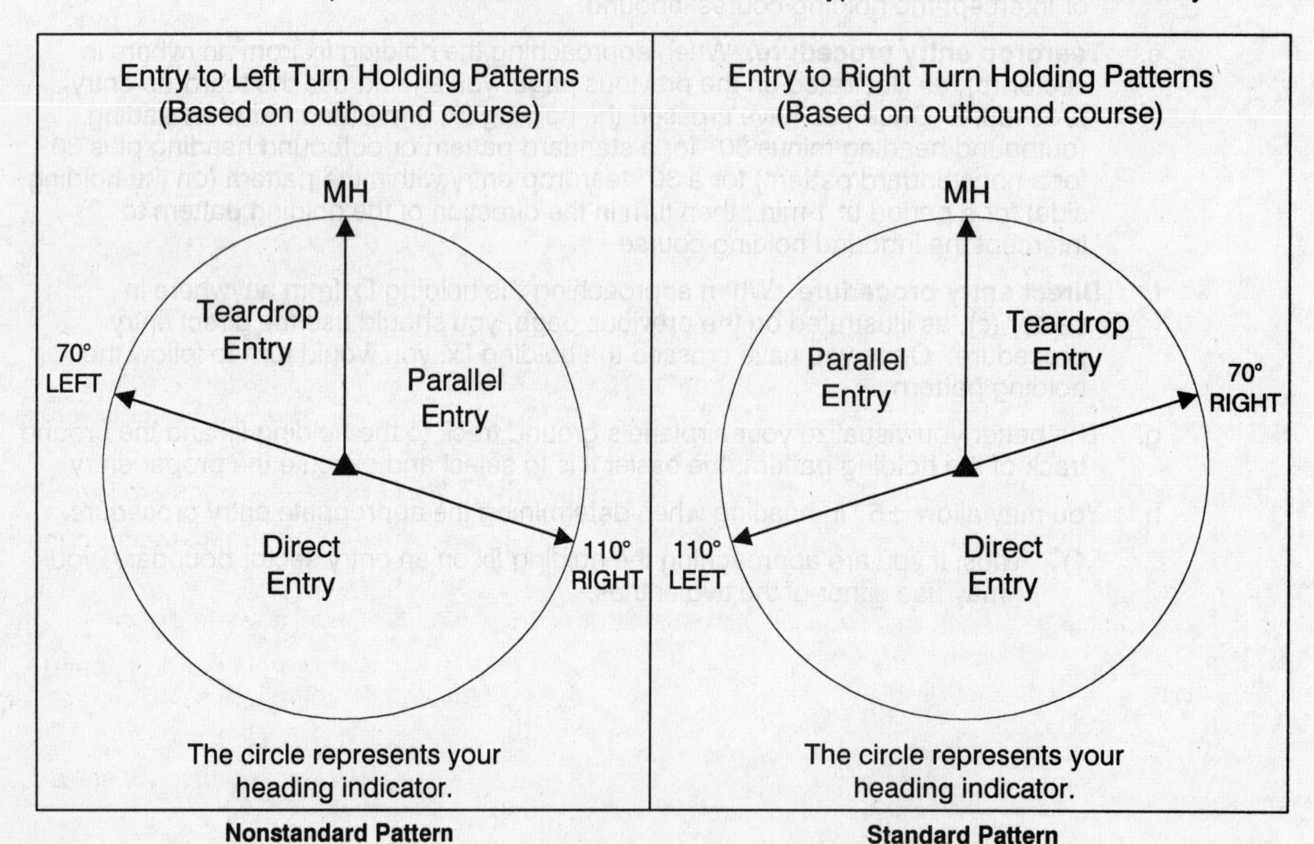

4. Recognize arrival at the holding fix and initiate prompt entry into the holding pattern.

a. Since a holding fix may be at a NAVAID site (VOR or NDB), an intersection, a DME fix, or a waypoint (GPS or other type of RNAV), you must be able to recognize when you are at the holding fix.

1) At a NAVAID site, station passage indicates you are at the holding fix.

2) While tracking on a radial or bearing, an intersection is identified by correctly identifying a cross radial or bearing (i.e., CDI centers or appropriate relative bearing on the ADF) that identifies that intersection.

3) Correctly reading the DME from the appropriate VORTAC will indicate arrival at the DME fix.

4) Understanding how your IFR-certified GPS or other type of RNAV system indicates arrival at a waypoint and the course information is essential. Manufacturers present this information in various styles.

b. Once you arrive at your holding fix, you should promptly execute the appropriate FAA-recommended entry into the holding pattern.

5. Comply with ATC reporting requirements.

a. You are required to report the time and altitude (or flight level) upon reaching the holding fix.

1) You are also required to report when leaving any assigned holding fix or point.

b. To aid you in remembering the necessary procedures and reports required upon entering the hold, you should learn and use the "five T's" each time you are at the fix. As soon as you start your entry, you should perform and confirm each one.

1) **Turn.** As soon as you cross the holding fix, you must turn to the required heading for the appropriate entry (i.e., parallel, direct, or teardrop).

a) Once established, turn to the outbound heading.

2) **Time.** Begin your outbound timing of 1 min. when you cross the fix for teardrop and parallel entries or when you complete the turn to the outbound heading or are abeam the fix for a direct entry.

a) When turning inbound, begin timing to the fix when the wings are level.

3) **Twist.** After you cross the fix, you should twist the OBS setting to the inbound course.

a) Check and adjust your HI, once the wings are level, after each turn.

4) **Throttle.** You are required to reduce speed to an appropriate holding speed for your particular make and model airplane within 3 min. prior to arriving at the fix.

a) Even if your airplane does not reach these maximum speeds, it is still a good idea to slow down to the approach airspeed.

b) It will make the holding pattern easier to fly, keep you within protected airspace, and conserve fuel.

5) **Talk.** You are required to report your entry into the hold, the altitude at which you are holding, and the time you arrived at the holding fix. This is very important to the controller and his/her handling of other aircraft around you.

a) Report leaving the holding fix.

6. **Use the proper timing criteria, where applicable, as required by altitude or ATC instructions.**
   a. The inbound leg should be 1 min. at or below 14,000 ft. MSL.
      1) Above 14,000 ft. MSL, the inbound leg should be 1.5 min.
   b. The initial outbound leg should be flown for 1 min. or 1.5 min. (appropriate to altitude). Timing for subsequent outbound legs should be adjusted, as necessary, to achieve proper inbound leg time.
      1) You may use any navigational means available (i.e., DME, RNAV, etc.) to ensure the appropriate inbound leg times.
   c. Outbound leg timing begins abeam the fix. If the abeam position cannot be determined, start timing when the turn to the outbound heading is completed; i.e., when wings are level.
      1) At a VOR, outbound timing starts when the TO/FROM indicator reverses (i.e., changes from FROM to TO if OBS is set to inbound course).
      2) At an intersection or waypoint, outbound timing starts at the completion of the outbound turn, since the abeam position cannot be determined.
      3) At an NDB, outbound timing starts when the ADF relative bearing is 90° (or 270° for nonstandard holds) plus or minus the drift correction angle.
      4) At a DME fix, outbound timing starts at the completion of the outbound turn, since the abeam position cannot be determined.
   d. You must adjust your outbound leg time to ensure your inbound leg is 1 min. (at or below 14,000 ft. MSL) or 1.5 min. (above 14,000 ft. MSL).
      1) One method to adjust your outbound leg timing is to double the inbound time deviation from the required time (e.g., 1 min.) and apply it to the outbound timing.
         a) If the inbound leg is 50 sec. (10 sec. short of the required 1 min.), double the deviation to 20 sec., and add it to the outbound time for a new time of 1 min. 20 sec.
         b) If the inbound leg is 1 min. 15 sec. (15 sec. more than the required 1 min.), double the deviation to 30 sec., and subtract it from the outbound time for a new time of 30 sec.
      2) For excessive deviations from the required inbound time (e.g., 1 min.), other adjustments may be appropriate (e.g., equal to or half of the deviation added to outbound time).
         a) If the inbound leg is 2 min. (1 min. more than the required 1 min.), halve the 1-min. excess and subtract 30 sec. for a new outbound time of 30 sec.
      3) Continue to make progressive adjustments until the correct outbound time is determined.
      4) This step is done simultaneously with item 8., on page 131. Changes in outbound timing will affect outbound heading and vice versa.
7. **Comply with pattern leg lengths when a DME distance is specified.**
   a. DME holding is subject to the same entry and holding procedures except that distances (expressed in NM) are used in lieu of time values.
   b. ATC specifies the outbound leg length.

c. The end of the outbound leg is determined by the DME reading.

1) EXAMPLE: When holding at a DME fix and the inbound course is toward the NAVAID, the fix distance is 10 NM, and the leg length is 5 NM, the end of the outbound leg will be reached when the DME reads 15 NM.

a) Sample clearance for the holding pattern depicted in the figure below is ". . . hold **east** of the 10 DME fix on the 090 radial, 5-mile legs. Expect further clearance at 1725. Time now is 1655."

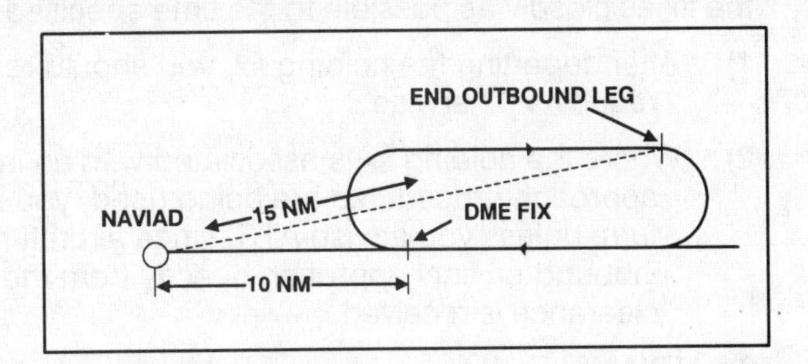

2) EXAMPLE: When holding at a DME fix and the inbound course is away from the NAVAID, the fix distance is 28 NM, and the leg length is 8 NM, the end of the outbound leg will be reached when the DME reads 20 NM.

a) Sample clearance for the holding pattern depicted in the figure below is ". . . hold **west** of the 28 DME fix on the 090 radial, 8-mile legs. Expect further clearance at 1725. Time now is 1655."

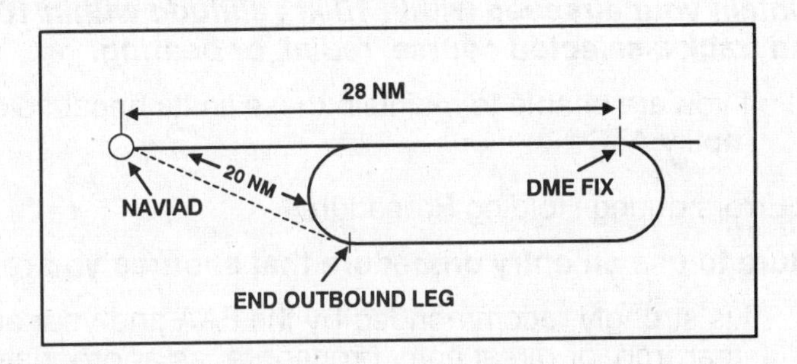

8. Use proper wind correction procedures to maintain the desired pattern and to arrive over the fix as close as possible to a specified time.

a. The symmetrical racetrack holding pattern cannot be tracked when a wind exists.

b. Compensate for wind effect by drift correction on the inbound and outbound legs.

1) First, determine the wind correction angle required while tracking on the inbound course. Once you have determined the wind correction angle inbound, triple it when on the outbound leg to compensate for wind drift during the turns and on the straight legs.

a) EXAMPLE: If correcting left by 8° when inbound, correct right by 24° when outbound.

b) Note that, while the FAA recommends tripling the inbound wind correction angle, large corrections on the outbound leg (i.e., over 30°) may result in overcompensation. To avoid overcompensation, some pilots will only double the inbound wind correction angle.

i) EXAMPLE: If correcting left by 20° when inbound, correct right by 40° when outbound.

2) Failure to use this technique may result in your flying into unprotected airspace (i.e., the non-holding side) and/or make it difficult to intercept and establish yourself on the inbound course, resulting in problems maintaining the pattern.

c. Normally it should take you no more than two or three complete circuits to have your wind corrections and the appropriate inbound leg time established.

d. While holding at a fix, when you receive instructions specifying the time of departure from the fix, you should adjust the timing of your holding pattern so you can leave the fix as closely as possible to the time specified.

1) After departing the holding fix, you should resume normal speed unless ATC requests otherwise.

2) Where the holding fix is associated with an instrument approach and timed approach procedures are being used, you should not execute a procedure turn, unless you advise ATC, since aircraft holding are expected to proceed inbound on final approach directly from the holding pattern when approach clearance is received.

3) EXAMPLE: You are cleared to leave the holding fix at 1207. As you arrive over the fix, you note the time is 1204. Since each turn requires 1 min. (total of 2 min.), 1 min. remains for both the outbound and the inbound leg. If there is no wind, you would proceed outbound for 30 sec. in order to arrive at the fix at 1207.

a) If you have a headwind or tailwind on the inbound course, you adjust the outbound time in order to arrive at the fix at the appropriate time.

9. *Maintain your airspeed within 10 kt.; altitude within 100 ft.; and headings within 10°; and track a selected course, radial, or bearing.*

a. If you are unable to maintain these limits because of turbulence or other situations, notify ATC.

C. Common Errors during Holding Procedures

1. Failure to use an entry procedure that ensures you remain in protected airspace.

a. It is strongly recommended by the FAA and your author that you use a parallel, teardrop, or direct entry procedure, as appropriate.

2. Failure to recognize holding fix passage.

a. You must divide your attention between your flight and navigation instruments.
b. Your holding fix will be at a VOR, an NDB, an intersection, a DME fix, or a waypoint.

3. Failure to comply with ATC instructions.

a. This error is normally due to not understanding the holding clearance.

1) Read back the instructions to ATC and ask for clarification of any part you do not understand.

4. Improper timing and wind drift corrections.

a. Outbound timing starts when you are abeam the holding fix.

1) If the abeam position cannot be identified, timing begins when you roll out (wings level) on the outbound heading.

b. The outbound time must be adjusted so that the inbound time is 1 min. (at or below 14,000 ft. MSL) or 1.5 min. (above 14,000 ft. MSL).

c. Outbound leg wind drift corrections should be triple the inbound leg wind drift correction.

D. VOR Intersection and ADF Holding Examples

1. VOR intersection holding

   a. Assume that you are tracking inbound on the 247° radial of VOR "A" (i.e., 067° course) and you are to hold at intersection "H."

   b. If your VOR equipment is limited to one receiver, it is especially important to establish your inbound heading for accurate course following, while orienting yourself to the 177° radial of VOR "B."

      1) To prevent overshooting the fix for holding entry, you must perform your position checks and tuning accurately and quickly.

      2) You can easily misinterpret your position with respect to the 177° radial if you fail to set the OBS correctly.

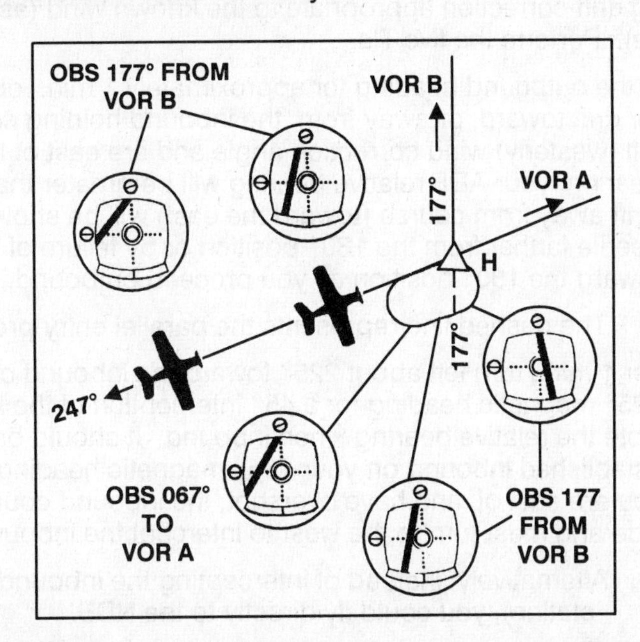

   c. When establishing a fix or an intersection by means of VOR stations on either side of your course, the TO/FROM indicator will read FROM if the OBS is set with the radial (not the reciprocal) under the index.

      1) Visualize flying outbound on the 177° radial of VOR "B." If you were west of your course, the CDI would be deflected toward the left as illustrated above in the two top CDI indications in the diagram.

      2) Note the opposite CDI indication is given if you set the OBS to the R-357 of VOR "B." The TO/FROM indicator will be TO, and you will receive right deflections when west of the 177° radial (visualize flying to the VOR on a 357° course/OBS).

   d. Roll into a standard-rate right turn as the CDI of VOR "B" centers. As you roll out on the outbound heading, check the CDI to determine the position of the 177° radial.

      1) Initially, you are east of the 177° radial with a right CDI deflection.

      2) Begin timing for the outbound leg as you roll out on the outbound heading.

      3) Then tune in VOR "A," setting the OBS on 067°, if you have only a single VOR.

      4) As illustrated, when you switch to the 067° course to VOR "A," you have a left deflection.

e. After 1 min. outbound, execute a standard-rate right turn to intercept the 247° radial of VOR "A" and track inbound to the holding fix (intersection H).

f. If you have two VORs, you do not need to switch back and forth from one VOR to the other.

  1) VOR "A" tuned in on VOR 1, OBS set to 067°
  2) VOR "B" tuned in on VOR 2, OBS set to 177°

2. NDB holding

a. Assume you are tracking inbound to an NDB on a 20° magnetic bearing TO the station using a fixed-card ADF. You have been cleared to hold north of the NDB on the 350° bearing FROM the station. See the diagram on the opposite page.

b. Turn to parallel the outbound course as the ADF needle indicates station passage, applying drift correction appropriate to the known wind (assume west for this example). Perform the five T's.

  1) Fly the outbound heading for approximately 1 min., observing the ADF needle for drift toward, or away from, the inbound holding course. If you apply a 10° left (westerly) wind correction angle and are east of the 350° magnetic bearing, your ADF relative bearing will be greater than 180° (needle left of tail). Drift away from course (toward the east) will be shown by movement of the needle farther from the 180° position or by failure of the needle to move toward the 180° position as you proceed outbound.

    a) The dashed line represents the parallel entry procedure.

  2) After 1 min., turn left about 225° toward the inbound course, rolling out on a 125° magnetic heading for a 45° interception of the inbound course of 170°. Note the relative bearing when inbound. It should be less than 45° when established inbound on your 125° magnetic heading. If it is greater than 045°, you are east of, and have overshot, the inbound course on the non-holding side and must turn to the west to intercept the inbound course.

    a) Alternatively, instead of intercepting the inbound course (170° TO the station), you could fly directly to the NDB.

  3) Lead the turn to the inbound course (170°) and roll out with drift correction. A 10° wind correction angle would be a 180° heading inbound. If you are on course, your wind correction angle will be equal and opposite to the ADF needle displacement from zero. In this example, your heading would be 10° right of course (180°), with the ADF needle pointing 10° left of the nose.

  4) Track inbound, using small corrections. The quicker you establish the desired track, the fewer your holding problems since both basic flight techniques and procedural details will keep you busy.

  5) Turn outbound on station passage. With the wind from the west, as shown, you would have a left (westerly) wind correction angle, and your track would end up closer to the inbound course than shown in the diagram on the opposite page.

6) Roll out on an outbound heading with a drift correction angle equal to triple the amount of inbound drift correction (assume 6° to 7° inbound for this example). With a 20° left (west) wind correction angle (330° heading), you will begin outbound timing (1 min.) when the ADF indicates a relative bearing of 110° (i.e., abeam the NDB station). As you maintain the outbound heading, the needle moves toward the 180° position. With experience, you will learn to recognize drift by rate of movement of the ADF needle -- rapidly toward the tail position if you drift inward or a strong tailwind exists, slowly toward the tail if you drift outward or a strong headwind exists.

7) With correct inbound and outbound wind correction angles, your ADF should read zero, plus or minus the appropriate wind correction angle, as you complete the turn to track inbound.

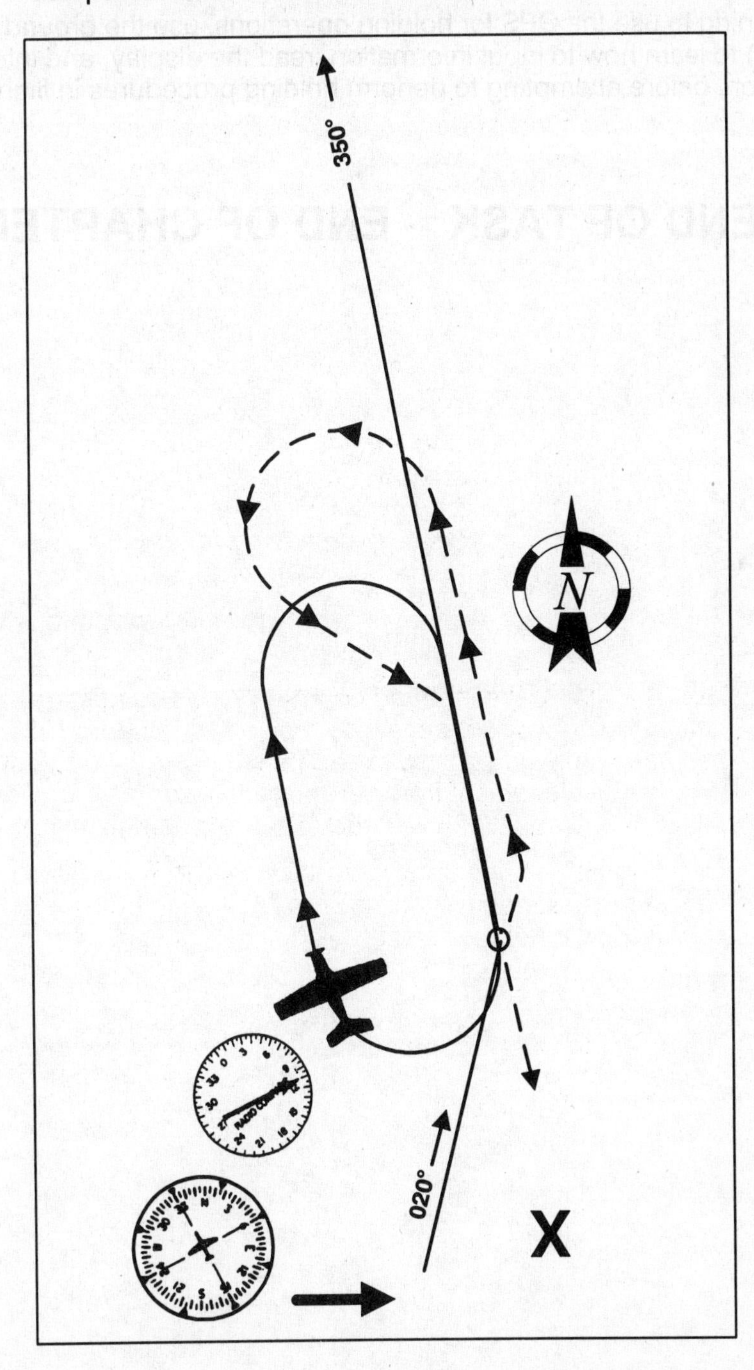

E. GPS Holding

1. No special knowledge about holding procedures is required when using a GPS.
2. The most important knowledge you must have is how to operate your GPS receiver in a holding pattern.
   a. Each manufacturer requires you to operate the GPS receiver in a specific manner (i.e., just because you know how to operate one make of GPS does not mean you can operate another make of GPS without training).
      1) Sometimes the GPS receiver is operated differently depending on whether you are at a published holding fix (i.e., the missed approach holding fix) or at an unpublished holding fix.
3. When learning to use the GPS for holding operations, use the ground training mode (if available) to learn how to input information, read the display, and interpret the displayed information, before attempting to perform holding procedures in flight.

END OF TASK -- END OF CHAPTER

CHAPTER IV
FLIGHT BY REFERENCE TO INSTRUMENTS

This chapter explains the seven tasks (A-G) of Flight by Reference to Instruments. These tasks include both knowledge and skill. Your examiner is required to test you on all seven tasks. Your examiner will select at least two of the tasks, A through E, to be performed without the use of the attitude and heading indicators. Task F, Steep Turns, shall be performed using all available instruments, and Task G, Recovery from Unusual Flight Attitudes, shall be performed without the use of the attitude indicator.

The FAA is concerned about numerous fatal accidents involving spatial disorientation of instrument-rated pilots who have attempted to control and maneuver their aircraft in instrument meteorological conditions (IMC) with inoperative gyroscopic heading and attitude indicators. Thus, the FAA has stressed that it is imperative for instrument pilots to acquire and maintain adequate partial panel instrument skills and not to be overly reliant upon the gyro-instrument systems.

Each task in this chapter requires a knowledge of attitude instrument flying procedures and a demonstration of the skills needed to perform the basic instrument maneuvers using both full and partial panel.

ATTITUDE INSTRUMENT FLYING

A. Attitude instrument flying may be defined in general terms as the control of an airplane's spatial position by use of instruments rather than by outside visual reference. Thus, proper interpretation of the flight instruments provides the same information as visual references outside the airplane.

B. There are two basic methods in use for learning attitude instrument flying -- **control and performance** and **primary and supporting**.

1. Both methods involve the use of the same instruments, and both use the same response for attitude control.

    a. They differ in their reliance on the attitude indicator (AI) and consequently on the use and interpretation of other instruments.

2. Your examiner is required to determine that you are competent in either the control and performance or the primary and supporting method of instrument flying.

    a. Your CFII will advise you on which method to pursue.

    b. The primary and supporting method of attitude instrument flying is described in AC 61-27, *Instrument Flying Handbook* (and this book), and is recommended by the FAA because it requires specific knowledge and interpretation of each individual instrument during training.

C. Control and Performance

1. Control instruments present the information necessary for changing and maintaining the power and attitude appropriate for the desired airplane performance.

   a. The AI exclusively controls pitch and bank.

   b. The tachometer (or manifold pressure gauge and tachometer in combination) controls power.

      NOTE: These are the only instruments that should be referred to when applying control inputs (pitch, bank, or power).

2. Performance instruments present the results from power and attitude control.

   a. Altimeter (ALT)
   b. Airspeed indicator (ASI)
   c. Vertical speed indicator (VSI)
   d. Heading indicator (HI)
   e. Turn coordinator (TC) or turn-and-slip indicator (T&SI)

D. Primary and Supporting

1. This method is recommended by the FAA and groups the instruments as they relate to control function as well as airplane performance.

   a. All maneuvers involve some degree of motion about the lateral, longitudinal, and/or vertical axis.

2. Attitude control is stressed in this book (and by the FAA) in terms of pitch control, bank control, power control, and trim control. Instruments are divided into the following three categories:

   a. Pitch instruments

      1) Attitude indicator (AI)
      2) Altimeter (ALT)
      3) Airspeed indicator (ASI)
      4) Vertical speed indicator (VSI)

   b. Bank instruments

      1) Attitude indicator (AI)
      2) Heading indicator (HI)
      3) Turn coordinator (TC) or turn-and-slip indicator (T&SI)
      4) Magnetic compass

   c. Power instruments

      1) Manifold pressure gauge (MP)
      2) Tachometer (RPM)
      3) Airspeed indicator (ASI)

3. For a particular maneuver or condition of flight, the pitch, bank, and power control requirements are most clearly indicated by certain key instruments.

   a. Those instruments which provide the most pertinent and essential information are referred to as primary instruments.

   b. Supporting instruments back up and supplement the information shown on the primary instruments.

   c. For each maneuver, one primary instrument from each of the above categories will be used. Several supporting instruments from each category may be used.

4. This concept of primary and supporting instruments in no way lessens the value of any particular instrument.
   a. The attitude indicator is the basic attitude reference. It is the only instrument that portrays instantly and directly the actual flight attitude.
      1) It should always be used, when available, in establishing and maintaining pitch and bank attitudes.
5. Remember, the primary instruments (for a given maneuver) are the ones that will show the greatest amount of change over time if the maneuver is being improperly controlled (pitch, bank, power).

E. During your attitude instrument training, you should develop three fundamental skills involved in all instrument flight maneuvers: instrument cross-check, instrument interpretation, and airplane control. Trim technique is a skill that should be refined for instrument flying.
   1. **Cross-checking** (also called scanning) is the continuous and logical observation of instruments for attitude and performance information.
      a. You will maintain your airplane's attitude by reference to instruments that will produce the desired result in performance.
      b. Since your AI is your reference instrument for airplane control and provides you with a quick reference to your pitch and bank attitude, it should be your start (or home-base) for your instrument scan. You should begin with the AI, scan one instrument (e.g., the HI), and then return to the AI before moving to a different instrument, as shown below.

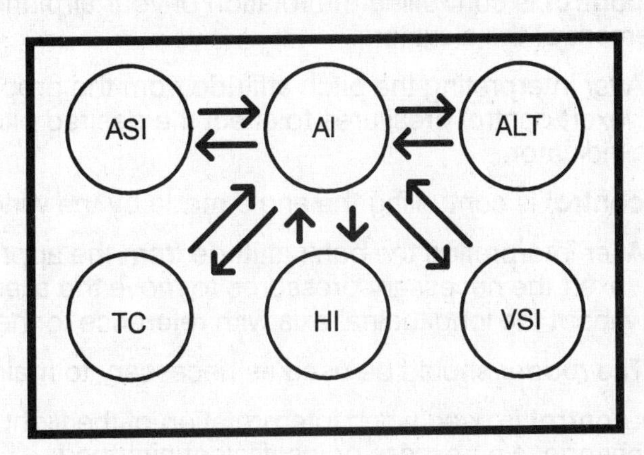

      c. Another type of instrument scan is called a "T" scan. Once again, the AI is the starting point. You should scan one instrument and then return to the AI, as shown below.

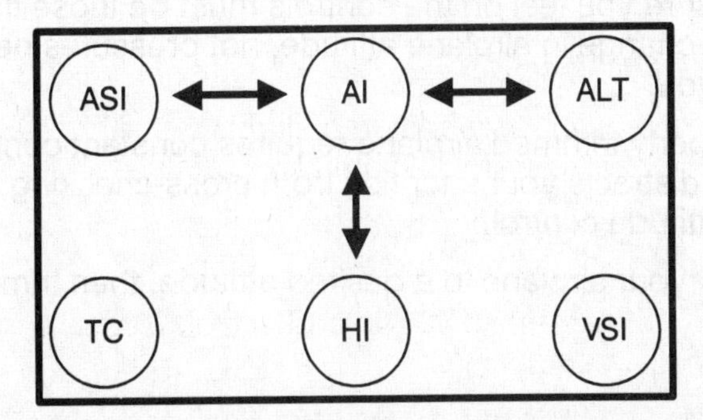

         1) The TC and VSI are checked as required by the flight maneuver (e.g., straight-and-level, climbing, descending, or turning).

d. Your instructor will show you various scanning techniques and discuss the pros and cons of each.

e. Frequent cross-check faults are

1) Fixation, or staring at a single instrument

2) Omission of an instrument from cross-check

3) Emphasis on a single instrument, instead of on a combination of instruments necessary for attitude information

f. The attitude indicator is at the center of the scan. Your cross-check pattern should include the attitude indicator as every second or third instrument scanned.

2. **Instrument interpretation** requires you to understand each instrument's construction, operating principle, and relationship to the performance of your airplane.

a. This understanding enables you to interpret the indication of each instrument during the cross-check.

b. Some instruments are quicker and more accurate than others.

1) EXAMPLE: The airspeed indicator tends to need time to settle after a pitch or power change before it portrays an accurate indication, while the attitude indicator gives almost instantaneous indication of pitch and bank changes.

3. **Airplane control** requires you to maintain your airplane's attitude or change it by interpretation of the instruments. It is composed of three elements.

a. **Pitch control** is controlling the rotation of your airplane about the lateral axis by movement of the elevators.

1) After interpreting the pitch attitude from the proper flight instruments, you will exert control pressures to effect the desired pitch with reference to the attitude indicator.

b. **Bank control** is controlling the angle made by the wing and the horizon.

1) After interpreting the bank attitude from the appropriate instruments, you will exert the necessary pressures to move the ailerons and roll your airplane about the longitudinal axis with reference to the attitude indicator.

2) The rudder should be used as necessary to maintain coordinated flight.

c. **Power control** is used when interpretation of the flight instruments indicates a need for a change, e.g., excess or insufficient airspeed.

4. **Trim** is used to relieve all possible control pressures held after a desired attitude has been attained.

a. The pressure you feel on the controls must be those that you apply while controlling a planned change in airplane attitude, not pressures held because you let the airplane control you.

b. An improperly trimmed airplane requires constant control pressures, produces tension, distracts your attention from cross-checking, and contributes to abrupt and erratic attitude control.

c. Always fly your airplane to a desired attitude, then trim. Do not try to fly your airplane with trim.

STRAIGHT-AND-LEVEL FLIGHT

IV.A. TASK: STRAIGHT-AND-LEVEL FLIGHT

REFERENCES: 14 CFR Part 61; AC 61-27.

Objective. To determine that the applicant:

1. Exhibits adequate knowledge of the elements related to attitude instrument flying during straight-and-level flight.
2. Maintains straight-and-level flight in the aircraft configuration specified by the examiner.
3. Maintains the heading within 10°, altitude within 100 ft. (30 meters) and airspeed within 10 kt.
4. Uses proper instrument cross-check and interpretation, and applies the appropriate pitch, bank, power, and trim corrections.

A. General Information

1. The objective of this task is for you to demonstrate your ability to perform attitude instrument flying during straight-and-level flight.
2. Our discussion of this maneuver is based on an airplane equipped with a turn coordinator (TC). If your airplane has a turn-and-slip indicator, the descriptions apply if "turn needle" is substituted for "miniature airplane of the TC."

B. Task Objectives

1. **Exhibit adequate knowledge of the elements related to attitude instrument flying during straight-and-level flight.**
   a. Flying straight means to maintain a constant heading on the heading indicator, which is done by keeping the wings level on the attitude indicator and the ball centered on the turn coordinator.
   b. Flying level means to maintain a constant altitude on the altimeter, which is done by holding the correct pitch attitude on the attitude indicator for a specific power setting and a level indication on the VSI, i.e., 0 feet per minute (fpm).
   c. Straight-and-level flight is very important as it is the beginning and end of most maneuvers. It is also important during cruise segments for comfort and fuel efficiency.
2. **Maintain straight-and-level flight in the airplane configuration specified by your examiner.**
   a. The configuration can be at any airspeed with landing gear extended or retracted (if applicable) and flaps extended or retracted.
   b. The attitude indicator will show different pitch attitudes for different configurations.
3. ***Maintain the heading within 10°, altitude within 100 ft., and airspeed within 10 kt.***
   a. Full panel
      1) The bank instruments are the heading indicator (HI), the turn coordinator (TC), and the attitude indicator (AI).
         a) The primary bank instrument for straight-and-level flight is the HI. If you deviate from the desired heading, you will correct by establishing a bank with reference to the AI and cross-check with the TC to ensure proper control is used to return your airplane to the desired heading.

2) The pitch instruments are the altimeter (ALT), the AI, and the vertical speed indicator (VSI).

   a) The primary pitch instrument for straight-and-level flight is the ALT. If you deviate from the assigned altitude, you should adjust the pitch with reference to the AI and cross-check with the VSI and ALT to ensure proper control was used to return your airplane to the assigned altitude.

   b) The secondary pitch instruments are the AI and VSI. Maintain cross-checking and interpretation to remain at the desired altitude.

      i) As a trend instrument, the VSI will show immediately, even before your ALT, the initial vertical movement of your airplane.

3) The power instruments are the airspeed indicator (ASI) and the manifold pressure (MP) and/or RPM (tachometer).

   a) The primary power instrument for straight-and-level flight is the ASI.

   b) Power control must be related to its effect on altitude and airspeed, since any change in power setting results in a change in the airspeed or the altitude of your airplane.

      i) At any given airspeed, the power setting determines whether your airplane is in level flight, in a climb, or in a descent.

   c) If the ASI is not showing the desired value, always check the ALT before deciding that a power change is necessary.

   d) If your altitude is higher than desired and your airspeed is low, or vice versa, a change in pitch alone may return your airplane to the desired altitude and airspeed.

      i) If both ALT and ASI are high, or low, a change in both pitch and power is necessary.

e) Refer to the MP and/or RPM gauges to ensure proper power control, and cross-check with the ASI until the desired value is maintained.

f) The secondary power instruments are the MP and/or RPM.

4) Primary and supporting instruments for straight-and-level flight (full panel) are presented below.

MANEUVER	PITCH	BANK	POWER
Straight-and-level flight			
Primary Supporting	ALT AI, VSI	HI AI, TC	ASI MP and/or RPM

b. Partial panel (inoperative attitude and heading indicators)

1) The bank instruments are the TC and the magnetic compass (MC).

a) The primary bank instrument with an inoperative HI is the TC. You will need to maintain wings level on the miniature airplane of the TC and cross-check with the MC to ensure that you are maintaining your desired heading.

2) The pitch instruments are the ALT and the VSI.

a) The only difference between full and partial panel operations is that you do not have the AI as a supporting instrument.

3) The power instruments are not affected by a loss of the attitude and heading indicators (i.e., a vacuum system failure).

4) Primary and supporting instruments for straight-and-level flight (partial panel) are presented below.

MANEUVER	PITCH	BANK	POWER
Straight-and-level flight			
Primary Supporting	ALT VSI	TC MC	ASI MP and/or RPM

4. Use proper instrument cross-check and interpretation, and apply the appropriate pitch, bank, power, and trim corrections.

a. Maintain level flight by adjusting your pitch as necessary to hold your assigned altitude.

1) Trim to relieve all elevator control pressures.

2) Small altitude deviations (i.e., 100 ft. or less) should be corrected with pitch only.

a) Rate of return to altitude should be approximately 200 fpm.

3) Large altitude deviations (i.e., greater than 100 ft.) may be more easily corrected by adjusting both pitch and power.

a) Use a correspondingly greater rate of return to altitude (approximately double your error in altitude).

4) The VSI becomes the primary pitch instrument while returning to altitude after a deviation has been noticed during straight-and-level flight.

b. Maintain straight flight by holding the wings level on the AI and maintaining your heading on the HI.

1) When you note a deviation from straight flight on your HI, make your correction to the desired heading by using an angle of bank no greater than the number of degrees to be turned.

a) Limit your bank corrections to a bank angle no greater than that required for a standard-rate turn.

b) Use coordinated aileron and rudder.

2) The ball of the TC should be centered. If not, your airplane is in uncoordinated flight either by being improperly trimmed (aileron) and/or by your holding rudder pressure.

a) Trim the airplane so the wings are level (if aileron trim is available) and the ball of the TC is centered (rudder trim).

c. Maintain airspeed with power.

C. Common Errors during Straight-and-Level Flight

1. Fixation, omission, and emphasis errors during instrument cross-check.

a. Fixation, or staring at a single instrument, usually occurs for a good reason but with poor results.

1) You may stare at (or fixate on) the ALT, which reads 200 ft. below assigned altitude, wondering how the needle got there. During that time, perhaps with increasing tension on the controls, a heading change occurs unnoticed, and more errors accumulate.

2) It may not be entirely a cross-checking error. It may be related to difficulties with one or both of the other fundamental skills (i.e., interpretation and control).

b. Omission of an instrument from the cross-check may be caused by failure to anticipate significant instrument indications following attitude changes.

1) Supporting instruments should be included in the scan.

c. Emphasis on a single instrument, instead of on the combination of instruments necessary for attitude information, is normal during the initial stages of instrument training.

1) You may tend to rely on the instrument that you understand the best, e.g., the ALT.

2) The VSI can give more immediate pitch information than the ALT.

2. Improper instrument interpretation.

a. This error may indicate that you do not fully understand each instrument's operating principle and relationship to the performance of your airplane.

b. You must be able to interpret even the slightest changes in your instrument indications from your cross-checking.

3. **Improper control applications.**
   a. This error normally occurs when you incorrectly interpret the instruments and then apply the improper controls to obtain a desired performance, e.g., using rudder pressure to correct for a heading error.
   b. It may also occur when you apply control inputs (pitch and bank) without referring to the AI.
4. **Failure to establish proper pitch, bank, or power adjustments during altitude, heading, or airspeed corrections.**
   a. You must understand which instruments provide information for pitch, bank, and power.
      1) The AI is the only instrument for pitch and bank control inputs.
   b. This error may indicate that you do not fully understand instrument cross-check, interpretation, and/or control.
5. **Faulty trim technique.**
   a. Trim should be used, not to substitute for control with the control yoke and rudder, but to relieve pressures already held to stabilize attitude.
   b. Excessive trim control induces control pressures that must be held until you retrim properly.
      1) Use trim frequently and in small amounts.
   c. Improper adjustment of seat or rudder pedals for comfortable positioning of legs and feet may contribute to trim errors.
      1) Tension in the ankles makes it difficult to relax rudder pressures.

END OF TASK

CHANGE OF AIRSPEED

IV.B. TASK: CHANGE OF AIRSPEED

REFERENCES: 14 CFR Part 61; AC 61-27.

Objective. To determine that the applicant:

1. Exhibits adequate knowledge of the elements relating to attitude instrument flying during change of airspeeds in straight-and-level flight and in turns.

2. Establishes a proper power setting when changing airspeed.

3. Maintains the heading within 10°, angle of bank within 5° when turning, altitude within 100 ft. (30 meters), and airspeed within 10 kt.

4. Uses proper instrument cross-check and interpretation, and applies the appropriate pitch, bank, power, and trim corrections.

A. General Information

1. The objective of this task is for you to demonstrate your ability to perform attitude instrument flying during change of airspeeds in straight-and-level flight and in turns.
2. Our discussion of this maneuver is based on an airplane equipped with a turn coordinator (TC). If your airplane has a turn-and-slip indicator, the descriptions apply if "turn needle" is substituted for "miniature airplane of the TC."
3. During your flight training, your instructor will have you perform slow flight and stalls to increase the proficiency of your instrument flying skills.

B. Task Objectives

1. **Exhibit adequate knowledge of the factors relating to attitude instrument flying during change of airspeeds in straight-and-level flight and in turns.**

   a. You will be required to fly at other than normal cruise airspeed, e.g., holding or approach speed.

      1) When power is changed to vary airspeed, a single-engine propeller-driven airplane tends to change attitude around all three axes of movement.

         a) Adding power will normally cause your airplane to yaw and roll left unless you apply counteracting aileron and rudder pressures. These left-turning tendencies are caused by one or more of the following:

            i) Torque reaction from the engine and propeller
            ii) Corkscrewing effect of the slipstream
            iii) Gyroscopic action of the propeller
            iv) Asymmetrical loading of the propeller (P-factor)

         b) As power and airspeed increase, slipstream velocity and downwash lift on the elevators increase. At the same time, lift on the wing increases, and the airplane has a nose-high tendency.

            i) If the airplane is trimmed for level flight, an increase in power and airspeed must be accompanied by forward pressure on the controls if you are to remain in level flight.

         c) As power and airspeed decrease, slipstream velocity and downwash lift on the elevators decrease. At the same time, lift on the wings decreases, and the airplane has a nose-low tendency.

            i) If the airplane is trimmed for level flight, a decrease in power and airspeed must be accompanied by back pressure on the controls if you are to remain in level flight.

b. This maneuver is effective for increasing your proficiency in all three basic instrument skills. It also contributes to your confidence in the instruments during attitude and power changes involved in more complex maneuvers.

2. Establish a proper power setting when changing airspeed.

a. Power control and airspeed changes are much easier when you know in advance the approximate power settings necessary to maintain various airspeeds in straight-and-level flight and in turns.

1) Your instructor will provide you with the power settings for your airplane.

3. Maintain heading within 10°, angle of bank within 5° when turning, altitude within 100 ft., and airspeed within 10 kt.

a. Change of airspeeds in straight-and-level flight (full panel)

1) The bank instruments are the heading indicator (HI), the turn coordinator (TC), and the attitude indicator (AI).

a) The primary bank instrument is the HI. If you deviate from the desired heading, you will correct by establishing a bank with reference to the AI and cross-check with the TC to ensure proper control is used to return your airplane to the desired heading.

2) The pitch instruments are the altimeter (ALT), the AI, and the vertical speed indicator (VSI).

a) The primary pitch instrument is the ALT. As your airspeed changes, you should adjust the pitch with reference to the AI and cross-check with the VSI and ALT to ensure proper control pressure is used to maintain your assigned altitude.

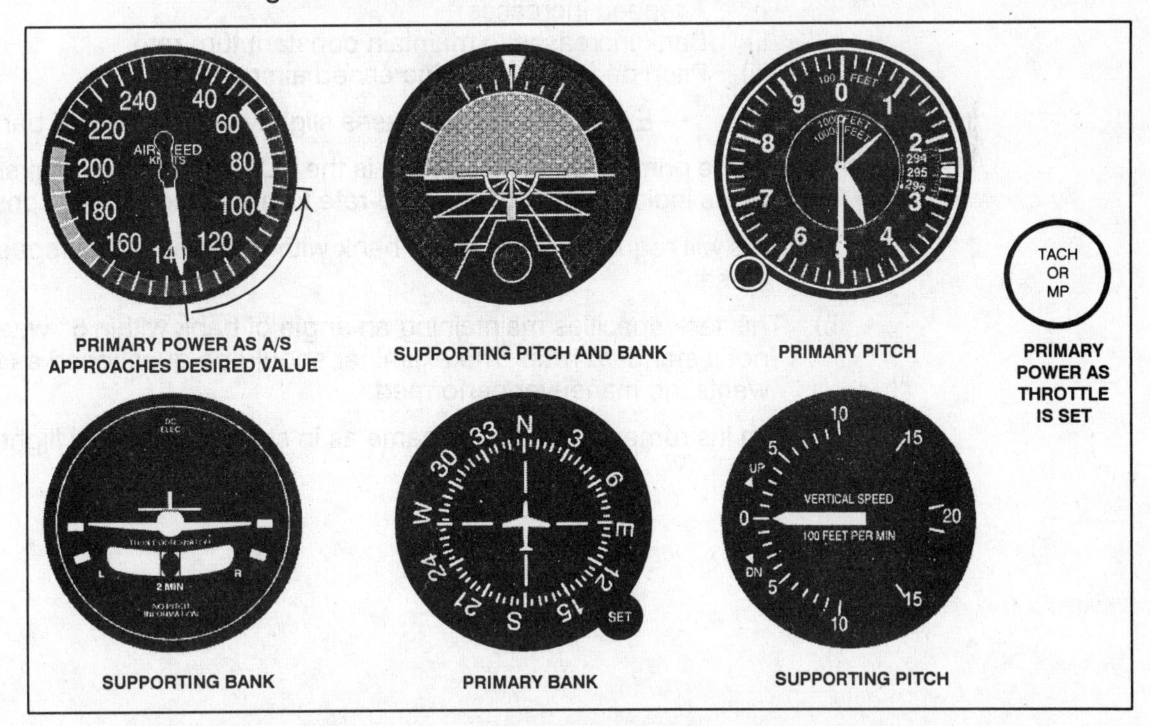

Straight-and-Level Flight (Airspeed Decreasing)

1. The ASI is indicating a decrease in airspeed from 140 kt. to 95 kt.
2. The AI should indicate a slightly nose-up attitude.
3. Altitude and heading remain constant.

3) The power instruments are the airspeed indicator (ASI) and the manifold pressure gauge (MP) and/or RPM (tachometer).

a) The MP and/or RPM is the primary power instrument because the throttle is set initially on the change of airspeed.

b) Power control must be related to its effect on altitude and airspeed since any change in power results in a change in the airspeed or the altitude of your airplane.

i) To maintain a constant altitude, you must decrease pitch as the airspeed increases and increase pitch as the airspeed decreases.

c) As the airspeed approaches the desired value, your ASI becomes the primary power instrument, and power is adjusted to maintain the selected airspeed.

b. Change of airspeeds in turns (full panel)

1) Changing airspeed in a turn requires simultaneous changes in all components of control. You must increase the rate of your instrument cross-check and interpretation.

a) Power reductions

i) Airspeed decreases.
ii) Bank decreases to maintain constant turn rate.
iii) Pitch increases with decreased airspeed.

- Even though it decreases slightly with decreased bank.

b) Power increases

i) Airspeed increases.
ii) Bank increases to maintain constant turn rate.
iii) Pitch decreases with increased airspeed.

- Even though it increases slightly with increased bank.

2) In a turn, the primary bank instrument is the TC, and, during an airspeed change, its indication (e.g., standard-rate turn) should remain constant.

a) This will require a decrease in bank with a decrease in airspeed and vice versa.

b) This task specifies maintaining an angle of bank within 5° when turning, not a standard rate. The examiner should be questioned as to how (s)he wants the maneuver performed.

3) The pitch instruments remain the same as in straight-and-level flight (i.e., ALT is primary).

Change of Airspeed in Turn

1. The ASI is indicating a decrease in airspeed from 140 kt. to 120 kt.
2. As airspeed decreases, the AI will indicate a decrease in bank angle to maintain standard-rate turn and a slightly nose-up attitude to maintain a constant altitude.
3. The HI is indicating a turn to the right from a heading of 280° toward 360°.
4. Altitude remains constant.

4) As in straight-and-level flight during a change of airspeed, MP and/or RPM is initially the primary power instrument. As the airspeed approaches the desired value, the ASI becomes the primary power instrument.

5) Two methods of changing airspeed in turns may be used.

   a) Change airspeed after your turn is established.

      i) This is the easiest method.

   b) Change airspeed simultaneously with the turn entry.

c. Primary and supporting instruments for a change in airspeed (full panel) are presented below.

MANEUVER	PITCH	BANK	POWER
Straight-and-level flight (full panel)			
Primary	ALT	HI	MP and/or RPM initially, ASI as desired airspeed is approached
Supporting	AI, VSI	AI, TC	Reverse of above
Level turn (full panel)			
Primary	ALT	TC	MP and/or RPM initially, ASI as desired airspeed is approached
Supporting	AI, VSI	AI	Reverse of above

d. Change of airspeeds in straight-and-level flight (partial panel--heading and attitude indicators inoperative)

1) The bank instruments are the TC and the magnetic compass (MC).

a) Changing airspeeds in straight-and-level flight will lead to acceleration/ deceleration errors in the MC.

i) On easterly and westerly headings, accelerating will indicate a turn to the north.

ii) Decelerating will indicate a turn to the south.

iii) Northerly or southerly headings do not experience acceleration/ deceleration errors.

b) The TC must be relied upon during airspeed changes until the MC settles back to the desired heading.

2) The pitch instruments are the ALT and VSI.

a) Changing airspeeds will cause the airplane to change pitch, i.e., rotate about its lateral axis.

b) You must apply forward control pressure to maintain level flight with an increase in airspeed.

i) The VSI will indicate a climb with an increase in airspeed.

ii) Keeping the VSI level (0 fpm) will decrease the airplane's pitch attitude and keep the airplane at a constant altitude, i.e., level flight.

c) You must apply back pressure to maintain level flight with a decrease in airspeed.

i) The VSI will indicate a descent with a decrease in airspeed.

ii) Keeping the VSI level (0 fpm) will increase the airplane's pitch attitude and keep the airplane at a constant altitude.

3) The power instruments are the same as those used during full-panel operations.

e. Change of airspeed in turns (partial panel--heading and attitude indicators inoperative)

1) Changing airspeeds while turning requires you to increase the rate of your instrument cross-check during full-panel operations, so partial panel requires a full understanding of the functional instruments.

2) The primary instruments (TC for bank, ALT for pitch, and MP and ASI for power) are the same as those used during full-panel operations.

a) Without the AI for supporting information, the TC and MC are your sole instruments for bank.

3) It is advisable to break this maneuver into two separate steps during partial-panel operations.

a) Turn to the desired heading.
b) Change to the desired airspeed.

NOTE: At times, you may be required to make changes in airspeed while turning.

f. Primary and supporting instruments for a change in airspeed (partial panel -- heading and attitude indicators inoperative) are presented below.

MANEUVER	PITCH	BANK	POWER
Straight-and-level flight (partial panel)			
Primary	ALT	TC initially, MC after airspeed is stabilized	MP and/or RPM initially, ASI as desired airspeed is approached
Supporting	AI, VSI	Reverse of above	Reverse of above
Level turn (partial panel)			
Primary	ALT	TC	MP and/or RPM initially, ASI as desired airspeed is approached
Supporting	VSI	MC	Reverse of above

4. Use proper instrument cross-check and interpretation, and apply the appropriate pitch, bank, power, and trim corrections.

a. Maintain straight and/or level flight as discussed in Task IV.A., Straight-and-Level Flight, beginning on page 141.

b. Changing airspeed in a turn requires you to adjust both pitch and bank.

1) Pitch will increase with a decrease in airspeed and vice versa.

2) Bank will decrease with a decrease in airspeed to maintain a standard-rate turn and vice versa.

c. You must know the required power settings for various airspeeds (e.g., cruise and approach speeds).

d. Trim is important throughout the maneuver to relieve control pressures.

C. Common Errors during a Change of Airspeed

1. **Fixation, omission, and emphasis errors during instrument cross-check.**

a. Fixation, or staring at a single instrument, usually occurs for a good reason but with poor results.

1) You may stare at the ASI, which reads 15 kt. below assigned airspeed, wondering how the needle got there. During that time, perhaps with increasing tension on the controls, an altitude change occurs unnoticed, and more errors accumulate.

2) It may not be entirely a cross-checking error. It may be related to difficulties with one or both of the other fundamental skills (i.e., interpretation and control).

b. Omission of an instrument from the cross-check may be caused by failure to anticipate significant instrument indications following attitude changes.

1) Supporting instruments should be included in the scan.

c. Emphasis on a single instrument, instead of on the combination of instruments necessary for attitude information, is normal during the initial stages of instrument training.

1) You may tend to rely on the instrument that you understand the best, e.g., the AI.

2) The ASI will be changing; however, the ALT is primary for pitch during this maneuver.

2. **Improper instrument interpretation.**

a. This error may indicate that you do not fully understand each instrument's operating principle and relationship to the performance of your airplane.

b. You must be able to interpret even the slightest changes in your instrument indications from your cross-checking.

3. **Improper control applications.**

a. This error normally occurs when you incorrectly interpret the instruments and/or apply the improper controls to obtain a desired performance, e.g., using pitch instead of power to correct an airspeed error.

4. **Failure to establish proper pitch, bank, or power adjustments during altitude, heading, or airspeed corrections.**

a. You must understand which instruments provide information for pitch, bank, and power.

1) The AI is the only instrument for pitch and bank control inputs.

b. This error may indicate that you do not fully understand instrument cross-check, interpretation, and/or control.

5. **Faulty trim technique.**

a. Trim should be used, not to substitute for control with the control yoke and rudder, but to relieve pressures already held to stabilize attitude.

b. Use trim frequently and in small amounts.

1) Trim changes should be expected during power and airspeed adjustments.

c. Improper adjustment of seat or rudder pedals for comfortable positioning of legs and feet may contribute to trim errors.

1) Tension in the ankles makes it difficult to relax rudder pressures.

D. Maneuvers Requiring Change of Airspeed

1. Airspeed changes are required for the basic flying skills like landing, which includes flap extension and gear extension (if applicable).

a. The nose tends to pitch down with gear extension.

b. As flaps are lowered, lift increases momentarily (at partial settings), followed by a noticeable increase in drag near maximum extension.

c. Lowering flaps will also cause the nose to pitch up.

2. Gleim's *Instrument Pilot Syllabus* will also require you to perform proficiency maneuvers like stalls and slow flight.

a. Both stalls and slow flight require airspeed changes along with gear and flap extension and retraction.

b. Knowledge of power settings and trim changes associated with the different combinations of airspeed, gear, and flap configurations will reduce your instrument cross-check and interpretation problems.

END OF TASK

CONSTANT AIRSPEED CLIMBS AND DESCENTS

IV.C. TASK: CONSTANT AIRSPEED CLIMBS AND DESCENTS

REFERENCES: 14 CFR Part 61; AC 61-27.

Objective. To determine that the applicant:

1. Exhibits adequate knowledge of the elements relating to attitude instrument flying during constant airspeed climbs and descents.
2. Demonstrates climbs and descents at a constant airspeed, between specific altitudes in straight or turning flight as specified by the examiner.
3. Enters constant airspeed climbs and descents from a specified altitude, airspeed, and heading.
4. Establishes the appropriate change of pitch and power to establish the desired climb and descent performance.
5. Maintains the airspeed within 10 kt., heading within 10° or, if in a turning maneuver, within 5° of the specified bank angle.
6. Performs the level-off within 100 ft. (30 meters) of the specified altitude.
7. Uses proper instrument cross-check and interpretation, and applies the appropriate pitch, bank, power, and trim corrections.

A. General Information

1. The objective of this task is for you to demonstrate your ability to perform constant airspeed climbs and descents, in straight or turning flight.
2. Our discussion of this maneuver is based on an airplane equipped with a turn coordinator (TC). If your airplane has a turn-and-slip indicator, the descriptions apply if "turn needle" is substituted for "miniature airplane of the TC."

B. Task Objectives

1. **Exhibit adequate knowledge of the elements relating to attitude instrument flying during constant airspeed climbs and descents.**
   a. For a constant airspeed climb with a given power setting, a single pitch attitude will maintain the desired airspeed.
      1) For some airspeeds, such as V_X or V_Y, the climb power setting and airspeed that will determine this climb attitude are given in the performance data found in your *POH*.
   b. A descent can be made at a variety of airspeeds and attitudes by reducing power, adding drag, and lowering the nose to a predetermined attitude. Sooner or later, the airspeed will stabilize at a constant value (i.e., a single pitch attitude will maintain the desired airspeed).
   c. You will be required to perform climbs and descents in turning flight.
      1) To perform this maneuver, you will combine the technique used in straight climbs and descents with the various turn techniques.
         a) Aerodynamic factors affecting lift and power control must be considered in determining power settings.
         b) Increase your rate of cross-check and interpretation to maintain the proper bank and pitch attitude.
2. **Demonstrate climbs and descents at a constant airspeed, between specific altitudes in straight or turning flight as specified by your examiner.**
   a. Your instructor and examiner may keep things simple by instructing you to climb/descend, then after you are established, instruct you to make a turn. (S)he may make it more complex by instructing you to turn and climb simultaneously or descend and possibly level off while still in a turn.
   b. Stay calm and be confident of your abilities.

3. Enter constant airspeed climbs and descents from a specified altitude, airspeed, and heading.

a. Normally your examiner will give you climb/descent instructions while you are in straight-and-level flight.

1) While receiving your instructions, you must maintain your altitude, airspeed, and heading, as discussed in Task IV.A., Straight-and-Level Flight, beginning on page 141.

4. Establish the appropriate change of pitch and power to establish the desired climb and descent performance.

a. Full panel

1) To enter a constant airspeed climb, use the attitude indicator (AI) to raise the nose to the approximate pitch attitude for the desired climb speed. Thus, during entry, the AI is primary for pitch.

a) Power may be advanced to the climb power setting simultaneously with the pitch change, or after the pitch is established and the airspeed approaches climb speed. Thus, the manifold pressure gauge (MP) and/or RPM (tachometer) is the primary power instrument.

b) As in straight flight, the primary instrument for bank is the heading indicator (HI).

Climb Entry for Constant Airspeed Climb (full panel)

1. The attitude indicator (AI) is indicating the nose-high pitch attitude for the climb.
2. The airspeed indicator (ASI) is indicating a decrease of airspeed from 140 kt. to the climb speed of 105 kt.
3. The vertical speed indicator (VSI) is indicating a positive rate of climb.
4. The altimeter (ALT) is indicating an increase in altitude.
5. The heading indicator (HI), turn coordinator (TC), and attitude indicator (AI) are indicating that straight flight is being maintained.

2) To enter a constant airspeed descent at an airspeed **lower** than cruise speed, the following method is effective either with or without an AI:

   a) Reduce airspeed to your selected descent airspeed while maintaining straight-and-level flight. Thus, MP and/or RPM is the primary power instrument.

   b) As the power is adjusted further to a predetermined setting for the descent, simultaneously lower the nose on the AI to maintain constant airspeed, and trim off control pressures. The airspeed indicator (ASI) is now the primary pitch instrument.

   c) As in all straight flight, the HI is the primary bank instrument.

3) To enter a constant airspeed descent at an airspeed **higher** than cruise speed, the following method should be used:

   a) Use the AI to lower the nose to the approximate pitch attitude for the desired descent speed. Thus, during entry, the AI is the primary pitch instrument.

   b) Power may be reduced to a descent power setting after the pitch is established and the airspeed approaches descent speed. Thus, MP and/or RPM is the primary power instrument.

      i) The amount of power reduced depends on the airspeed increase and rate of descent desired.

         - For a large airspeed increase, no power reduction may be necessary.
         - For a small airspeed increase, a larger power reduction may be required.

4) Primary and supporting instruments for entering a constant airspeed climb or descent are presented below.

MANEUVER	PITCH	BANK	POWER
Transitioning from straight and level to constant airspeed climb/descent (full panel)			
Primary	AI	HI	MP and/or RPM
Supporting	VSI	AI, TC	

b. Partial panel (inoperative attitude and heading indicators)

1) Climb entries on partial panel (i.e., no AI or HI) are more easily and accurately controlled if you enter the maneuver from climbing speed.

a) Slow to climb speed while maintaining altitude as discussed in Task IV.B., Change of Airspeed, beginning on page 146.

b) When the desired climb speed is reached, apply climb power while simultaneously raising the nose to maintain the airspeed.

2) To enter a constant airspeed descent on partial panel, it is best to keep the maneuver as simple as possible by using an airspeed lower than cruise.

a) During a partial panel descent, the ASI, ALT, and VSI will be showing varying rates of change until your airplane decelerates to a constant airspeed at a constant attitude.

3) Primary and supporting instruments for entering a constant airspeed climb or descent are presented below.

MANEUVER	PITCH	BANK	POWER
Transitioning from straight and level to constant airspeed climb/descent (partial panel)			
Primary	ASI	TC	MP and/or RPM
Supporting	VSI, ALT	MC	

5. *Maintain the airspeed within 10 kt., heading within 10°, or if in a turning maneuver, within 5° of the desired bank angle.*

a. Full panel

1) During the straight, constant airspeed climb or descent the ASI becomes the primary pitch instrument. Make small pitch changes to maintain the desired airspeed.

a) Never exceed any airspeed limitations of your aircraft (e.g., in a high-speed descent).

b) The MP and/or RPM remains the primary power instrument, which is used to ensure that the proper climb or descent power is maintained.

i) Watch for excessive RPM on fixed-pitch props during descents.

ii) MP will have to be increased during climbs and reduced during descents (approximately 1" Hg/1,000 ft.).

2) The HI remains the primary instrument for bank control.

Stabilized Climb at Constant Airspeed

1. ASI is the primary pitch instrument.
2. HI is the primary bank instrument (straight flight).
   a. During a turn, the TC is the primary bank instrument.

3) During the turning, constant airspeed climb/descent, the primary instruments for pitch, bank, and power are ASI, TC, and MP and/or RPM, respectively.

a) The ball should be checked throughout the turn for coordination.

b) The airplane's nose will tend to rise as the wings are being returned to level attitude. Sufficient elevator pressure must be applied to maintain the appropriate airspeed.

4) Primary and supporting instruments for a constant airspeed climb or descent are presented below.

MANEUVER	PITCH	BANK	POWER
Straight constant airspeed climb/descent (full panel)			
Primary	ASI	HI	MP and/or RPM
Supporting	AI, VSI	AI, TC	ASI
Turning constant airspeed climb/descent (full panel)			
Primary	ASI	TC	MP and/or RPM
Supporting	AI, VSI	TC	ASI

b. Partial panel (inoperative attitude and heading indicators)

1) The pitch instruments are the ASI, the ALT, and the VSI.

a) The primary pitch instrument for a constant airspeed climb or descent is the ASI. Since the ASI indications lag, it is vitally important that you make small corrections to maintain a constant airspeed.

b) The supporting instruments are the ALT and the VSI. These are cross-checked to ensure that the airplane is continuing the climb or descent to the desired altitude.

2) In a straight or turning constant airspeed climb or descent, the bank instruments are the TC and the MC.

a) The primary bank instrument is the TC.

i) In a straight climb or descent, check to ensure that the wings remain level on the miniature airplane of the TC.

b) In a stabilized straight climb or descent, the MC should be cross-checked to ensure that the heading is maintained.

3) The power instruments are the ASI, the MP, and/or the RPM.

a) The primary power instrument is the MP and/or the RPM. The desired power setting should be set for the climb or descent.

b) The supporting power instrument is the ASI.

4) Primary and supporting instruments for a constant airspeed climb or descent are presented below.

MANEUVER	PITCH	BANK	POWER
Straight constant airspeed climb/descent (partial panel)			
Primary	ASI	TC	MP and/or RPM
Supporting	ALT, VSI	MC	ASI
Turning constant airspeed climb/descent (partial panel)			
Primary	ASI	TC	MP and/or RPM
Supporting	ALT, VSI	MC	ASI

6. ***Perform the level-off within 100 ft. of the specified altitude.***

a. Full panel

1) To level off from a climb, it is necessary to start the level-off before reaching the desired altitude. An effective practice is to lead the altitude by 10% of the vertical speed (e.g., at 500 fpm, the lead would be 50 ft.).

a) Apply smooth, steady forward elevator pressure toward level flight attitude for the speed desired. As the AI shows the pitch change, the VSI will move slowly toward zero, the ALT will move more slowly, and the ASI will increase.

b) Once the ALT, AI, and VSI show level flight, constant changes in pitch and application of nose-down trim will be required as the airspeed increases.

c) As the ASI approaches cruise speed, reduce power to cruise setting.

2) The level-off from a descent must be started before you reach the desired altitude. Assuming a 500-fpm rate of descent, lead the altitude by 100 to 150 ft. for a level-off at an airspeed higher than descending speed.

a) At the lead point, add power to the appropriate level flight cruise setting. Since the nose will tend to rise as the airspeed increases, hold forward elevator pressure to maintain the descent until approximately 50 ft. above the altitude; then smoothly adjust pitch to the level flight attitude.

3) To level off from a descent at descent airspeed, lead the desired altitude by approximately 50 ft. (or 10% of the vertical speed), simultaneously adjusting the pitch attitude to level flight and adding power to a setting that will hold the airspeed constant.

a) Trim off the control pressures, and continue with the normal straight-and-level flight cross-check.

Level-off at Cruise Airspeed

1. The ASI is indicating an increase in airspeed from climb to cruise airspeed.
2. The AI is indicating a decrease in the nose-high pitch attitude to level flight.
3. The ALT is indicating a level-off at 5,000 ft.
4. The VSI is indicating a decrease in vertical speed.
5. The TC and HI are indicating a constant heading for straight flight.

4) To level off at an airspeed lower than descent airspeed (i.e., descent airspeed greater than cruise airspeed), use the AI to raise the nose to the pitch attitude for level cruise (on or near the horizon).

   a) As the airspeed approaches cruise speed, adjust power to the cruise power setting.

   b) Trim may be adjusted before and after power adjustment.

5) Primary and supporting instruments for a constant airspeed climb or descent are presented below.

MANEUVER	PITCH	BANK	POWER
Level-off at cruising airspeed (full panel)			
Primary	ALT	HI	MP and/or RPM initially, ASI as desired airspeed is approached
Supporting	AI, VSI	AI, TC	

b. Partial panel (inoperative attitude and heading indicators)

1) To level off from a constant airspeed climb, the procedures are similar to those for full-panel operations.

a) Lead the desired altitude by 10% of the rate shown on the VSI.

b) At that time, the ALT becomes the primary pitch instrument.

i) Gradually add forward pressure on the control and note a decreasing rate on the VSI and the ALT needle moving more slowly.

c) Maintain straight flight by keeping the wings level on the miniature airplane of the TC and the ball centered.

d) As the airspeed increases to cruise, scan the ALT and VSI to maintain your altitude and trim as often as necessary.

e) As the airspeed approaches cruise speed, reduce power to the cruise setting and retrim the airplane.

2) To level off from a constant airspeed descent, the procedures are similar to those for full-panel operations.

a) Lead the desired altitude by 10% of the rate shown on the VSI.

b) At this time, smoothly add power to the cruise setting. The ALT becomes the primary pitch instrument.

i) Gradually add back control pressure and note a decreasing rate on the VSI and the ALT needle moving more slowly.

c) Maintain straight flight by keeping the wings level on the miniature of the TC and the ball centered.

d) Trim the airplane as necessary.

3) Primary and supporting instruments for a constant airspeed climb or descent are presented below.

MANEUVER	PITCH	BANK	POWER
Level-off at cruising airspeed (partial panel)			
Primary	ALT	TC	MP and/or RPM initially, ASI as desired airspeed is approached
Supporting	VSI	MC	

7. **Use proper instrument cross-check and interpretation, and apply the appropriate pitch, bank, power, and trim corrections.**

   a. Maintain airspeed by adjusting your pitch as necessary.

      1) Trim to relieve all elevator control pressures.

      2) Lead the level-off as appropriate by smoothly returning the nose to a level pitch attitude.

         a) After leveling off, adjust pitch as necessary to hold the desired altitude.

   b. Maintain straight flight by holding the wings level on the AI and maintaining your heading on the HI.

      1) When you note a deviation from straight flight on your HI, make your correction to the desired heading by using an angle of bank no greater than the number of degrees to be turned.

         a) Limit your bank corrections to a bank angle no greater than that required for a standard-rate turn.

         b) Use coordinated aileron and rudder.

      2) The ball of the TC should be centered. If not, your airplane is in uncoordinated flight either by being improperly trimmed (aileron) and/or by your holding rudder pressure.

         a) Trim the airplane so the wings are level (if aileron trim is available) and the ball of the TC is centered (rudder trim).

   c. If turning, adjust bank as necessary to maintain the desired rate of turn (e.g., standard rate) on the TC (or angle of bank on the AI), and use the rudder to keep the ball centered.

      1) This task specifies maintaining an angle of bank within 5° when turning, not a standard rate. The examiner should be questioned as to how (s)he wants the maneuver performed.

   d. You must know the required power settings for your desired climb and descent performance and for level-off at your desired airspeed.

C. Common Errors during Constant Airspeed Climbs and Descents

1. **Fixation, omission, and emphasis errors during instrument cross-check.**

   a. Fixation, or staring at a single instrument, usually occurs for a good reason but with poor results.

      1) You may stare at the ASI, which reads 20 kt. below assigned airspeed, wondering how the needle got there. During that time, perhaps with increasing tension on the controls, a heading change occurs unnoticed, and more errors accumulate.

      2) It may not be entirely a cross-checking error. It may be related to difficulties with one or both of the other fundamental skills (i.e., interpretation and control).

b. Omission of an instrument from the cross-check may be caused by failure to anticipate significant instrument indications following attitude changes.

1) Supporting instruments should be included in the scan.

c. Emphasis on a single instrument, instead of on the combination of instruments necessary for attitude information, is normal during the initial stages of instrument training.

1) You may tend to rely on the instrument that you understand the best, e.g., the AI.

2) The ALT will be changing; however, the ASI is primary for pitch during this maneuver.

2. Improper instrument interpretation.

a. This error may indicate that you do not fully understand each instrument's operating principle and relationship to the performance of the airplane.

b. You must be able to interpret even the slightest changes in your instrument indications from your cross-checking.

3. Improper control applications.

a. This occurs when you incorrectly interpret the instruments and/or apply the improper controls to obtain a desired performance, e.g., using power instead of pitch to correct a minor airspeed error.

4. Failure to establish proper pitch, bank, or power adjustments during heading and airspeed corrections.

a. You must understand which instruments provide information for pitch, bank, and power.

1) The AI is the only instrument for pitch and bank control inputs.

b. This error may indicate that you do not fully understand instrument cross-check, interpretation, and/or control.

5. Improper entry or level-off technique.

a. Until you learn and use the proper power setting and pitch attitudes in climbs and descents, you may tend to make larger than necessary pitch adjustments.

1) You must restrain the impulse to change a flight attitude until you know what the result will be.

a) Do not chase the needles.

b) The rate of cross-check must be varied during speed, power, or attitude changes on climbs and descents.

c) During the level-off, you must note the rate of climb or descent to determine the proper lead.

i) Failure to do this will result in overshooting or undershooting the desired altitude.

2) "Ballooning" (allowing the nose to pitch up) on level-off results when descent attitude with forward elevator pressure is not maintained as power is increased.

3) You must maintain an accelerated cross-check until straight-and-level flight is positively established.

6. **Faulty trim technique.**

a. Trim should be used, not to substitute for control with the control yoke and rudder, but to relieve pressures already held to stabilize attitude.

b. Use trim frequently and in small amounts.

1) Trim should be expected during any pitch, power, or airspeed change.

c. Improper adjustment of seat or rudder pedals for comfortable positioning of legs and feet may contribute to trim errors.

1) Tension in the ankles makes it difficult to relax rudder pressures.

END OF TASK

RATE CLIMBS AND DESCENTS

IV.D. TASK: RATE CLIMBS AND DESCENTS

REFERENCES: 14 CFR Part 61; AC 61-27.

Objective. To determine that the applicant:

1. Exhibits adequate knowledge of the elements relating to attitude instrument flying during rate climbs and descents.
2. Demonstrates climbs and descents at a constant rate between specific altitudes in straight or turning flight as specified by the examiner.
3. Enters rate climbs and descents from a specified altitude, airspeed, and heading.
4. Establishes the appropriate change of pitch, bank, and power to establish the specified rate of climb or descent.
5. Maintains the specified rate of climb and descent within 100 feet per minute, airspeed within 10 kt., heading within 10°, or if in a turning maneuver, within 5° of the specified bank angle.
6. Performs the level-off within 100 ft. (30 meters) of the specified altitude.
7. Uses proper instrument cross-check and interpretation, and applies the appropriate pitch, bank, power, and trim corrections.

A. General Information

1. The objective of this task is for you to demonstrate your ability to perform rate climbs and descents in straight or turning flight.
2. Our discussion of this maneuver is based on an airplane equipped with a turn coordinator (TC). If your airplane has a turn-and-slip indicator, the descriptions apply if "turn needle" is substituted for "miniature airplane of the TC."
3. Your instructor will provide you with power settings in various configurations that can be used to establish a certain rate of descent at a specific airspeed.
    a. These settings are important to know especially when conducting instrument approaches.
4. Section 5, Performance, in your airplane's *POH* will contain information on rate climb performance numbers.

B. Task Objectives

1. **Exhibit adequate knowledge of the elements relating to attitude instrument flying during rate climbs and descents.**
    a. A rate climb and descent is similar to a constant airspeed climb and descent (see Task IV.C. on page 154, except that your primary pitch instrument is now the VSI and your primary power instrument is the ASI.
    b. Various situations may require rate climbs and descents.
        1) Rate climbs are necessary for those departure procedures that require a specified climb gradient.
        2) The last 1,000 ft. of a climb or descent to a specified altitude should be made at 500 fpm.
        3) On precision approaches (i.e., ILS), you must use a rate descent to maintain the glide slope.

c. You may also be required to perform this task in turning flight.

1) You will combine the techniques used in straight climbs and descents with the various turn techniques.

a) Aerodynamic factors affecting lift and power control must be considered in determining power settings.

b) Increase your rate of cross-check and interpretation to maintain the proper bank and pitch attitude.

2. Demonstrate climbs and descents at a constant rate between specific altitudes in straight or turning flight, as specified by your examiner.

a. Your instructor and examiner may keep things simple by instructing you to climb/descend, then after you are established, instruct you to make a turn. Normally, (s)he may make it more complex by instructing you to turn and climb simultaneously or descend and possibly level off while still in a turn.

b. Stay calm and be confident of your abilities.

3. Enter rate climbs and descents from a specified altitude, airspeed, and heading.

a. Normally your instructor and examiner will give you climb/descent instructions while you are in straight-and-level flight.

1) While receiving your instructions, you must maintain your altitude, airspeed, and heading, as discussed in Task IV.A., Straight-and-Level Flight, beginning on page 141.

4. Establish the appropriate change of pitch, bank, and power to establish the desired rate of climb or descent.

a. Full panel

1) To enter a constant-rate climb, use your attitude indicator (AI) to increase the pitch to the approximate nose-high indication appropriate to the airspeed and rate of climb.

a) As the power is increased, the airspeed indicator (ASI) is primary for pitch control until the vertical speed approaches the desired value.

i) As the vertical speed indicator (VSI) stabilizes, it becomes primary for pitch control, and the ASI becomes primary for power control.

ii) Trim as airspeed decreases.

b) In straight flight, the primary instrument for bank is the heading indicator (HI).

2) To enter a constant-rate descent, the following method is effective either with or without an AI:

a) Reduce airspeed to your selected descent airspeed while maintaining straight-and-level flight. Thus, MP and/or RPM is the primary power instrument initially.

b) As the power is adjusted further to a predetermined setting for the descent, simultaneously lower the nose on the AI to establish a constant rate, and trim off control pressures. The VSI is now the primary pitch instrument.

c) As in all straight flight, the HI is the primary bank instrument.

3) Primary and supporting instruments for rate climbs and descents are presented below.

MANEUVER	PITCH	BANK	POWER
Transitioning from straight and level to a climb/descent (full panel)			
Primary	ASI	HI	MP and/or RPM
Supporting	AI, VSI	TC	ASI

b. Partial panel (inoperative attitude and heading indicators)

1) Climb entries on partial panel (i.e., no AI and HI) are more easily and accurately controlled if you enter the maneuver from climbing speed.

a) Slow to climb speed while maintaining altitude, as discussed in Task IV.B., Change of Airspeed, beginning on page 146.

b) When the desired climb speed is reached, apply climb power while simultaneously raising the nose to maintain the airspeed until the VSI approaches the desired value.

2) To enter a constant rate descent, use the same procedures as those used for full-panel operations. The only difference is that you will not have the AI or HI.

a) The TC will be your primary bank instrument.

b) During a partial-panel descent, the ASI, ALT, and VSI will be showing varying rates of change until your airplane descends at a constant rate and at a constant attitude. Then small pitch variations can be made to maintain a constant rate descent.

3) If you know the power setting and airspeed to attain a certain rate of descent, partial-panel operations will be easier for you.

4) Primary and supporting instruments for rate climbs and descents are presented below.

MANEUVER	PITCH	BANK	POWER
Transitioning from straight and level to a climb/descent (partial panel)			
Primary	ASI	TC	MP and/or RPM
Supporting	ALT, VSI	MC	ASI

5. ***Maintain the desired rate of climb and descent within 100 feet per minute (fpm), airspeed within 10 kt., heading within 10°, or if in a turning maneuver, within 5° of the desired bank angle.***

   a. Full panel

      1) During the constant-rate climb, the VSI becomes the primary pitch instrument, and the ASI becomes primary for power.

         a) Pitch and power corrections must be quickly and closely coordinated.

            i) If the VSI is correct but the ASI is low, add power.

               - As power is increased, lower the pitch attitude using the AI to maintain constant vertical speed.

            ii) If the VSI is high and the ASI is low, lower the pitch using the AI and note the increase in the ASI to determine if a power change is necessary.

Stabilized Climb at Constant Rate

1. ASI is the primary power instrument.
2. VSI is the primary pitch instrument.
3. HI is the primary bank instrument (straight flight).
   a. During a turn, the TC is the primary bank instrument.

2) During the constant-rate descent, the VSI becomes the primary pitch instrument, and the ASI becomes primary for power.
   a) Pitch and power must be closely coordinated, as in climbs, when corrections are made.
3) During the turning, constant-rate climb/descent, the primary instruments for pitch, bank, and power are the VSI, the TC, and the ASI, respectively.
   a) The ball of the TC should be checked throughout the turn for coordination.
   b) Your airplane's nose will tend to rise as the wings are being returned to level flight.
      i) Sufficient elevator pressure must be applied to maintain the appropriate vertical speed (rate).
4) Primary and supporting instruments for rate climbs and descents are presented below.

MANEUVER	PITCH	BANK	POWER
Straight constant rate stabilized climb/descent (full panel)			
Primary	VSI	HI	ASI
Supporting	AI	AI, TC	MP and/or RPM

b. Partial panel (inoperative attitude and heading indicators)
   1) Maintaining a constant rate descent or climb under partial panel is very similar to performing the maneuver under full-panel operations.
   2) The primary pitch, bank, and power instruments are the VSI, the TC, and the ASI, respectively.
      a) The supporting pitch instrument is the ALT, and the supporting bank instrument is the MC.
   3) Performing turns while climbing or descending uses the same primary and supporting instruments.
   4) Primary and supporting instruments for rate climbs and descents are presented below.

MANEUVER	PITCH	BANK	POWER
Straight constant rate stabilized climb/descent (partial panel)			
Primary	VSI	TC	ASI
Supporting	ALT	MC	MP and/or RPM

6. ***Perform the level-off within 100 ft. of the specified altitude.***

a. Full panel

1) To level off from a climb, it is necessary to start the level-off before reaching the desired altitude. An effective practice is to lead the altitude by 10% of the vertical speed (e.g., at 500 fpm, the lead would be 50 ft.).

a) Apply smooth, steady forward elevator pressure toward level flight attitude for the speed desired. As the AI shows the pitch change, the VSI will move slowly toward zero, the ALT will move more slowly, and the ASI will increase.

b) Once the ALT, AI, and VSI show level flight, constant changes in pitch and application of nose-down trim will be required as the airspeed increases.

c) As the ASI approaches cruise speed, reduce power to cruise setting.

2) The level-off from a descent must be started before you reach the desired altitude. Assuming a 500-fpm rate of descent, lead the altitude by 100 to 150 ft. for a level-off at an airspeed higher than descending speed.

a) At the lead point, add power to the appropriate level flight cruise setting. Since the nose will tend to rise as the airspeed increases, hold forward elevator pressure to maintain the descent until approximately 50 ft. above the altitude; then smoothly adjust pitch to the level flight attitude.

3) To level off from a descent at descent airspeed, lead the desired altitude by approximately 50 ft. (or 10% of vertical speed), simultaneously adjusting the pitch attitude to level flight and adding power to a setting that will hold the airspeed constant.

4) Trim off the control pressures, and continue with the normal straight-and-level flight cross-check.

5) Primary and supporting instruments for rate climbs and descents are presented below.

MANEUVER	PITCH	BANK	POWER
Level-off at cruising airspeed			
Primary	ALT	HI	MP and/or RPM initially, ASI as desired airspeed is approached
Supporting	AI, VSI	AI, TC	

b. Partial panel (inoperative attitude and heading indicators)

1) To level off from a constant rate climb, the procedures are similar to those for full-panel operations.

a) Lead the desired altitude by 10% of the rate shown on the VSI.

b) At that time, the ALT becomes the primary pitch instrument.

i) Gradually add forward pressure on the control and note a decreasing rate on the VSI and the ALT needle moving more slowly.

c) Maintain straight flight by keeping the wings level on the miniature airplane of the TC and the ball centered.

d) As the airspeed increases to cruise, scan the ALT and VSI to maintain your altitude and trim as often as necessary.

e) As the airspeed approaches cruise speed, reduce power to the cruise setting and retrim the airplane.

2) To level off from a constant airspeed descent, the procedures are similar to those for full-panel operations.

a) Lead the desired altitude by 10% of the rate shown on the VSI.

b) At this time, smoothly add power to the cruise setting. The ALT becomes the primary pitch instruments.

i) Gradually add back control pressure and note a decreasing rate on VSI and the ALT needle moving more slowly.

c) Maintain straight flight by keeping the wings level on the miniature of the TC and the ball centered.

d) Trim the airplane as necessary.

3) Primary and supporting instruments for rate climbs and descents are presented below.

MANEUVER	**PITCH**	**BANK**	**POWER**
Level-off at cruising airspeed (partial panel)			
Primary	ALT	TC	MP and/or RPM initially, ASI as desired airspeed is approached
Supporting	VSI	MC	

7. Use proper instrument cross-check and interpretation, and apply the appropriate pitch, bank, power, and trim corrections.

a. Maintain rate of climb or descent by adjusting your pitch as necessary.

1) Trim to relieve all elevator control pressures.

2) Lead the level-off as appropriate by smoothly returning the nose to a level pitch attitude.

a) After leveling off, adjust pitch as necessary to hold the desired altitude.

b. Maintain straight flight by holding the wings level on the AI and maintaining your heading on the HI.

1) When you note a deviation from straight flight on your HI, make your correction to the desired heading by using an angle of bank no greater than the number of degrees to be turned.

a) Limit your bank corrections to a bank angle no greater than that required for a standard-rate turn.

b) Use coordinated aileron and rudder.

2) The ball of the TC should be centered. If not, your airplane is in uncoordinated flight either by being improperly trimmed (aileron) and/or by your holding rudder pressure.

a) Trim the airplane so the wings are level (if aileron trim is available) and the ball of the TC is centered (rudder trim).

c. If turning, adjust bank as necessary to maintain the desired rate of turn (e.g., standard rate) on the TC (or angle of bank on the AI), and use the rudder to keep the ball centered.

1) Adjust pitch to maintain the desired rate of climb or descent.

2) This task specifies maintaining an angle of bank within 5° when turning, not a standard rate. The examiner should be questioned as to how (s)he wants the maneuver performed.

d. You must know the required power settings for your desired climb and descent performance and for level-off at your desired airspeed.

C. Common Errors during Rate Climbs and Descents

1. **Fixation, omission, and emphasis errors during instrument cross-check.**

a. Fixation, or staring at a single instrument, usually occurs for a good reason but with poor results.

1) You may stare at the VSI, which reads 200 fpm below assigned rate, wondering how the needle got there. During that time, perhaps with increasing tension on the controls, a heading change occurs unnoticed, and more errors accumulate.

2) It may not be entirely a cross-checking error. It may be related to difficulties with one or both of the other fundamental skills (i.e., interpretation and control).

b. Omission of an instrument from the cross-check may be caused by failure to anticipate significant instrument indications following attitude changes.

1) Supporting instruments should be included in the scan.

c. Emphasis on a single instrument, instead of on the combination of instruments necessary for attitude information, is normal during the initial stages of instrument training.

1) You may tend to rely on the instrument that you understand the best, e.g., the attitude indicator.

2) The ALT will be changing; however, the VSI is primary for pitch during this maneuver.

2. **Improper instrument interpretation.**

a. This error may indicate that you do not fully understand each instrument's operating principle and relationship to the performance of your airplane.

b. You must be able to interpret even the slightest changes in your instrument indications from your cross-checking.

3. Improper control applications.

a. This error normally occurs when you incorrectly interpret the instruments and/or apply the improper controls to obtain a desired performance, e.g., using power instead of pitch to correct a minor vertical speed error.

4. Failure to establish proper pitch, bank, or power adjustments during altitude, heading, or airspeed corrections.

a. You must understand which instruments provide information for pitch, bank, and power.

1) The AI is the only instrument for pitch and bank control inputs.

b. This error may indicate that you do not fully understand instrument cross-check, interpretation, and/or control.

5. Faulty trim technique.

a. Trim should be used, not to substitute for control with the control yoke and rudder, but to relieve pressures already held to stabilize attitude.

b. Use trim frequently and in small amounts.

1) Trim should be expected during any pitch, power, or airspeed changes.

c. Improper adjustment of seat or rudder pedals for comfortable positioning of legs and feet may contribute to trim errors.

1) Tension in the ankles makes it difficult to relax rudder pressures.

END OF TASK

TIMED TURNS TO MAGNETIC COMPASS HEADINGS

IV.E. TASK: TIMED TURNS TO MAGNETIC COMPASS HEADINGS

REFERENCES: 14 CFR Part 61; AC 61-27.

NOTE: If the aircraft has a turn-and-slip indicator, the phrase "miniature aircraft of the turn coordinator" applies to the turn needle.

Objective. To determine that the applicant:

1. Exhibits adequate knowledge of elements and procedures relating to calibrating the miniature aircraft of the turn coordinator, the operating characteristics and errors of the magnetic compass, and the performance of timed turns to specified compass headings.
2. Establishes indicated standard-rate turns, both right and left.
3. Applies the clock correctly to the calibration procedure.
4. Changes the miniature aircraft position, as necessary, to produce a standard-rate turn.
5. Makes timed turns to specified compass headings.
6. Maintains the altitude within 100 ft. (30 meters), airspeed within 10 kt., bank angle 5° of a standard or half-standard-rate turn, and rolls out on specified headings within 10°.

A. General Information

1. The objective of this task is for you to demonstrate your ability to perform timed turns to magnetic compass headings.
   a. If your aircraft has a turn-and-slip indicator, the phrase "miniature airplane of the turn coordinator" applies to the turn needle.
2. For additional reading, see Chapter 2, Airplane Instruments, Engines, and Systems, in *Pilot Handbook* for a five-page detailed discussion on the magnetic compass construction and operation.

B. Task Objectives

1. **Exhibit adequate knowledge of elements and procedures relating to calibrating the miniature aircraft of the turn coordinator, the operating characteristics and errors of the magnetic compass, and the performance of timed turns to specified compass headings.**
   a. A timed turn is a turn in which the clock and the turn coordinator (TC) are used to change heading a definite number of degrees in a given time. For example, using a standard-rate turn (3° per second), your airplane should turn 45° in 15 sec.
      1) Prior to performing timed turns, the TC should be calibrated to determine the accuracy of its indications. This should be done in both left and right turns.
   b. In your airplane, the magnetic compass is probably the only direction-indicating instrument independent of other airplane instruments and power sources. Because of its operating characteristics, called compass errors, you are most likely to use it only as a reference for setting the heading indicator (HI).
      1) A knowledge of magnetic compass characteristics will enable you to use the instrument to turn your airplane to correct headings and maintain them.

2) Compass errors

a) **Variation** is the angular difference between true and magnetic north.

i) This difference varies around the earth and is indicated on charts.

b) **Deviation** is the magnetic disturbance within your airplane that deflects the compass needles from alignment with magnetic north.

i) Deviation varies with the electrical components in use, and the magnetism changes with jolts from hard landings and installation of additional radio equipment.

ii) Deviation is compensated by adjustment of the N-S/E-W magnets, and the errors remaining are recorded on the compass card.

- To fly compass headings, you must refer to the compass correction card for corrected headings to steer.

c) **Magnetic dip** is the tendency of the compass needles to point down as well as to the magnetic pole. The dip causes significant errors when the airplane is turning or accelerating. This error is the greatest at the poles and zero at the magnetic equator.

i) If you are on a northerly heading and you start a turn to the east or west, the indication of the compass lags, or shows a turn in the opposite direction.

ii) If you are on a southerly heading and you start a turn toward the east or west, the compass indication precedes the turn, showing a greater amount of turn than is actually occurring.

iii) When you are on an easterly or westerly heading, the compass correctly indicates when you start a turn to either the north or the south.

iv) If you are on an easterly or westerly heading, acceleration results in a northerly turn indication.

v) If you are on an easterly or westerly heading, deceleration results in a southerly turn indication.

vi) If you maintain a northerly or southerly heading, no error results from diving, climbing, or changing airspeed.

3) With an angle of bank between 15° and 18°, the amount of lead or lag to be used when turning to northerly or southerly headings varies with, and is approximately equal to, the latitude of the locality over which the turn is being made.

a) When turning to a heading of north, the lead for rollout must include the number of degrees of your latitude, plus your normal lead used in your turn recovery.

b) During a turn to a southerly heading, maintain the turn until the compass passes south the number of degrees of your latitude, minus your normal rollout lead.

c) When turning to a heading of east or west, use only your normal rollout lead.

d) For other than cardinal headings, apply proportionally less correction than used for north or south rollouts.

4) Abrupt changes in attitude or airspeed and the resulting erratic movements of the compass card make accurate interpretations of the instrument very difficult.
   a) Proficiency in compass turns depends on knowledge of the compass characteristics, smooth control technique, and accurate bank and pitch control.

c. Timed turns to specified compass headings will assist you in heading changes with the HI inoperative.
   1) Use the magnetic compass at the completion of the turn to check turn accuracy, taking compass deviation errors into consideration.

d. Primary and supporting instruments for timed turns are presented below.

MANEUVER	PITCH	BANK	POWER
Establishing a level standard-rate turn			
Primary	ALT	AI	ASI
Supporting	AI, VSI	TC	MP and/or RPM
Stabilized standard-rate turn			
Primary	ALT	TC	ASI
Supporting	AI, VSI	AI	MP and/or RPM
Recovery from a level standard-rate turn			
Primary	ALT	AI	ASI
Supporting	AI, VSI	TC, HI	MP and/or RPM

2. Establish indicated standard-rate turns, both right and left.

a. The standard-rate turn is indicated by the mark below the wings-level mark on the TC.
   1) The bank required for a standard-rate turn varies with airspeed.
      a) Bank in degrees will be approximately 15% of airspeed.
      b) EXAMPLE: At 100 kt., a bank of approximately 15 is required.

b. This calibration procedure must have been accomplished using both right and left turns.

Turn Coordinator Calibration

1. ASI is the primary power instrument.
2. ALT is the primary pitch instrument.
3. TC is the primary bank instrument.
4. AI and HI indicate a left turn (from 360° toward 270°).

c. Roll into the turn by using coordinated aileron and rudder pressure in the proper direction.

   1) Use the attitude indicator (AI) to establish the amount and direction of bank desired. If using a standard-rate (3° per second) turn, check the miniature airplane of the TC for the standard rate indication.

      a) If using a turn-and-slip indicator, a standard-rate turn is normally shown when the needle is deflected one needle's width to the doghouse mark.

      b) The ball should be checked throughout the turn for coordination.

   2) As long as the airplane is in a coordinated bank, it normally continues to turn with neutral ailerons.

3. Apply the clock correctly to the calibration procedure.

a. While holding the indicated standard rate of turn, note the heading on the HI as the sweep second hand of the clock passes a cardinal point (12, 3, 6, 9).

b. If you are not comfortable with a 10-sec. interval, use a 30-sec. (90°) or 1-min. (180°) interval to calibrate your TC.

1) The flight conditions will affect your decision.
   a) In smooth conditions in which you are maintaining a constant rate of turn, a 10-sec. interval is all you need.
   b) In bumpy conditions (convective turbulence) in which you are having problems holding a constant rate on your TC, a 30-sec. interval may be appropriate.

4. Change the miniature aircraft position, as necessary, to produce a standard-rate turn.

a. While holding a constant rate of turn, note the heading changes at the appropriate time intervals you are using.
   1) EXAMPLE: If using a 10-sec. interval, you will note the heading changes every 10 sec., which at a standard-rate turn is 30°.

b. Make a larger or smaller deflection of the miniature airplane of the TC to produce a standard-rate turn.

c. When you have calibrated the TC during turns in each direction, note the corrected deflections, if any, and apply them during all timed turns.

5. Make timed turns to specified compass headings.

a. In a timed turn, the clock and TC are used to change the heading by a certain number of degrees in a given time.
   1) Using a standard-rate turn (3° per second), an airplane turns 45° in 15 sec.
   2) Using a half-standard-rate turn (1.5° per second), the airplane turns 45° in 30 sec.

b. In making timed turns, use the same cross-check and control technique that you use for any turn, except substitute the clock for the HI.
   1) The miniature aircraft of the TC is primary for bank control.
   2) The altimeter (ALT) is primary for pitch control.
   3) The airspeed indicator (ASI) is primary for power control.

c. Start the roll-in when the clock's second hand passes a cardinal point, hold the turn at the calibrated standard-rate indication (or half-standard rate for small changes in heading).
   1) If using a digital timer, start the roll-in after starting the timer.

d. Begin the rollout when the computed number of seconds has elapsed.

e. If the rates of roll-in and rollout are the same, the time taken during entry and recovery need not be considered in the time computation.

f. Use your magnetic compass at the completion of the turn to check turn accuracy.
   1) Take compass deviation errors into consideration by referring to your airplane's compass correction card.

g. If you have an operative HI, refer to it to determine your progress toward the desired heading and the time to begin the rollout. During a timed turn, the clock will indicate the appropriate time to begin the rollout, and the HI should be checked for accuracy of the turn.
   1) The rollout to a desired heading must be started before the heading is reached.
   2) Practice with a lead of one-half the angle of bank until you determine the lead suitable for your technique.

3) Apply coordinated aileron and rudder opposite the direction of turn. As you initiate the turn recovery, the AI becomes the primary bank instrument.

6. ***You are required to maintain altitude, ±100 ft., airspeed, ±10 kt., and bank angle, ±5° of a standard or half-standard-rate turn, and to roll out on specified headings, ±10°.***

C. Common Errors during Timed Turns

1. **Incorrect calibration procedures.**
   a. Establish and maintain a constant rate of turn while you are calibrating the TC.
   b. Ensure that you know the number of degrees of heading change desired with the time interval you are using, e.g., 30° in 10 sec.
   c. Make necessary corrections to obtain a standard-rate turn, and note these deflections on the TC.
   d. These steps must be followed for both left and right turns.

2. **Failure to maintain altitude, airspeed, and desired rate of turn.**
   a. Use the same cross-check and control techniques as used in other level turns, except substitute the clock for the HI.
   b. Do not limit your instrument scan to only the clock and TC. If you do, you will have trouble in controlling your airplane.

3. **Improper timing.**
   a. Start the roll-in when the clock second hand passes a cardinal point. Timing does NOT begin after you have rolled into your turn.
   b. The rollout starts after the computed time has elapsed.
   c. Practice calculating times for various heading changes (standard rate).
      1) 360° = 2 min.
      2) 180° = 1 min.
      3) 90° = 30 sec.
      4) 45° = 15 sec.
      5) 30° = 10 sec.
      6) 10° = approximately 3 sec.

4. **Faulty technique when making small changes of heading.**
   a. When making small changes of heading, you should use a half-standard-rate turn (i.e., 1.5° per second rate of turn).

5. **Uncoordinated use of controls.**
   a. Maintain coordinated flight throughout this maneuver.
   b. Do not use just the rudder.

6. **Faulty trim technique.**
   a. Trim should be used, not to substitute for control with the control yoke and rudder, but to relieve pressures already held to stabilize attitude.
   b. Use trim frequently and in small amounts.
   c. Improper adjustment of seat or rudder pedals for comfortable positioning of legs and feet may contribute to trim errors.
      1) Tension in the ankles makes it difficult to relax rudder pressures.

END OF TASK

STEEP TURNS

IV.F. TASK: STEEP TURNS

REFERENCES: 14 CFR Part 61; AC 61-27.

Objective. To determine that the applicant:

1. Exhibits adequate knowledge of the factors relating to attitude instrument flying during steep turns.
2. Enters a turn using a bank of approximately 45° for an airplane.
3. Maintains the specified angle of bank for either 180° or 360° of turn, both left and right.
4. Maintains altitude within 100 ft. (30 meters), airspeed within 10 kt., 5° of specified bank angle, and rolls out within 10° of the specified heading.
5. Uses proper instrument cross-check and interpretation, and applies the appropriate pitch, bank, power, and trim corrections.

A. General Information

1. The objective of this task is for you to demonstrate your ability to perform steep turns.
2. During your practical test, this test shall be performed using all available instruments.
3. For IFR, any turn above a standard-rate turn is steep because standard-rate turns are the maximum used for IFR.
   a. Steep turns are a safety training maneuver for showing how to deal with the unexpected.
   b. They also assist your skill in recovery from unusual attitudes.

B. Task Objectives

1. **Exhibit adequate knowledge of the factors relating to attitude instrument flying during steep turns.**
   a. For the purposes of instrument flight training, any turn greater than a standard rate may be considered steep. The exact angle of bank at which a normal turn becomes steep is unimportant.
      1) What is important is that you maintain control of your airplane with bank angles in excess of those you normally use on instruments.
      2) Practice in steep turns not only will increase your proficiency in the basic instrument skills but also will enable you to react smoothly, quickly, and confidently to unexpected abnormal flight attitudes under IFR conditions.
   b. Pronounced changes occur in the effects of aerodynamic forces on airplane control at progressively steepening bank attitudes.
      1) Since the vertical lift component is greatly reduced, pitch control is an important aspect of this maneuver.
         a) Unless immediately noted and corrected with a pitch increase, the loss of vertical lift results in rapid movement of the altimeter (ALT), the vertical speed indicator (VSI), and the airspeed indicator (ASI).
         b) The faster the rate of bank change, the faster the lift changes occur.
      2) Increased speed in cross-check, interpretation, and control is necessary in proportion to the amount of these changes.
      3) The techniques for entering, maintaining, and recovering from steep turns are the same in principle as those for standard-rate turns.

2. **Enter a turn using a bank of approximately 45°.**
    a. Enter a steep turn as you normally enter a turn, but prepare to cross-check rapidly as you increase the bank to approximately 45° on the attitude indicator (AI).
    b. As the bank changes, apply smooth, steady back elevator pressure to maintain constant altitude.
        1) If this is not done, the loss of vertical lift and increased load factor reach a point where increased back elevator pressure tightens the turn without raising the nose.
            a) This is a diving spiral, and an accelerated stall may develop.
        2) Do not use excessive pitch attitudes.
    c. Use coordinated aileron and rudder to keep the ball centered on the turn coordinator (TC).
3. **Maintain the specified angle of bank for either 180° or 360° of turn, both left and right.**
    a. Your examiner will instruct you to perform a 180° or 360° turn. This turn will be performed both to the left and to the right.
    b. You may be required to roll immediately from a left turn to a right (or vice versa) without any period of level flight.
4. ***Maintain altitude within 100 ft., airspeed within 10 kt., and 5° of specified bank angle, and roll out within 10° of the specified heading.***
    a. If you observe a rapid downward movement of the ALT or VSI needles, together with an increase in airspeed, despite your application of back elevator pressure, you are in a diving spiral.
        1) Immediately shallow the bank with smooth and coordinated aileron and rudder pressures, hold or slightly relax elevator pressure, and increase your cross-check of the AI, ALT, and VSI.
            a) Reduce power if the airspeed increase is rapid.
        2) When you note that the elevator is effective in raising the nose, hold the bank angle shown on the AI, and adjust elevator control pressures smoothly for the nose-high attitude appropriate to the bank maintained.
        3) The AI is the primary instrument for bank.
    b. The power necessary to maintain constant airspeed increases as the bank, angle of attack, and drag increase.
        1) You should learn the power settings appropriate to specific bank attitudes and make adjustments to power without undue attention to airspeed and power instruments when establishing the bank.
        2) If you keep pitch relatively constant, you have more time to cross-check, interpret, and control for accurate airspeed and bank.
    c. Use the AI to maintain the bank specified by your examiner (i.e., approximately 45°).
    d. During the maneuver, you should trim your airplane to relieve the control pressures you are holding.

e. During recovery from steep turns, the elevator and power control must be coordinated with bank control in proportion to the increase of the vertical lift component.

1) Forward elevator pressure may be required initially, due to nose-up trim, and power decreased (if used) to maintain airspeed.

2) Lead your rollout at approximately one-half your bank angle.

a) EXAMPLE: While maintaining a 45° bank, lead your rollout to the desired heading by 20° to 25° (45 ÷ 2).

3) Retrim if necessary.

4) Use coordinated aileron and rudder to keep the ball centered on the TC.

5. Use proper instrument cross-check and interpretation, and apply the appropriate pitch, bank, power, and trim corrections.

a. Maintain level flight by adjusting your pitch as necessary to hold your assigned altitude.

1) As bank increases, more back pressure will be needed to maintain altitude.

a) Trim to relieve control pressure.

2) Coordinate pitch attitude with bank angle to maintain level flight.

3) As you roll out of your turn, use forward pressure to maintain altitude.

4) The ALT is your primary pitch instrument, but you will need to cross-check against the supporting instruments (i.e., the AI and the VSI).

a) As a trend instrument, your VSI will show immediately the initial vertical movement of your airplane, even before your ALT.

b) Avoid excessive vertical speeds, i.e., greater than 500 fpm.

i) Vertical speeds less than 500 fpm can usually be corrected with pitch only (maintain 45° bank).

ii) Vertical speeds greater than 500 fpm may require bank and/or power adjustments in addition to pitch corrections.

b. Maintain the desired bank angle by using your AI. Cross-check the TC to ensure that the turn is coordinated (i.e., the ball is centered).

1) In a steep bank, aileron pressure opposite the direction of turn may be needed to avoid overbanking.

2) Monitor the progress of your turn with the HI.

c. Use power, as necessary, to assist you in maintaining airspeed.

1) Monitor your ASI to prevent too fast, or too slow, an airspeed.

2) If you maintain a relatively constant pitch attitude, you can add power to a known setting upon initial roll-in and not reset it until rollout.

C. Common Errors during Steep Turns

1. **Errors for steep turns are the same as for standard-rate turns, except they are more exaggerated and more difficult to correct.**
2. **Failure to recognize and make proper corrections for pitch, bank, or power errors.**
   a. Your cross-check must increase.
      1) Include the VSI often in your scan.
   b. As you note the need for pitch, immediately apply smooth, steady back elevator pressure.
      1) Do not pitch excessively.
      2) Refer to the AI for pitch and bank adjustments.
   c. Frequently, due to insufficient back pressure, the airplane descends and may enter a diving spiral (increasing airspeed).
      1) DO NOT exert additional back pressure UNTIL you have decreased the bank.
   d. During recovery, release any back elevator pressure you may be holding.
3. **Uncoordinated use of controls.**
   a. Maintain coordinated flight throughout this maneuver.
4. **Faulty trim technique.**
   a. Trim should be used, not to substitute for control with the control yoke and rudder, but to relieve pressures already held to stabilize attitude.
   b. Use trim frequently and in small amounts.
   c. Improper adjustment of seat or rudder pedals for comfortable positioning of legs and feet may contribute to trim errors.
      1) Tension in the ankles makes it difficult to relax rudder pressures.

END OF TASK

RECOVERY FROM UNUSUAL FLIGHT ATTITUDES

IV.G. TASK: RECOVERY FROM UNUSUAL FLIGHT ATTITUDES

REFERENCES: FAR Part 61; AC 61-27.

NOTE: Any intervention by the examiner to prevent the aircraft from exceeding any operating limitations, or entering an unsafe flight condition, shall be disqualifying.

Objective. To determine that the applicant:

1. Exhibits adequate knowledge of the elements relating to attitude instrument flying during recovery from unusual flight attitudes (both nose-high and nose-low).
2. Uses proper instrument cross-check and interpretation, and applies the appropriate pitch, bank, and power corrections in the correct sequence to return the aircraft to a stabilized level flight attitude.

A. General Information

1. The objective of this task is for you to demonstrate your ability to recover from unusual flight attitudes (both nose-high and nose-low).
    a. Any intervention by your examiner to prevent your airplane from exceeding any operating limitations or entering an unsafe flight condition shall be disqualifying.
2. Unusual attitudes result from loss of positive control.
    a. Unless corrected immediately, they precede a crash, destruction of the aircraft, and almost certain death or serious injury.
    b. Unusual attitudes are EMERGENCY SITUATIONS.
    c. A proficiency in recovering from unusual attitudes cannot be overemphasized.
3. It is imperative to recover from an unusual attitude promptly, not only to avoid a crash, but also to avoid the following:
    a. Stall/spin
    b. Airspeed limits: V_{NO}, V_A, V_{NE}
    c. Bank or pitch limits
    d. Engine limits (RPM)
    e. Altitude limits (e.g., obstacles, other aircraft)
    f. Load factor (or "G") limits

4. Various factors contribute to an unusual attitude while under IFR.
   a. Presumably you have already encountered an unusual attitude when flying VFR. How did it happen?
      1) Were you looking at a chart?
      2) Were you distracted by a passenger?
      3) Did it result from excessive attention to an equipment problem?
   b. Such scenarios can also result in an unusual attitude when flying in IMC, only more so.
      1) You must be in control and STAY in control. Fly your airplane first.
      2) Keep your cockpit well organized with everything you may possibly need easily available.
      3) You must be organized.
      4) Plan ahead.
      5) Always keep the airplane trimmed so it will not stray into unusual attitudes as easily.
         a) If an unusual attitude does occur, a properly trimmed airplane facilitates recovery to straight-and-level flight.
5. The failure of flight instruments (especially the attitude indicator) while in IMC is a major reason for the instrument cross-check.
   a. Usually when an attitude indicator (AI) fails and the horizon falls over, you follow it because it appears you are banking in the opposite direction.
      1) Think about this and visualize it.
      2) When the AI horizon falls one way, you appear to be turning in the opposite direction.
      3) Remember, when the AI does NOT respond to a significant control movement (pitch or bank), it has probably failed.
         a) Compare bank to the turn coordinator (TC).
         b) Compare pitch to the airspeed indicator (ASI), the vertical speed indicator (VSI), and the altimeter (ALT).
6. During your flight training, your instructor will have your perform a recovery from an unusual attitude using both full and partial panel.
   a. During your practical test, this task shall be performed without the use of the attitude indicator.

B. Task Objectives

1. **Exhibit adequate knowledge of the elements relating to attitude instrument flying during recovery from unusual flight attitudes (both nose-high and nose-low).**

   a. As a general rule, anytime there is an instrument rate of movement or indication other than those associated with basic instrument flight maneuvers, assume an unusual attitude and increase the speed of cross-check to confirm the attitude, instrument error, or instrument malfunction.

   b. When a critical attitude is noted on the flight instruments, the immediate priority is to recognize what your airplane is doing and decide how to return it to straight-and-level flight as quickly as possible.

   c. To avoid aggravating the critical attitude with a control application in the wrong direction, the initial interpretation of the instruments must be accurate.

   d. Nose-high attitudes are shown by the rate and direction of movement of the ALT, VSI, and ASI needles, as shown in the figure below.

Unusual Attitude -- Nose High

1. ASI is decreasing from 140 kt. down to 75 kt.
2. ALT is increasing from 4,500 ft. toward 5,000 ft.
3. TC indicates a right turn.
4. HI indicates a right turn from 270° toward 360°.
5. VSI indicates a positive rate of climb.
6. This task requires you to perform this maneuver without an AI, which your examiner will cover.

e. Nose-low attitudes are shown by the same instruments, but in the opposite direction, as shown in the figure below.

Unusual Attitude -- Nose Low

1. ASI is increasing from 140 kt. to 190 kt.
2. ALT is decreasing from 6,500 ft. to 6,000 ft.
3. TC indicates a right turn.
4. HI indicates a right turn from 270° toward 360°.
5. VSI indicates a negative vertical speed (i.e., descent).
6. This task requires you to perform this maneuver without an AI, which your examiner will cover.

2. Use proper instrument cross-check and interpretation, and apply the appropriate pitch, bank, and power corrections in the correct sequence to return your airplane to a stabilized level flight attitude.

a. Recovery from a nose-high attitude

1) Nose-high unusual attitude is indicated by

a) Decreasing airspeed
b) Increasing altitude
c) A turn on the TC

2) Take action in the following sequence.

a) Add power. If the airspeed is decreasing or below the desired airspeed, increase power (as necessary in proportion to the observed deceleration).

b) Reduce pitch. Apply forward elevator pressure to lower the nose and prevent a stall.

i) Deflecting ailerons to level the wings before the angle of attack is reduced could result in a spin.

c) Level the wings. Correct the bank (if any) by applying coordinated aileron and rudder pressure to level the miniature airplane and center the ball of the TC.

3) The corrective control applications should be made almost simultaneously but in the sequence beginning on the previous page.

4) After initial control has been applied, continue with a fast cross-check for possible overcontrolling, since the necessary initial control pressures may be large.

a) As the rate of movement of the ALT and VSI needles decreases, the attitude is approaching level flight. When the needles stop and reverse direction, your airplane is passing through level flight.

5) When airspeed increases to normal speed, set cruise power.

b. Recovery from a nose-low attitude

1) Nose-low unusual attitude is indicated by

a) Increasing airspeed
b) Decreasing altitude
c) A turn on the TC

2) Take action in the following sequence.

a) Reduce power. If the airspeed is increasing, or is above the desired speed, reduce power to prevent excessive airspeed and loss of altitude.

b) Level the wings. Correct the bank attitude with coordinated aileron and rudder pressure to straight flight by referring to the TC.

i) Increasing elevator back pressure before the wings are leveled will tend to increase the bank and make the situation worse.

ii) Excessive G-loads may be imposed, resulting in structural failure.

c) Raise the nose. Smoothly apply back elevator pressure to raise the nose to level flight.

i) With the higher-than-normal airspeed, it is vital to raise the nose very smoothly to avoid overstressing the airplane.

3) The corrective control applications should be made almost simultaneously but in the sequence above.

4) After initial control has been applied, continue with a fast cross-check for possible overcontrolling, since the necessary initial control pressures may be large.

a) As the rate of movement of the ALT and VSI needles decreases, the attitude is approaching level flight. When the needles stop and reverse direction, your airplane is passing through level flight.

5) When airspeed decreases to normal speed, set cruise power.

c. As the indications of the ALT, TC, and ASI stabilize, the TC should be checked to determine that you are returning to straight flight and that the ball is centered.

1) Slipping or skidding sensations can easily aggravate disorientation and retard recovery.

2) You should return to your last assigned altitude after stabilizing in straight-and-level flight.

d. Unlike the control applications in normal maneuvers, larger control movements in recoveries from unusual attitudes may be necessary to bring the airplane under control.

1) Nevertheless, such control applications must be smooth, positive, prompt, and coordinated.

2) Once the airplane is returned to approximately straight-and-level flight, control movements should be limited to small adjustments.

e. Flight instruments in straight-and-level flight (without the AI) are presented below.

MANEUVER	PITCH	BANK	POWER
Straight-and-level flight (without the AI)			
Primary	ALT	HI	ASI
Supporting	ASI, VSI	TC	MP, RPM

1) In the discussion on recovery, we stressed the use of instruments as if the attitude and heading indicators were inoperative. During your practical test you will have the use of the HI; thus it is included in the above table.

C. Common Errors during Recovery from Unusual Flight Attitudes

1. **Failure to recognize an unusual flight attitude.**

a. This error is due to poor instrument cross-check and interpretation.

b. Once you are in an unusual attitude, determine how to return to straight-and-level flight, NOT how your airplane got there.

c. Unusually loud or soft engine and wind noise may provide an indication.

2. **Attempting to recover from an unusual flight attitude by "feel" rather than by instrument indications.**

a. The most hazardous illusions that lead to spatial disorientation are created by information received by our motion-sensing system, located in each inner ear.

b. The system is not capable of detecting a constant velocity or small changes in velocity, nor can it distinguish between centrifugal force and gravity.

c. The motion-sensing system, functioning normally in flight, can produce false sensations.

d. During unusual flight attitudes, you must learn to believe and interpret the flight instruments because spatial disorientation is normal in unusual flight attitudes.

3. **Inappropriate control applications during recovery.**

a. Accurately interpret the initial instruments before recovery is started.

b. Follow the recovery steps in the proper sequence.

c. Control movements may be larger than normal but must be smooth, positive, prompt, and coordinated.

4. **Failure to recognize from instrument indications when the airplane is passing through a level flight attitude.**

a. Without an attitude indicator, a level pitch attitude is indicated by the reversal and stabilization of the airspeed indicator and altimeter needles.

END OF TASK -- END OF CHAPTER

CHAPTER V
NAVIGATION SYSTEMS

This chapter explains the one task of Navigation Systems. This task includes both knowledge and skill.

You will be required to be able to use any IFR-approved navigation system and to intercept and track desired courses. If your airplane is equipped with a VOR, an ADF, and a GPS, your examiner may choose any of these systems to test you; thus you must be conversant in whatever system(s) is (are) installed in your airplane.

Once en route under IFR, you must always keep track of your EXACT location, direction, and speed. Remember: PLAN AHEAD AND BE PREPARED. Navigation systems form the basis upon which you will always know where you are; they also assist you in executing your desired course.

INTERCEPTING AND TRACKING NAVIGATIONAL SYSTEMS AND DME ARCS

V.A. TASK: INTERCEPTING AND TRACKING NAVIGATIONAL SYSTEMS AND DME ARCS

REFERENCES: 14 CFR Parts 61, 91; AC 61-27; AIM.

NOTE: Any reference to DME arcs, ADF, or GPS shall be disregarded if the aircraft is not equipped with these specified navigational systems.

Objective. To determine that the applicant:

1. Exhibits adequate knowledge of the elements related to intercepting and tracking navigational systems and DME arcs.
2. Tunes and correctly identifies the navigation facility.
3. Sets and correctly orients the radial to be intercepted into the course selector or correctly identifies the radial on the RMI.
4. Intercepts the specified radial at a predetermined angle, inbound or outbound from a navigational facility.
5. Maintains the airspeed within 10 kt., altitude within 100 ft. (30 meters), and selected headings within 5°.
6. Applies proper correction to maintain a radial, allowing no more than three-quarter-scale deflection of the CDI or within 10° in case of an RMI.
7. Determines the aircraft position relative to the navigational facility or from a waypoint in case of GPS.
8. Intercepts a DME arc and maintains that arc within 1 NM.
9. Recognizes navigational receiver or facility failure, and, when required, reports the failure to ATC.

A. General Information

1. The objective of this task is for you to demonstrate your ability to intercept and track radials, bearings, courses, and DME arcs using various navigational systems.
   a. Any reference to DME arcs, ADF, or GPS can be disregarded if your airplane is not equipped with these specified navigational systems.
2. For additional reading, see *Pilot Handbook*, Chapter 10, Navigation Systems, for a 26-page discussion on the operating principles, components, and orientation and tracking procedures of various navigation systems.
3. Since the displays vary on the IFR-approved GPS systems, consult your GPS operating handbook for detailed information.
4. We have divided the discussion of intercepting and tracking radials, DME arcs, bearings, and courses into separate topics to simplify your studying. Each topic is listed below with the page on which the discussion of the topic begins:

B. VOR and DME Arc Task Objectives

1. **Exhibit adequate knowledge of the elements related to intercepting and tracking VOR radials and DME arcs.**

   a. Interception of VOR radials and DME arcs requires that you orient (visualize) your position with respect to your desired radial or DME arc. Then you must determine an interception angle.

   b. Tracking a radial or DME arc requires that you make any necessary wind-drift correction to remain on the radial or arc.

      1) This involves bracketing to determine the correct amount of drift correction.

2. **Tune and correctly identify the VOR facility.**

   a. Tune the VOR receiver to the frequency of the selected VOR station. (We will use the term VOR to include VOR, VORTAC, or VOR/DME facilities.)

      1) The VOR frequency will be on your en route chart.

   b. You should positively identify the VOR transmitting station by its Morse code identification or by a recorded voice identification which states the name of the station followed by the letters "V-O-R." This positive identification is necessary because many FSSs transmit voice messages on the same frequency as VORs.

      1) At sufficiently high altitudes in certain parts of the country, it is possible to receive two VORs, each broadcasting on the same frequency.

      2) During periods of maintenance, the facility may radiate a T-E-S-T code (– . . . . –), or the code may be removed.

      3) Voice identification has been added to numerous VORs. The transmission consists of a voice announcement, "Airville VOR," alternating with the usual Morse code identification.

3. **Set and correctly orient the radial to be intercepted into the course selector or correctly identify the radial on the RMI.**

   a. It is important that you understand your clearance as to which radial you are to intercept and whether you are to track inbound or outbound.

   b. Before setting the desired course, you should orient yourself by centering the CDI needle and then determine the difference between the radial to be intercepted and the radial on which you are located.

   c. You must ensure that you correctly set the desired radial or inbound course to be intercepted. Remember, radials are FROM the station.

      1) If you are tracking outbound FROM the station, rotate the OBS until the desired radial is under the index at the top of the instrument.

      2) If you are tracking inbound TO the station, rotate the OBS until the desired radial is at the bottom of the instrument. The reciprocal (i.e., inbound course) will be under the index at the top of the instrument.

         a) If you cannot see the numbers at the bottom of the instrument, you can use the HI or perform the calculation to determine the inbound course.

   d. When using a radio magnetic indicator (RMI), the head of the needle will point to the VOR station.

      1) The tail of the needle indicates the VOR radial on which you are presently located.

4. Intercept the specified radial at a predetermined angle, inbound or outbound from a VOR facility.

a. The following steps may be used to intercept a predetermined VOR radial, either outbound or inbound:

1) Turn to a heading to parallel your desired course, in the same direction as the course to be flown.

2) Rotate the OBS knob until the CDI needle centers with a TO indication if you are to track toward the VOR or a FROM indication if you are to track outbound from the VOR. This procedure will help ensure that you remain oriented.

a) Note the course indicated under the index at the top of your VOR instrument.

3) Rotate the OBS to your desired course (which is also your present heading) and determine the difference between the course you are on [from item 2) above] and your desired course.

4) Double the difference to determine the intercept angle (not less than 20° or greater than 90°).

5) Turn toward the CDI to an intercept heading that will establish an angle equal to the amount determined in 4) above.

6) Hold this heading until the CDI begins to center, indicating that you are approaching your desired course.

7) You must learn to judge the rate of closure with the course centerline so you can lead your turn to prevent overshooting the desired radial.

8) As you turn onto the desired radial, check to ensure that you are flying inbound or outbound as directed in your clearance.

b. Remember, as you approach the VOR, the CDI becomes more sensitive.

1) Avoid large intercept angles when close to the VOR.

a) At 30 NM from the VOR, 1 dot = 1 mile. At 60 NM from the VOR, 1 dot = 2 miles

c. When using an RMI, the orientation and intercept angle computation is similar to when using a CDI.

1) Turn to a heading to parallel the desired course, in the same direction as the course to be flown.

a) In the example on the opposite page, you are to intercept and track inbound on the 180-R.

2) Note the bearing pointer. It indicates your position as being on the 150-R, a 30° difference from your desired radial.

3) Double the deflection to determine an interception angle that is not less than 20° or greater than 90°.

a) In this example, your interception angle is 60° (30° x 2).

4) Turn to intercept heading.

5) Hold this heading until the needle points to the desired course (360° in this example); then turn inbound.

a) To intercept a radial outbound, hold intercept heading until the tail of the needle points to the desired radial.

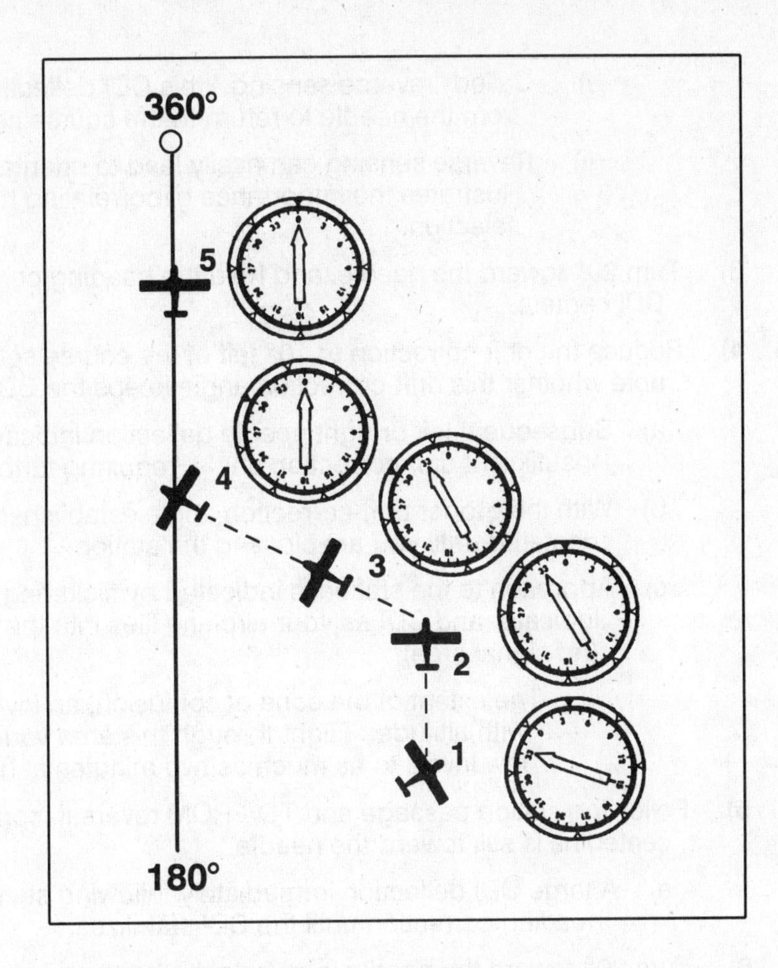

5. ***Maintain the airspeed within 10 kt., altitude within 100 ft., and selected headings within 5°.***

   a. Do not fixate on your CDI or RMI. You must cross-check and interpret all of your instruments to maintain airspeed, altitude, and headings.

   b. Do not make a lot of heading changes to get on course more quickly. Select headings for intercepting and tracking until the navigation system you are using indicates a change is needed.

6. ***Apply proper correction to maintain a radial, allowing no more than three-quarter-scale deflection of the CDI or within 10° in the case of an RMI.***

   a. To track inbound when using a CDI with the wind unknown, proceed using the following steps (corresponding to the figure on the left side of page 197). Outbound tracking procedures are the same.

      1) With the CDI centered, maintain the heading corresponding to the desired course.

      2) As you hold the heading, observe the CDI for deflection to left or right. The direction of CDI deflection from the centerline shows the direction of the crosswind component.

         a) The left VOR head (OBS correctly set to 360°) shows a left deflection, thus a left crosswind.

         b) Note the indications with the reciprocal of the inbound course set on the OBS. The right VOR head (OBS incorrectly set to 180°) shows right CDI deflection. The indicator correctly shows FROM and the airplane to the left of the centerline with reference to the selected course (i.e., 180° FROM the station).

i) Called "reverse sensing," this CDI deflection indicates a turn **away** from the needle to return to the course centerline.

ii) Reverse sensing can easily lead to orientation problems and illustrates the importance of correlating heading and course selection.

3) Turn 20° toward the needle, and hold the heading correction (340°) until the CDI centers.

4) Reduce the drift correction to 10° left of the course setting (heading 350°), and note whether this drift-correction angle keeps the CDI centered.

a) Subsequent left or right needle deflection indicates an excessive or insufficient drift-correction angle, requiring further bracketing.

b) With the proper drift-correction angle established, the CDI will remain centered until you are close to the station.

c) Approach to the station is indicated by flickering of the TO/FROM indicator and CDI as your airplane flies into the "cone of confusion" (no-signal area).

i) The extent of the cone of confusion, an inverted cone, increases with altitude. Flight through this area varies from a few seconds at low levels to as much as two minutes at high altitudes.

5) Following station passage and TO/FROM reversal, correction to the course centerline is still toward the needle.

a) A large CDI deflection immediately following station passage calls for no heading correction until the CDI stabilizes.

6) Turn 20° toward the needle, and hold the heading correction (010°) until the CDI centers.

b. To track inbound when using an RMI with the wind unknown, proceed using the following steps (corresponding to the figure on the right side of page 197). Outbound tracking procedures are the same.

1) Your heading is 360°, you are inbound on the 180-R, and the bearing pointer indicates 360°.

2) As you maintain a constant heading, the bearing pointer indicates that you have drifted off course approximately 10° to the right. This drift indicates a left crosswind component.

3) Turn left 20° to a heading of 340°.

4) Your airplane has returned to course, and the bearing pointer again indicates 360°.

5) Reduce the drift correction to 10° left, or a heading of 350°. If the course is not maintained, further bracketing will be required.

7. Determine your airplane's position relative to the VOR facility.

a. To determine your position relative to a VOR station, you should rotate the OBS until the CDI needle centers with a FROM indication.

1) Read the radial that you are on under the index at the top of the instrument. This is your position relative to the VOR station.

b. With an RMI, the tail of the needle indicates which radial you are on.

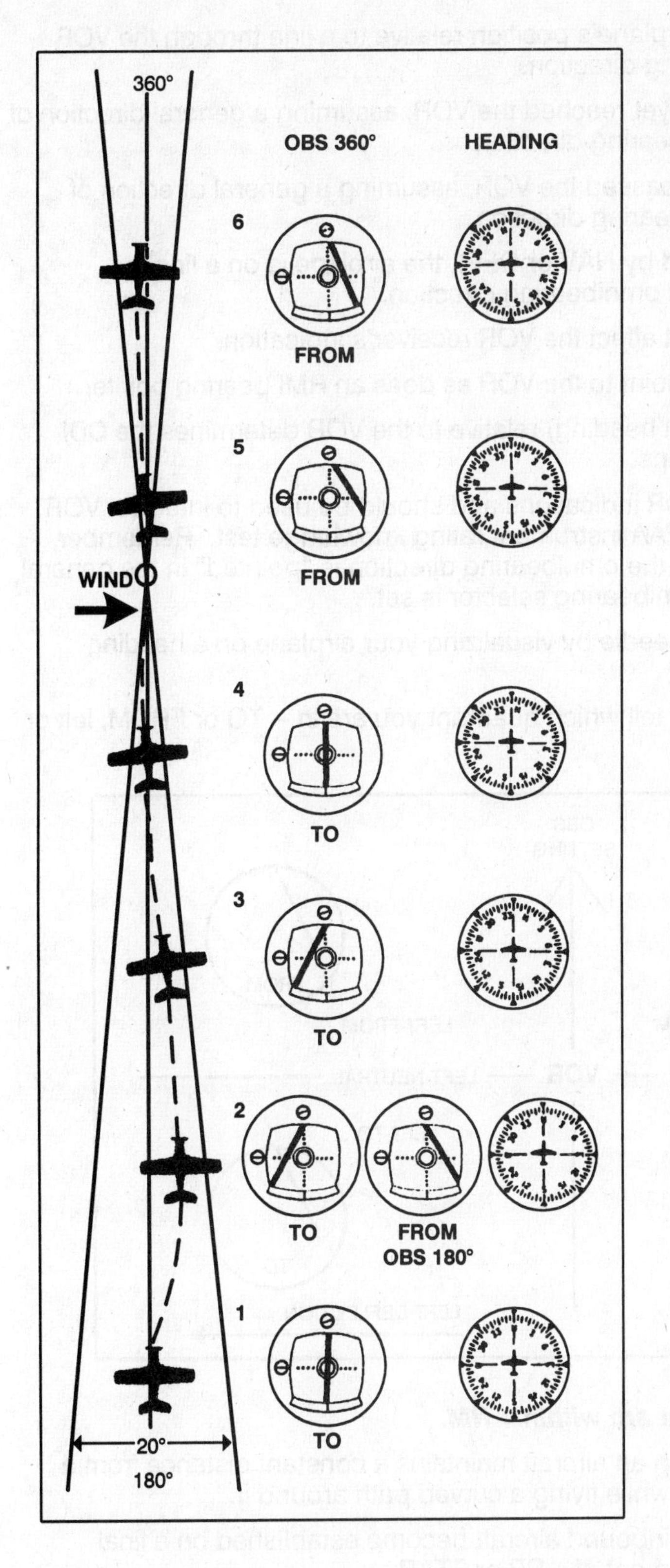
360°
OBS 360°
HEADING
6
FROM
5
FROM
WIND
4
TO
3
TO
2
TO
FROM
OBS 180°
1
TO
20°
180°

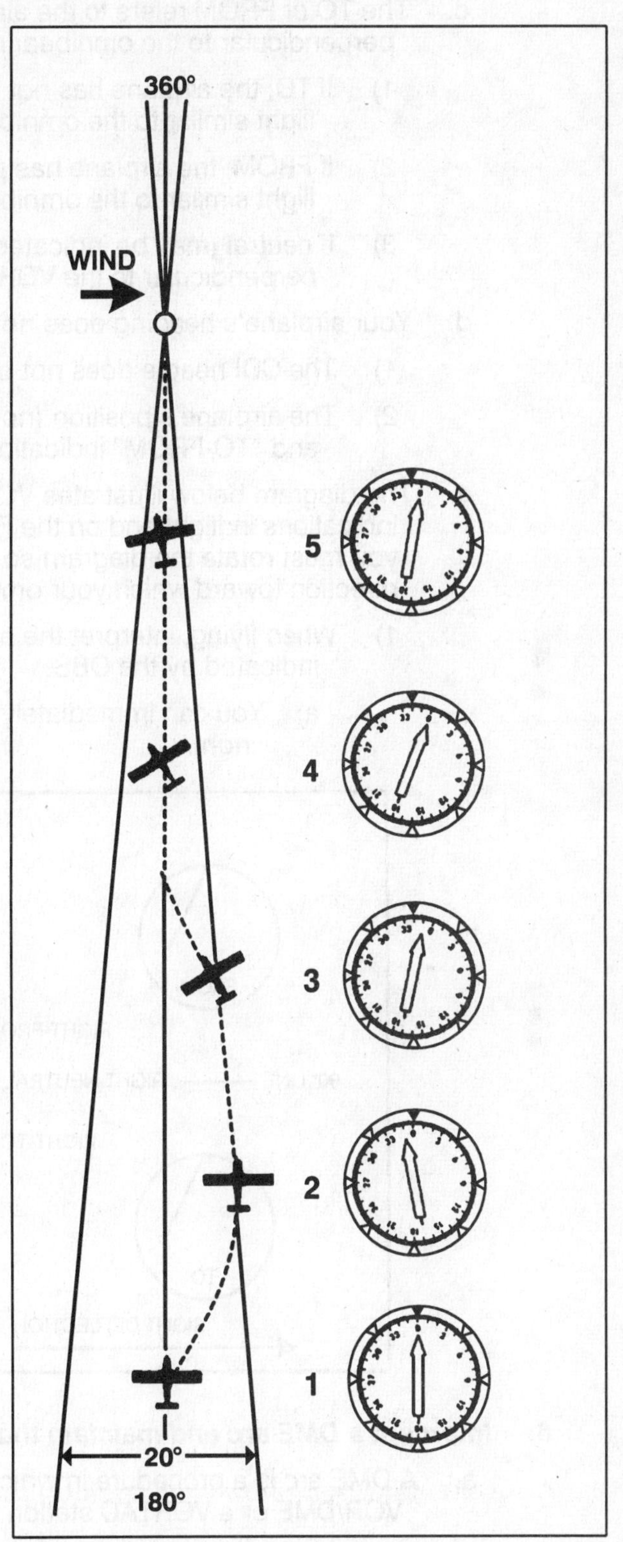
360°
WIND
5
4
3
2
1
20°
180°

c. The TO or FROM refers to the airplane's position relative to a line through the VOR perpendicular to the omnibearing direction.

  1) If TO, the airplane has not yet reached the VOR, assuming a general direction of flight similar to the omnibearing direction.

  2) If FROM, the airplane has passed the VOR, assuming a general direction of flight similar to the omnibearing direction.

  3) If neutral (may be indicated by NAV or OFF), the airplane is on a line perpendicular to the VOR omnibearing direction.

d. Your airplane's heading does not affect the VOR receiver's indication.

  1) The CDI needle does not point to the VOR as does an RMI bearing pointer.

  2) The airplane's position (not heading) relative to the VOR determines the CDI and "TO-FROM" indications.

e. The diagram below illustrates VOR indications and should be used to interpret VOR indications in flight and on the FAA instrument rating knowledge test. Remember, you must rotate the diagram so the omnibearing direction is "pointed" in the general direction toward which your omnibearing selector is set.

  1) When flying, interpret the needle by visualizing your airplane on a heading indicated by the OBS.

    a) You can immediately tell which quadrant you are in -- TO or FROM, left or right.

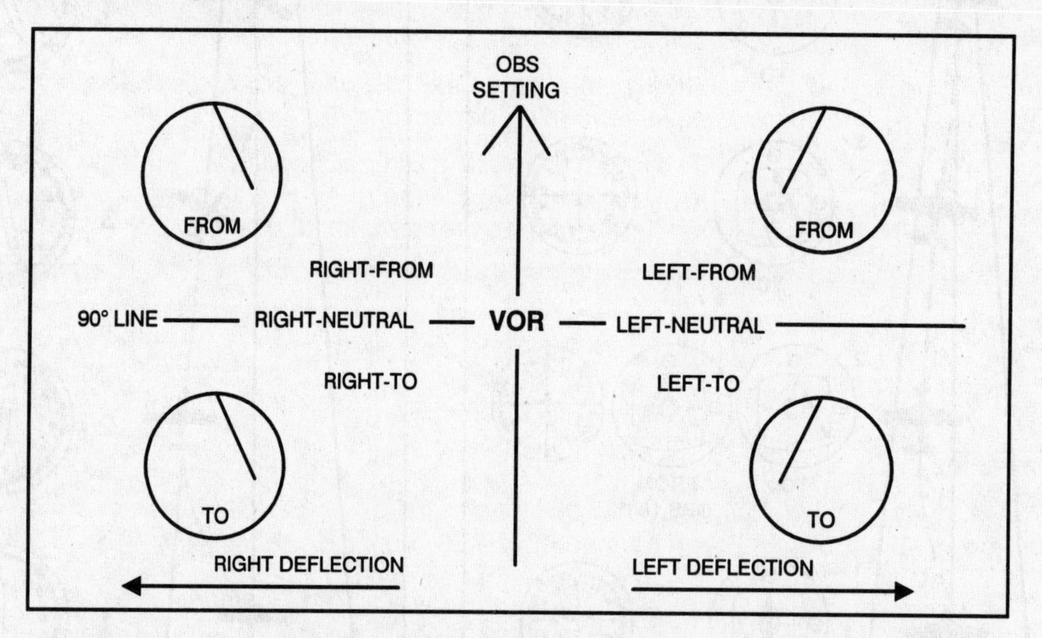

8. *Intercept a DME arc and maintain that arc within 1 NM.*

a. A DME arc is a procedure in which an aircraft maintains a constant distance from a VOR/DME or a VORTAC station while flying a curved path around it.

  1) DME arcs are used to help inbound aircraft become established on a final approach course or a segment of a DP or STAR.

b. Refer to the figure on page 199, and follow these steps to intercept the 10 DME arc when inbound or outbound on the 325-R of OKT VORTAC:

  1) Track inbound/outbound on the 325-R, frequently checking the DME mileage readout.

2) A 0.5 NM lead is satisfactory for groundspeeds of 150 kt. or less. Start your turn to the arc at 10.5 NM inbound or 9.5 NM outbound.
   a) Higher groundspeeds use a proportionately greater lead.
3) Continue the turn for 90°. In this example, the rollout heading will be 055° in no-wind conditions.
4) During the last part of the intercepting turn, monitor the DME closely. If it appears that you are overshooting the arc, continue past the originally planned rollout point.
   a) If the arc is being undershot, roll out of your turn early.

c. Flying the DME arc requires that you apply proper wind correction and maintain positional awareness at all times.
   1) Use the VOR receiver to maintain constant awareness of the radials you are passing through.
      a) Always operate with the FROM flag so that the radial is under the index at the top of the instrument.
      b) Set the OBS to 10° ahead of the radial you are on (initially 335° in the example).
      c) When the needle centers (from tracking the arc), reset OBS to 10° ahead again.
   2) Use the DME readout as you would a CDI to determine whether you are left or right of your desired ground track.
      a) When resetting the OBS to a new radial, check the DME readout, and make a heading correction.
         i) In the example, if DME approximately equals 10.0 NM, turn 10°.
         ii) If the DME is less than 10.0 NM, maintain your heading.
         iii) If the DME is greater than 10.0 NM, turn 20°.

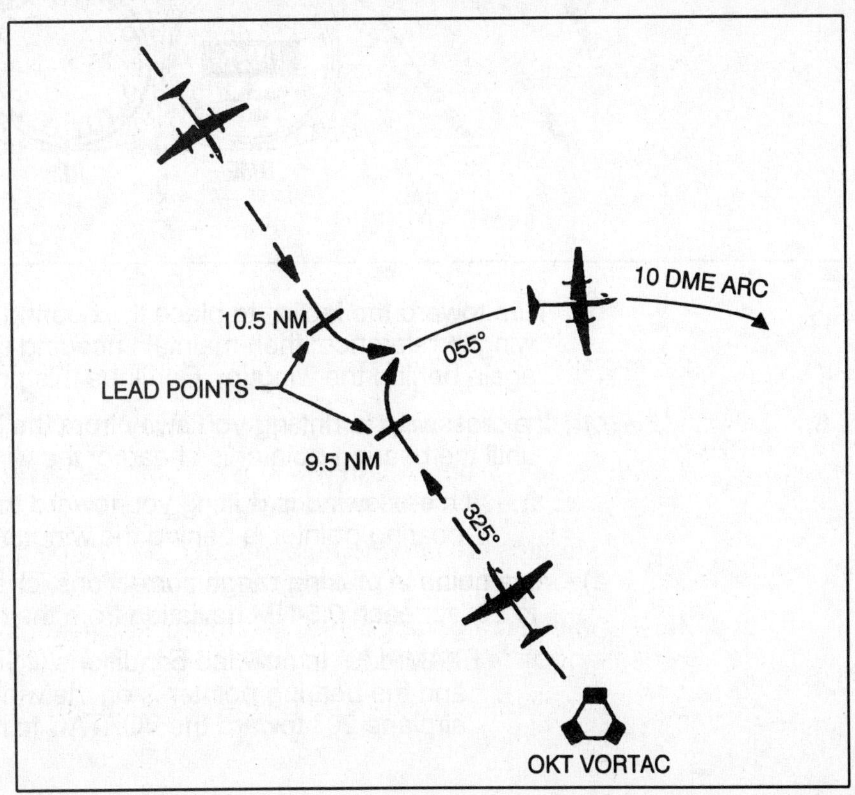

3) Maintain the DME arc within 1 NM.

a) In actual flight, a DME arc has the same width (protected airspace) as a federal airway (i.e., 4 NM on either side of the centerline).

d. With an RMI in a no-wind condition, you should be able to fly an exact circle around the VORTAC by maintaining a relative bearing of 90° or 270°. This may sound more complicated than using a CDI. Why not also use a DME as a "CDI," as presented in c. on the previous page? It works really well and is easy.

1) In actual practice, a series of short legs are flown. To maintain the arc (see the figure below), proceed as follows:

a) With the RMI bearing pointer on the wingtip reference (90° or 270° position) and your airplane at the desired DME range, maintain a constant heading, and allow the bearing pointer to move 5° to 10° behind the wingtip, thus increasing the range slightly.

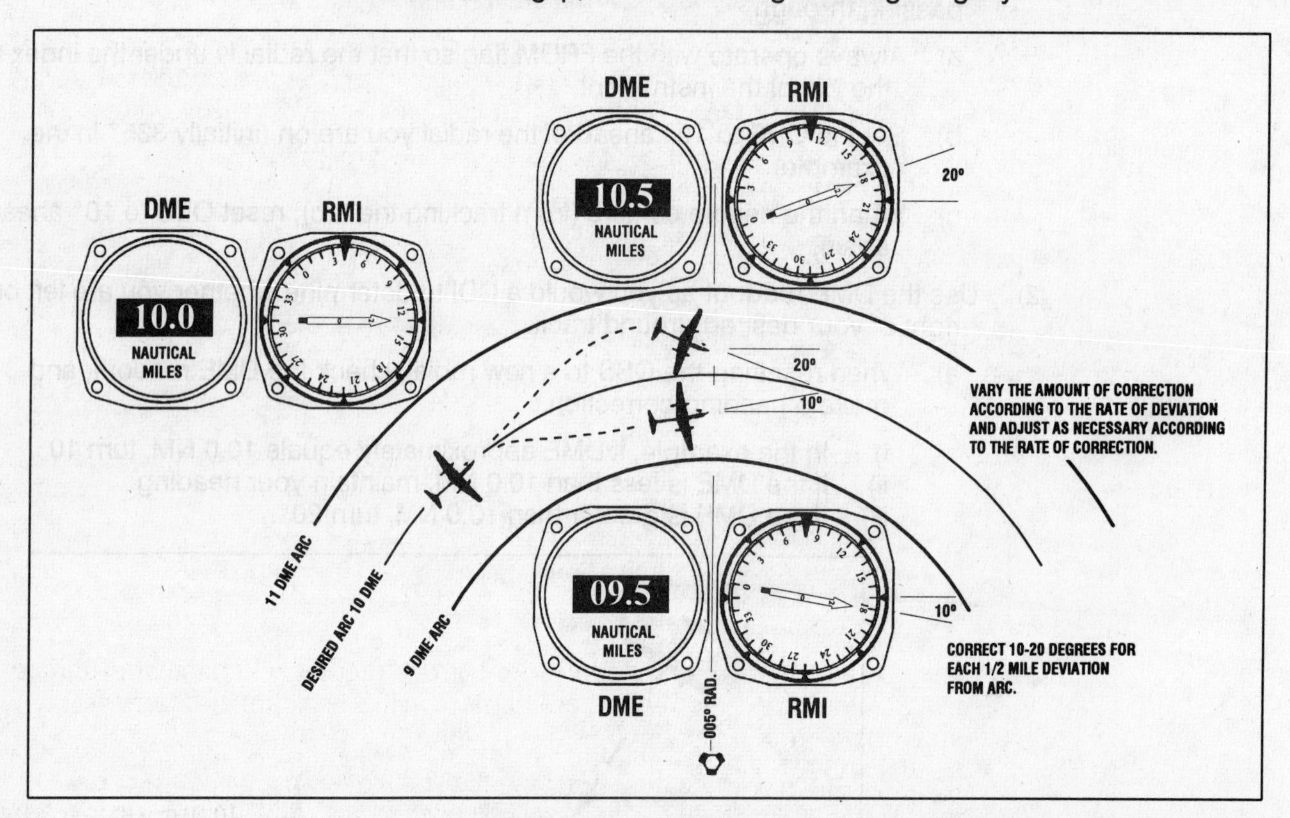

b) Turn toward the facility to place the bearing pointer 5° to 10° ahead of the wingtip reference; then maintain heading until the bearing pointer is again behind the wingtip. Continue this procedure to maintain the arc.

c) If a crosswind is drifting you away from the VORTAC, turn your airplane until the bearing pointer is ahead of the wingtip reference.

i) If a crosswind is drifting you toward the VORTAC, turn until the bearing pointer is behind the wingtip.

d) As a guide in making range corrections, change the relative bearing 10° to 20° for each 0.5-NM deviation from the desired arc.

i) EXAMPLE: In no-wind conditions, if you are 0.5 NM outside the arc and the bearing pointer is on the wingtip reference, turn your airplane 20° toward the VORTAC to return to the arc.

e. While flying a DME arc, you must keep a continuous mental picture of your position relative to the VORTAC. Since the wind-drift-correction angle is constantly changing throughout the arc, wind orientation is important.

f. Keep track of the radial you are on and the radial you are going to intercept for an approach.

1) When within 10° of the approach radial, set OBS for approach course and prepare for course intercept.

a) Remember, you are holding a 90° intercept to the course; the CDI needle will move quickly.

9. Recognize a VOR receiver or facility failure, and, when required, report the failure to ATC.

a. If the VOR facility is out of service, the Morse code identification is not transmitted. This serves to alert pilots that this VOR station should not be used for navigation.

b. VOR receivers have an alarm flag (frequently a red NAV or OFF flag between the TO and FROM) to indicate when the received signal strength is inadequate to operate the navigational equipment. The alarm flag appears if one of the following applies:

1) The VOR facility is out of service.

2) Your airplane is too far from the VOR.

3) Your airplane is too low to be in line-of-sight of the VOR facility transmitting signals.

c. Erratic movement of the CDI needle also indicates one of the following:

1) An unreliable signal from a distant VOR
2) Receiver problems

d. When operating under IFR in controlled airspace, you are required to report to ATC any malfunction of navigational, approach, or communication equipment occurring in flight (FAR 91.187).

C. Common Errors during Intercepting and Tracking of VOR Radials and DME Arcs

1. Incorrect tuning and identification procedures.

a. Select the desired VOR facility, and tune the correct frequency into the appropriate VOR navigation radio. Frequencies are printed on IFR en route and terminal procedure charts.

b. Before attempting to navigate, you must identify the station by its Morse code identifier. Some VORs use a voice identifier which states the station's name followed by the letters V-O-R.

c. If your DME is separate from the VOR receiver, you must ensure that you properly tune and identify the desired DME facility.

2. Failure to properly set the course selector on the radial to be intercepted.

a. You must know whether you are to track inbound or outbound on the specified radial.

1) When tracking inbound (TO) on a radial, you should set the course selector (OBS) to the reciprocal of the radial.

2) When tracking outbound (FROM) on a radial, set the OBS to that radial.

3. **Failure to use proper procedures for radial or DME arc interception and tracking.**
   a. The key to radial or DME arc interception and tracking is the ability to visualize where you are in relation to the VOR and the radial or arc you want to be on.
   b. Follow the procedures as you would a checklist. Following them in sequence will enable you to perform these maneuvers.
   c. Remember, the CDI gives relevant information only if the airplane is oriented near the same heading as in the OBS.
   d. You should parallel the desired radial on an intercept problem.
      1) Without this step, orientation to the desired radial can be confusing.
      2) Since you think in left/right terms, aligning your airplane's heading to the desired radial is essential.
   e. Do not chase the CDI needle.
      1) Fly by headings on your heading indicator.
      2) Careless heading control and failure to bracket wind corrections make this a common error.

D. ADF Task Objectives

1. **Exhibit adequate knowledge of the elements related to intercepting and tracking NDB bearings.**
   a. Interception of NDB bearings requires you to orient yourself to the NDB and the desired bearing and compute an interception angle. Then you must be able to recognize when you arrive at the desired bearing.
   b. Tracking requires you to make any necessary wind-drift corrections to remain on the desired bearing, either inbound or outbound from the NDB.
      1) This involves interpretation of the heading indicator (HI) and ADF needle to intercept and hold a desired magnetic bearing.
   c. Some airplanes may be equipped with an RMI to present that navigational information from the ADF receiver. If your airplane is so equipped, review the discussion on the RMI throughout item B., VOR and DME Arc Task Objectives, beginning on page 193.
      1) The RMI display is the same, whether it is showing information from a VOR, an NDB, or both.
2. **Tune and correctly identify the NDB facility.**
   a. Tune the ADF receiver to the frequency of the selected NDB facility, with the function switch in the "receive" or antenna (ANT) position.
      1) This position selects the sense antenna only and allows maximum sensitivity to radio signal. It also disables the ADF needle.
   b. Next, positively identify the NDB facility by its Morse code identifier.
      1) Once this is done, switch the function switch to ADF; the needle should point toward the NDB facility.
      2) Use the test function to swing the ADF needle away from the station, and check to determine that it returns to the proper position.
   c. Since ADF receivers do not have a "flag" to warn you when erroneous bearing information is being displayed, you should set the volume to a level that allows constant monitoring of the NDB's identification.

3. **Correctly orient yourself by accurately determining the relative bearing of the NDB facility.**

   a. The relative bearing is an angle measured clockwise from the nose of your airplane to the NDB facility. On an ADF with a fixed dial, the direction in which the needle points is the relative bearing.

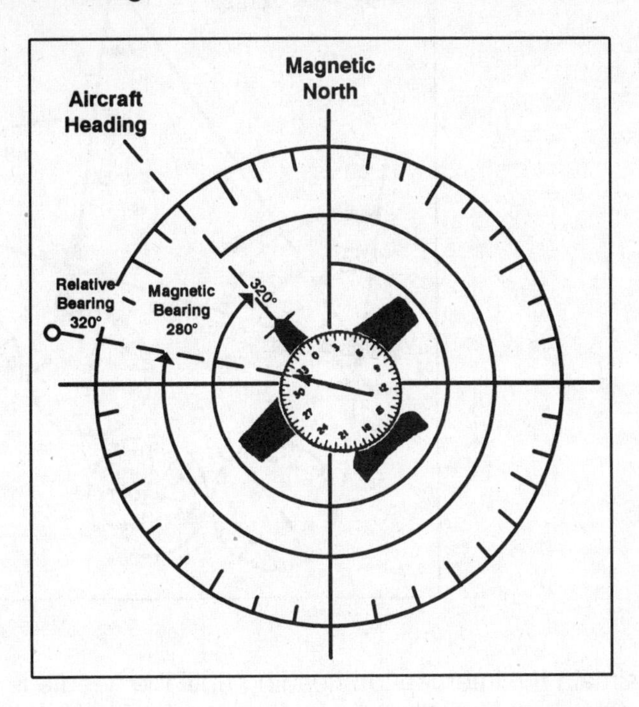

      1) In the figure above, the ADF needle is pointing to 320°; thus the relative bearing of the NDB is 320°.

4. **Intercept the specified bearing at a predetermined angle, inbound or outbound from a navigational facility.**

   a. To intercept an inbound (TO) magnetic bearing of an NDB facility, the following steps may be used (see the figure on the following page for example):

      1) Determine your position in relation to the NDB facility by turning to the magnetic heading of the bearing to be intercepted.

         a) You must understand your clearance as to whether the bearing given to you is TO or FROM the NDB facility.

         b) In this example, you are to intercept the 355° bearing TO the NDB facility, so you should turn your airplane to a heading of 355°.

      2) Note whether the station is to the right or left of the nose position. Determine the number of degrees of needle deflection from the zero position, and double this amount for the interception angle.

         a) In this example, the needle is 40° to the right of the nose, so the interception angle will be 80° (40° x 2).

         b) Use an interception angle of not less than 20° or greater than 90°.

      3) Turn your airplane the number of degrees determined for the interception angle in the same direction as the needle is pointing from the nose (i.e., left or right).

a) In this example, you will turn right 80° to a heading of 075°.

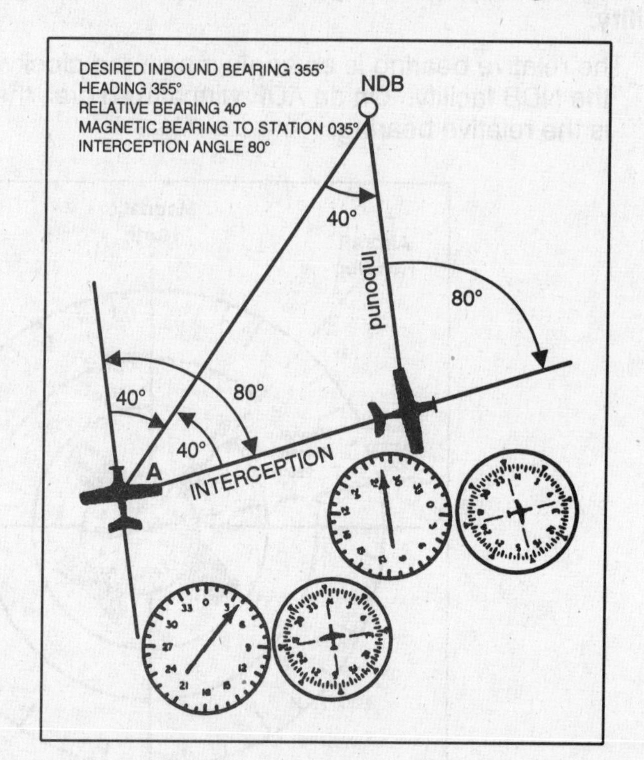

4) Maintain the interception heading until the needle is deflected the same number of degrees from the zero position as the angle of interception (minus lead appropriate to the rate of bearing change).

a) In this example, as you are heading 075°, the interception angle is 80°, and the needle deflection is to the left less than 80°; thus you have not reached the desired bearing.

i) If the needle deflection is greater than 80° to the left, you have flown through the desired bearing.

b) Once the needle indicates 80° to the left of the nose, you are on the 355° bearing to the NDB facility.

5) Turn inbound (with an appropriate lead) and track the bearing to the NDB facility.

b. Interception of an outbound (FROM) NDB bearing can be accomplished by the procedures used to intercept the inbound bearing, with some differences.

1) Substitute the 180° position for the zero position on the ADF needle.

2) Once you make your turn to intercept, the needle will be deflected, greater than your interception angle, from the 180° position.

a) If the number of degrees from the 180° position is less than your interception angle, you have flown through your desired bearing.

5. ***Maintain the airspeed within 10 kt., altitude within 100 ft., and selected headings within 5°.***

a. Do not fixate on your ADF or RMI. You must cross-check and interpret all of your instruments to maintain airspeed, altitude, and headings.

b. Do not make a lot of heading changes to get on course more quickly. Select headings for intercepting and tracking until the navigation system you are using indicates a change is needed.

c. When using the ADF for navigation, check your HI against the magnetic compass and make corrections as necessary. This procedure is important to follow after every turn.

6. *Apply proper correction to maintain a bearing within 10°.*

a. To track inbound to an NDB facility with the wind unknown, proceed using the following steps (see the figure on the left side of page 206):

1) As you intercept the desired bearing, turn to the heading that corresponds with the desired course. Maintain this heading until off-course drift is indicated by left or right needle deflection.

a) Deflection of the ADF needle shows a crosswind: needle left/wind from left; needle right/wind from right.

2) When a 5° change in needle deflection is observed, turn 20° in the direction of needle deflection.

a) Some pilots wait for a 10° change before making a heading change.

3) When the needle is deflected 20° (deflection = interception angle), the bearing has been intercepted. Turn 10° back toward the inbound course. You are now inbound with a 10° drift-correction angle.

4) If you observe off-course deflection in the original direction, turn again to the original interception heading.

5) When you have reintercepted the desired course, turn 5° toward the inbound course, proceeding inbound with a 15° drift correction.

6) If the initial 10° drift correction is excessive, as shown by needle deflection away from the wind, turn to parallel the desired course and let the wind drift you back on course. When the needle is again zeroed, turn into the wind with a reduced drift-correction angle.

7) Station passage is indicated when the needle points to either wingtip position or settles at or near the 180° position.

b. Procedures for tracking outbound are identical to those for tracking inbound. However, the directions of the ADF needle deflections are different from those noted during inbound tracking, as shown in the figure on the right side of page 206.

1) When tracking inbound, a change of heading toward the desired bearing results in movement of the ADF needle toward zero.

2) When tracking outbound, a change of heading toward the desired bearing results in needle movement further away from the 180° position.

7. Determine your airplane's position relative to the NDB facility.

a. As discussed earlier, the relative bearing (RB) is the angle measured clockwise from the nose of your airplane to the NDB facility. On an ADF with a fixed dial, the direction in which the needle points is the relative bearing.

b. The magnetic bearing (MB) is the magnetic course from your airplane to the NDB facility. Use the MB to visualize your position relative to the NDB facility.

1) The MB to the NDB facility can be determined by adding the RB to the magnetic heading (MH) of your airplane, or RB + MH = MB(TO).

a) EXAMPLE: If the RB is 060° and the MH is 130°, the MB to the station is 190° (060° + 130°).

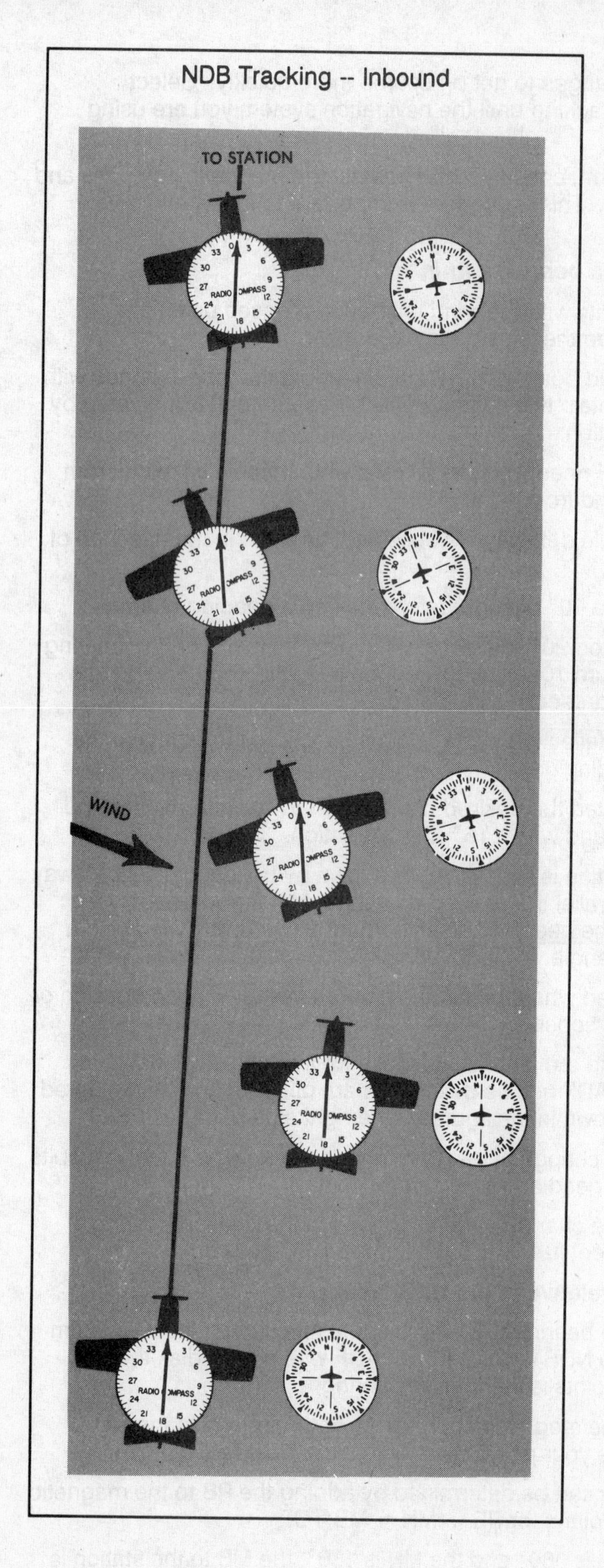
NDB Tracking -- Inbound
TO STATION
WIND
RADIO COMPASS

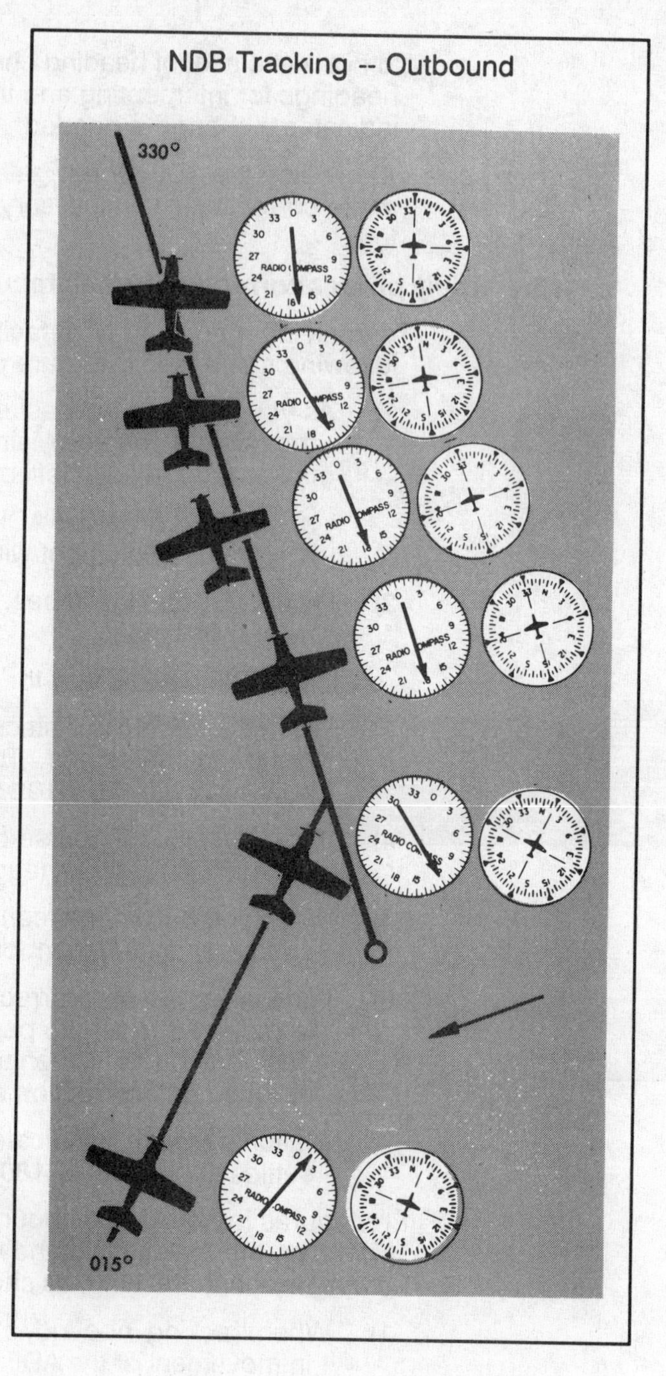
NDB Tracking -- Outbound
330°
015°
RADIO COMPASS

2) If the total exceeds 360°, subtract 360° to obtain the magnetic bearing to the station.
   a) EXAMPLE: If the RB is 270° and the MH is 300°, 360° is subtracted from the total (270° + 300°), or 570° – 360° = 210°, which is the MB to the station.
3) To determine the magnetic bearing FROM the station, 180° is added to (or subtracted from) the magnetic bearing TO the station. This is a reciprocal bearing used when plotting position fixes.
4) This method is good for calculating MB on tests, but a simpler approach to the MB problem can be used in the cockpit.

c. You will be able to orient yourself more readily if you think in terms of nose/tail and left/right needle indications, visualizing the ADF dial in terms of the longitudinal axis of your airplane.
   1) When the needle points to zero (0°), the nose of your airplane points directly to the NDB.
   2) When the needle points to 090°, the NDB facility is off the right wingtip. Thus, to turn directly to the NDB, turn right 90°.

d. You must remember that the RB shown on the ADF dial does not, by itself, indicate your airplane's position, and the RB must be related to your airplane's MH to determine direction to or from the NDB facility.

8. Recognize an ADF receiver or NDB facility failure, and, when required, report the failure to ATC.

a. Any time that you can no longer hear the NDB Morse code identifier due to static or silence, you should assume that the signal is unreliable or a failure has occurred.
   1) Some NDB stations may broadcast weather reports or other items, which will block the Morse code identifier but still be usable.

b. Look for erratic needle movements or needle rotation of 360°, especially if a good indication was being displayed and you know that you are not close to the NDB facility.
   1) If the needle has held position for a relatively long time, use the test function to ensure that the needle swings back to its original position.

c. When operating under IFR in controlled airspace, you are required to report to ATC any malfunction of navigational, approach, or communication equipment occurring in flight (FAR 91.187).

E. Common Errors during Intercepting and Tracking of NDB Bearings

1. Incorrect tuning and identification procedures.

a. Select the proper NDB facility, and tune the correct frequency into the appropriate navigation radio. Frequencies are printed on IFR en route and terminal procedure charts.

b. Before attempting to navigate, you must identify the NDB facility by its Morse code identifier.
   1) Many pilots have navigated to the wrong NDB facility.

c. Since the ADF has no warning flag that the signal is unreliable, continuously monitor the NDB facility's Morse code identifier to confirm signal reliability.

2. Failure to follow the procedure for the determination of a magnetic bearing to or from an NDB facility.

a. Listen to your clearance carefully and repeat it back to ATC (or your examiner). An NDB bearing is TO the station unless you are told to track a bearing FROM the station.

b. Apply the proper orientation procedures so you know exactly where you are relative to the NDB facility.

3. Failure to follow procedure in the interception and tracking of a magnetic bearing.

a. Follow the procedures as you would a checklist. Following them in sequence will enable you to perform this maneuver.

b. Errors associated with failing to follow procedures are discussed below.

1) Careless interception angles are very likely if you rush the initial orientation procedure.

2) Overshooting and undershooting predetermined magnetic bearings are often due to forgetting the course interception angles used.

3) Selected headings are not maintained. Any heading change is accompanied by an ADF needle change. The instruments must be read in combination before any interpretation is made.

4) Track corrections close to the station are overcontrolled (chasing the ADF needle) due to failure to understand or recognize station approach.

5) Heading indicator is not consistently set with magnetic compass.

F. GPS Task Objectives

1. Exhibit adequate knowledge of the elements related to intercepting and tracking GPS courses.

a. Intercepting courses is much easier using a GPS rather than a VOR or an NDB.

1) You can plot a course from your present position to any preprogrammed waypoint or user-defined waypoint.

b. Tracking the course between waypoints requires you to maintain the bearing from each waypoint to the next waypoint. The bearing is maintained by keeping the CDI indicator centered.

1) Additionally, most GPS receivers provide both bearing and track information.

a) If you keep your bearing and track the same, you will be on course.

2. Correctly enter the desired waypoint.

a. You must enter the proper waypoint or waypoints into the GPS receiver.

1) Most GPS receivers require that you enter a "K" prior to the three-letter airport identifier.

a) Gainesville would be entered as "KGNV."

b) This allows the unit to differentiate airports and VORs sharing the same name.

b. GPS receivers are regularly updated and must be kept current to ensure safe navigation.

c. GPS receivers are made by many different manufacturers and have different operating characteristics.

1) You should be proficient with the operation of the GPS receiver that you are using.

a) Most GPS receivers have detailed operating manuals, and some have simulation modes or tutorials.

d. Garmin International offers a free PC simulator for its 400- and 500-series of panel-mounted GPS receivers. The programs accurately reproduce the functions of both series of GPS receivers, and you can use "virtual" buttons on the receiver's face in order to practice entering waypoints and flight plans, flying GPS approaches, and more.

1) Garmin's web site is at www.garmin.com
2) The 400-series simulator is at www.garmin.com/software/train430.zip
3) The 500-series simulator is at www.garmin.com/software/train 530.zip

3. Set the desired course to be intercepted into the GPS receiver.

a. It is important that you understand your clearance as to which course, radial, or bearing you are to intercept and whether you are to track inbound or outbound.

b. GPS receivers will allow you to specify a course (radial or bearing) to or from a waypoint.

1) ATC may instruct you to maintain your present heading to intercept a specific radial to or from a VOR (or a bearing to or from an NDB).

c. Each GPS has different procedures for setting the course to be intercepted, and you will need to follow the procedures for your airplane's GPS.

1) You will also need to know what effect this has on other functions of the GPS. For example, in some receivers, automatic leg sequencing or turn anticipation is not available.

4. Intercept the specified course inbound or outbound from a navigational facility or waypoint.

a. Follow the GPS receiver indications when intercepting the selected course.

1) In some GPS receivers, this is accomplished by using the CDI as you would when intercepting a VOR radial using your VOR instrument.

b. Follow the manufacturer's procedures for your airplane's GPS receiver.

5. Maintain the airspeed within 10 kt., altitude within 100 ft., and selected headings within 5°.

a. Do not fixate on your GPS indications. You must cross-check and interpret all of your instruments to maintain airspeed, altitude, and headings.

b. Do not make a lot of heading changes to get on course more quickly. Select headings for intercepting and tracking until the navigation system you are using indicates a change is needed.

6. Apply proper correction to maintain a radial, allowing no more than three-quarter-scale deflection of the CDI.

a. Tracking a course with a GPS is easy, given the information available to you.

b. Once you are on course, the GPS receiver can supply you with the bearing or desired track to the next waypoint and the actual track you are flying.

1) By keeping these two numbers equal, you will remain on course.

c. The GPS receiver can also provide you information on how far you are from the centerline of your desired course or how many degrees you need to turn to parallel the desired course.

d. Ensure that you fully understand all of the functions of your airplane's GPS receiver.

7. **Determine your airplane's position relative to a waypoint.**
   a. GPS receivers provide magnetic course, bearing track, and distance to selected waypoints.
      1) Some GPS receivers also provide a moving map display, which helps you maintain situational awareness.
8. ***Intercept a DME arc and maintain that arc within 1 NM.***
   a. A GPS can be used to track a DME arc.
      1) Follow the manufacturer's procedures on loading the DME arc procedure into the GPS receiver.
   b. To intercept and track the DME arc, most GPS receivers require you to keep the CDI centered.
      1) Use the GPS distance information to ensure that you are maintaining the proper DME arc.
9. **Recognize a GPS receiver failure, and, when required, report the failure to ATC.**
   a. Most GPS receivers will provide an audible signal or text message to alert the user that the GPS is unreliable for position information.
      1) The GPS is unable to provide position information with less than four satellites.
   b. Loss of satellite reception can occur due to aircraft attitude, antenna location, or satellite positions relative to the horizon.
      1) The pilot should always be prepared for a GPS failure.
         a) The aircraft regularly changes attitude during flight.
         b) The relative positions of satellites are constantly changing.
   c. When operating under IFR in controlled airspace, you are required to report to ATC any malfunction of navigational, approach, or communication equipment occurring in flight (FAR 91.187).

G. Common Errors during Intercepting and Tracking GPS Courses

1. **Incorrect waypoint selection procedures.**
   a. Select the proper waypoint that you will be using. VORs, NDBs, and intersections are printed on IFR en route and terminal procedure charts.
2. **Failure to properly set the GPS to the radial to be intercepted.**
   a. You must know whether you are to track inbound or outbound on the specified radial.
      1) GPS receivers require that you input the desired course.
         a) EXAMPLE: If you are to track inbound on the 253° radial to the VOR, you will input 073° into the GPS.
   b. Failure to understand the operating procedures of the GPS can also be a factor.
3. **Failure to use proper procedures for course or DME arc interception and tracking.**
   a. Follow the GPS procedures as you would a checklist. Following them in sequence will enable you to perform these maneuvers.

END OF TASK -- END OF CHAPTER

CHAPTER VI
INSTRUMENT APPROACH PROCEDURES

This chapter explains the five tasks (A-E) of Instrument Approach Procedures. These tasks include both knowledge and skill. Your examiner is required to test you on two nonprecision approaches using two different approach systems (VOR, LOC, NDB, LDA, SDF, or GPS) and one precision approach (ILS). For example, you may do a VOR and a LOC approach, but you may not do a VOR and a VOR/DME approach. At least one instrument approach procedure must be performed in your airplane. At least one nonprecision approach must be performed without the use of gyroscopic heading and attitude indicator. At the discretion of your examiner, the instrument approach(es) and missed approach(es) not selected for actual flight may be performed in a flight simulator that meets the requirements of Appendix 1 of the FAA's Instrument Rating Practical Test Standards (see page 351).

Since Task A, Nonprecision Instrument Approach, is a general task, we have used the task to explain VOR, NDB, GPS (RNAV), LOC, LDA, and SDF approaches. General information about GPS (RNAV) approaches begins on page 251.

ARRIVAL PROCEDURES

A. Approved Instrument Approach Procedures

1. Instrument approach procedures (IAPs) are a series of predetermined maneuvers for the orderly transfer of an airplane under instrument flight conditions from the beginning of the initial approach to a landing or to a point from which a landing may be made visually.
   a. U.S. civil standard instrument approach procedures are approved by the FAA as prescribed under FAR Part 97 and are available for public use.
   b. U.S. military standard instrument approach procedures are approved and published by the Department of Defense.
   c. Special instrument approach procedures are approved by the FAA for individual operators but are not published in FAR Part 97 for public use; i.e., these IAPs are for private airports.

B. Segments of an Instrument Approach

1. An IAP may have as many as four separate segments depending on how the IAP is structured.
   a. The **initial approach segment** begins at the initial approach fix (IAF) and is used to provide a method for you to align your airplane with the approach course. This alignment may be accomplished by flying a procedure turn, holding pattern, DME arc, radial, or radar vector. The **initial approach fix** is marked on the plan view of the IAP charts as **IAF**. A given IAP may have more than one IAF.

b. The **intermediate approach segment** begins at the intermediate fix (IF) or point and ends upon arrival at the final approach fix. Most approaches do not have an IF; thus, the intermediate segment begins at the point where you begin proceeding inbound to the final approach fix.

1) This point could be the end of the procedure or holding pattern turn inbound, or the end of the DME arc.

2) A nonprecision approach with no depicted final approach fix (such as an on-airport VOR or NDB) does not have an intermediate approach segment.

c. The **final approach segment** begins at the final approach fix (or final approach point) and ends at the runway, airport, or missed approach point.

1) The **final approach fix (FAF)** is designated on the plan view by a Maltese cross symbol (NACO and JEPP charts) and the lightning bolt symbol (NACO charts) or glide slope intercept (JEPP charts) for precision approaches.

a) When ATC directs a lower-than-published glide slope (or glide path) altitude, the FAF is the actual point of the glide slope intercept.

2) The **final approach point** is applicable only to a VOR or NDB IAP with no depicted FAF (such as an on-airport VOR or NDB). It is the point where the aircraft is established inbound on the final approach course from the procedure turn and where the final approach descent to the MDA may be commenced.

d. The **missed approach segment** begins at the missed approach point (MAP) or point of arrival at the decision height (DH) and ends at the missed approach fix at the prescribed altitude.

1) The **missed approach fix** is specified in the missed approach procedure and is normally a fix for holding.

2. **Feeder (transition) routes**, while not technically an approach segment, may be depicted on an IAP chart to designate routes on which to proceed from the en route structure to the IAF.

C. **Radar Approach Control**

1. Radar approach control not only is used for radar approaches but also is used to provide vectors in conjunction with published nonradar approaches.

a. Radar vectors can provide course guidance and expedite traffic to the final approach course of any established IAP or to the traffic pattern for a visual approach.

2. Approach control facilities that provide radar service will operate in the following manner:

a. Arriving aircraft are cleared to an outer fix most appropriate to the route being flown and, if required, given holding information.

1) If radar handoffs are used, aircraft are cleared to the airport or to a fix so located that the handoff will be completed prior to the time at which the aircraft reaches the fix.

b. After release to approach control, you will be vectored to the appropriate instrument final approach course.

1) Radar vectors and altitude will be issued as required for spacing and separating aircraft.

a) Thus, you must NOT deviate from the headings issued by approach control.

2) You will normally be informed when it is necessary to vector across the final approach course for spacing or other reasons.

a) You should question the controller if crossing the final approach course is imminent but you have not been informed that you are being vectored across the final approach course.

c. You are not expected to turn inbound on the final approach course unless you have received an approach clearance.

1) The approach clearance will normally be issued with the final vector for interception of the final approach course, and the vector will be such as to enable you to establish your airplane on the final approach course prior to reaching the final approach fix.

d. If you are already established on the final approach course, you will be issued an approach clearance before you reach the final approach fix (FAF).

1) When you are established inbound on the final approach course, radar separation will be maintained, and you are expected to complete the approach using the approach aid designated in the clearance (e.g., VOR, ILS) as the primary means of navigation.

a) Thus, once established on the final approach course, do not deviate from it unless a clearance to do so is received from ATC.

e. After passing the FAF on final approach, you are expected to continue inbound on the final approach course and to complete the approach or execute the published missed approach procedure.

3. ARTCCs are approved and may provide approach control services to specific airports.

a. The radar systems used by an ARTCC do not provide the same precision as airport surveillance radar (ASR)/precision approach radar (PAR) used by approach control facilities, and their update rate is not as fast as the ASR/PAR update rate.

1) Thus, you may be requested to report that you are established on the final approach course.

4. Whether you are vectored to the appropriate final approach course or navigate on your own on published routes to the final approach course, radar service is automatically terminated when the landing is completed or when you are instructed to change to advisory frequency at airports without an operating control tower, whichever occurs first.

D. Advance Information on Instrument Approach

1. When landing at an airport where there are approach control services and where two or more IAPs are published, you will be provided in advance of your arrival with the type of approach to expect or you will be informed that you will be vectored for a visual approach.

a. This information will be broadcast by either a controller or an ATIS.

b. It will not be furnished when the visibility is 3 SM or better and the ceiling is at or above the highest initial approach altitude established for any IAP for the airport.

1) In this case, you will probably be told to expect a visual approach.

2. The purpose of this information is to aid in your planning; however, it is not an ATC clearance and is subject to change.

a. Keep in mind that fluctuating weather, shifting winds, blocked runway, etc., are conditions that may result in changes to approach information previously received.

b. You must advise ATC immediately if you are unable to execute the approach ATC previously advised or if you prefer another type of approach.

1) Do not accept a visual approach if you cannot remain in VFR conditions throughout the approach.

3. When your destination is an airport that does not have an operating control tower but does have automated weather data and broadcast capability, you are responsible to monitor the ASOS/AWOS frequency to determine the current weather conditions at the airport.

a. You shall advise ATC when you have received the ASOS/AWOS information and state your intentions.

4. When making an IFR approach to an airport without an operating control tower or FSS, after ATC advises, "CHANGE TO ADVISORY FREQUENCY APPROVED," you should broadcast your intentions, including the type of approach you are executing; your position; and your arrival over the FAF inbound (nonprecision approach) or over the outer marker (ILS approach).

a. Continue to monitor the CTAF for reports from other pilots.

E. Instrument Approach Procedure (IAP) Charts

1. FAR 91.175, Takeoff and Landing under IFR, requires the use of standard IAPs prescribed for the airport in FAR Part 97, Standard Instrument Approach Procedures, unless otherwise authorized by the FAA.

a. FAR 91.175 also requires civil pilots flying into or out of military airports to comply with the IAPs and takeoff and landing minimums prescribed by the authority having jurisdiction over those airports.

2. All IAPs are based on joint civil and military criteria contained in the U.S. Standard for Terminal Instrument Procedures (TERPS).

a. The design of an IAP takes into account the interrelationship between airports, facilities, and the surrounding environment, terrain, obstacles, noise sensitivity, etc.

1) Appropriate altitudes, courses, headings, distances, and other limitations are specified, and, once approved, the procedures are published and distributed by government (e.g., NACO) and commercial (e.g., JEPP) cartographers as IAP charts.

b. Radar IAPs are established where requirements and facilities exist, but they are printed in tabular form by the NACO.

1) JEPP radar approaches are printed in chart form.

c. Approach minimums are based on the local altimeter setting for that airport, unless annotated otherwise.

1) When a different altimeter source is required, or more than one source is authorized, it will be annotated on the approach chart.

a) EXAMPLE: Use Sidney altimeter setting, if not received, use Scottsbluff altimeter setting.

2) Approach minimums may be raised when a nonlocal altimeter source is authorized.

3) When more than one altimeter source is authorized, and the minima are different, they will be shown by separate lines in the approach minima box or a note.

a) EXAMPLE: Use Manhattan altimeter setting; when not available, use Salina altimeter setting and increase all MDAs 40 feet.

4) When the altimeter must be obtained from a source other than ATC, a note will indicate the source.

a) EXAMPLE: Obtain local altimeter setting on CTAF.

5) When the altimeter setting(s) on which the IAP is based is not available, the approach is not authorized.

d. By adhering to the altitudes, flight paths, and weather minimums depicted on the IAP chart or vectors and altitudes assigned by the radar controller, you are assured of terrain and obstruction clearance and runway or airport alignment during approach for landing.

e. IAPs are designed to provide an IFR descent from the en route environment to a point where a safe landing can be made.

1) IAPs are prescribed and approved to ensure a safe descent during instrument flight conditions at a specific airport.

2) You must understand these procedures and their use prior to attempting to fly instrument approaches.

f. TERPS criteria are provided for the following types of IAPs:

1) Precision approaches when an electronic glide slope is provided, such as an ILS approach

2) Nonprecision approaches when glide slope information is not provided, such as VOR or GPS

3. The methods used to depict prescribed altitudes on instrument approach charts differ according to the techniques used by different chart publishers.

a. Altitude is depicted on NACO charts in the profile view with underscore, overscore, or both to identify it as minimum, maximum, or mandatory.

1) Minimum altitude is depicted with the altitude value underscored, i.e., $\underline{2000}$.

a) You are required to maintain altitude at or above the depicted value.

2) Maximum altitude is depicted with the altitude value overscored, i.e., $\overline{2000}$.

a) You are required to maintain altitude at or below the depicted value.

3) Mandatory altitude is depicted with the altitude value both underscored and overscored, i.e., $\overline{\underline{2000}}$.

a) You are required to maintain altitude at the depicted level.

b. With very few exceptions, NACO charts use only the underscore to identify minimum altitudes.

1) The depiction of maximum and minimum altitudes is used almost exclusively for military charts.

c. You are cautioned to adhere to altitudes as prescribed because, in certain instances, they may be used as the basis for vertical separation of aircraft by ATC.

1) When a depicted altitude is specified in an ATC clearance, that altitude becomes mandatory as defined in item a.3) above.

4. **Instrument approach procedure naming**

a. Straight-in IAPs (those which the final approach course is within 30° of the centerline of the runway) are identified by the navigation system providing the final approach guidance and the runway to which the approach is aligned (e.g., VOR RWY 13).

b. Circling only IAPs (those which the final approach course is more than 30° than the runway heading) are identified by the navigational system providing final approach guidance and a letter (e.g., VOR-A).

c. More than one navigational system separated by a slash indicates that more than one type of equipment must be used to execute the final approach (e.g., VOR/DME RWY 31).

d. More than one navigational system separated by the word "or" indicates either type of system may be used to execute the final approach (e.g., VOR or GPS RWY 15).

e. In some cases, other types of navigation systems may be required to execute other portions of the approach, such as an NDB in the missed approach procedure.

   1) You must ensure that your airplane is equipped with the required navigational system(s) in order to execute the approach, including the missed approach.

f. In the future, NACO will add a new notation for localizer (LOC) approaches when charted on an ILS approach requiring other navigational systems to fly the approach course.

   1) The LOC minimums will be annotated with the navigational system required (e.g., "DME Required" or "RADAR Required").

   2) During the transition period, ILS approach charts will still exist without the annotation.

g. The naming of multiple approaches of the same type to the same runway is also changing.

   1) New approaches with the same navigational system for guidance will be annotated with an alphabetical suffix beginning at the end of the alphabet and working backwards for subsequent procedures (e.g., ILS Z RWY 28, ILS Y RWY 28, etc.).

   2) The existing annotations (e.g., ILS 2 RWY 28 or Silver ILS RWY 28) will be phased out and eventually replaced with the new designation.

   3) Category II and III ILS procedures are not subject to this naming convention.

h. GPS, Wide Area Augmentation System (WAAS), and lateral navigation (LNAV)/vertical navigation (VNAV) approach procedures will be identified by the term RNAV and the runway (e.g., RNAV RWY 21).

   1) VOR/DME RNAV approaches will continue to be identified as VOR/DME RNAV and the runway (e.g., VOR/DME RNAV RWY 21).

5. **Minimum safe/sector altitudes (MSA)** are published for emergency use on IAP charts.

   a. For conventional navigation systems, the MSA is normally based on the primary omnidirectional facility (VOR or NDB) on which the IAP is predicated.

      1) The MSA depiction on the approach chart contains the facility identifier of the NAVAID used to determine the MSA.

   b. For RNAV approaches, the MSA is based on the runway waypoint (RWY WP), for straight-in approaches, or the airport waypoint (APT WP), for circling approaches.

   c. For GPS approaches, the MSA center will be the missed approach waypoint (MAWP).

   d. MSAs are expressed in feet above mean sea level and normally have a 25-NM radius; however, this radius may be expanded to 30 NM if necessary to encompass the airport landing surfaces.

e. Ideally, a single sector altitude is established and depicted on the plan view of approach charts.

1) When necessary, as many as four MSAs may be established, due to obstructions.

a) Each MSA sector will have no less than a 90° spread.

f. MSAs provide a 1,000-ft. clearance over all obstructions but do not necessarily assure acceptable navigation signal coverage.

6. **Minimum vectoring altitudes (MVA)** are established for use by ATC when radar vectoring is used but is not shown on your charts.

a. The MVA provides a 1,000-ft. clearance above the highest obstacle in nonmountainous areas and a 2,000-ft. clearance above the highest obstacle in designated mountainous areas.

1) Where lower MVAs are required in designated mountainous areas to be compatible with terminal routes or to permit vectoring to an IAP, a 1,000-ft. obstacle clearance may be authorized with the use of airport surveillance radar (ASR).

b. It is important for you to understand MVAs. Because of differences in the areas considered for MVA and the ability to isolate specific obstacles, some MVAs may be lower than the nonradar MEA, MOCA, or other minimum altitudes depicted on charts for a given location.

1) While being radar vectored, IFR altitude assignments by ATC will be at or above the MVA.

7. **Visual descent points (VDP)** are being incorporated into selected nonprecision IAPs.

a. The VDP is a defined point on the final approach course on a nonprecision straight-in approach procedure from which normal descent from the MDA to the runway touchdown point may begin, provided the required visual reference(s) is established.

b. The VDP will normally be identified by DME on VOR and LOC procedures and by along track distance (ATD) to the next waypoint for RNAV procedures.

c. The VDP is identified on the profile view of the IAP chart by the symbol: **V**.

d. VDPs are intended to provide additional guidance where they are implemented.

1) No special technique is required to fly a procedure with a VDP.

2) You should not descend below the MDA prior to reaching the VDP and acquiring the necessary visual reference(s).

e. If your airplane is not equipped to identify the VDP (e.g., you do not have a DME, but the VDP is identified by a DME fix), you should fly the approach procedure as though no VDP had been provided.

8. **Visual portion of the final segment**

a. Instrument procedure designers perform a visual area obstruction evaluation off the approach end of each runway authorized for instrument landing, straight-in, or circling.

b. Restrictions to instrument operations are imposed if penetrations of the obstruction clearance surface exists.

1) These restrictions vary based on the severity of the penetrations and may include increasing visibility, denying VDPs, and prohibiting night instrument operations to the runway.

9. **Vertical descent angle (VDA) on nonprecision approaches**
   a. Descent angles are currently published on some nonprecision approaches and the FAA's intent is to publish VDAs on all nonprecision approaches.
      1) Published along with the VDA is the threshold crossing height (TCH).
   b. The VDA describes a computed path from the altitude at the FAF to the runway at the TCH.
      1) The optimum descent angle is 3.00°; and, whenever possible, the approach will be designed to accommodate this angle.
   c. The VDA will provide you with a means to establish a stabilized approach from the FAF, or stepdown fix, to the TCH.
      1) Stabilized descent along this path is a key factor in the reduction of controlled flight into terrain (CFIT) incidents.
      2) You can use the published VDA and estimated/actual groundspeed to find a target rate of descent from the rate of descent table published with the IAPs.
   d. You must be aware that the published VDA is for information only and it is strictly advisory in nature.
      1) There is no implicit additional obstacle clearance protection below the MDA.
      2) You must still respect the published MDA unless the necessary visual clues are present.
   e. In rare cases, the published VDA will not be the same as the visual glide slope indicator (VGSI); VASI or PAPI.
      1) In these cases, the procedure will be annotated: "VGSI and descent angle not coincident."

F. **Approach Clearance**

1. If you have been cleared to a holding fix and subsequently "cleared ... approach," you have not received new routing instructions.
   a. Even though clearance for the approach may be issued prior to your reaching the holding fix, ATC expects you to proceed by way of the holding fix (your last assigned route) and the feeder route associated with that fix (if a feeder route is published on the IAP chart) to the initial approach fix (IAF) to begin the approach.
   b. When cleared for the approach, the published off-airway (feeder) routes that lead from the en route structure to the IAF are part of the clearance.
2. If a feeder route to an IAF begins at a fix located along the route of flight prior to reaching the holding fix, and clearance for the approach is issued, you should begin the approach via the published feeder route; i.e., you are not expected to overfly the feeder route and return to it.
   a. You are expected to begin the approach in a similar manner at the IAF, if the IAF for the procedure is located along the route of flight to the holding fix.
3. If the controller desires a route of flight directly to the IAF, the controller should say so with phraseology that includes the words "direct...," "proceed direct," or a similar phrase that you can understand without question.
   a. When you are uncertain about the clearance, immediately query ATC as to the desired route of flight.

4. The name of an instrument approach, as published, is used to identify the approach, even though a component of the approach aid, such as the glide slope on an ILS, is inoperative or unreliable.
    a. The controller will use the name of the approach as published but must advise you, at the time that an approach clearance is issued, that the inoperative or unreliable approach aid component is unusable.

G. **Instrument Approach Procedures**

1. Instrument approach minimums are specified for various aircraft approach categories.
    a. The following aircraft approach categories are based on an airspeed of 1.3 V_{SO} (1.3 times the stalling speed of the aircraft in the landing configuration at maximum certificated gross landing weight). An aircraft will fit only one category.
        1) Category A: 90 kt. or less
        2) Category B: 91 kt. to 120 kt.
        3) Category C: 121 kt. to 140 kt.
        4) Category D: 141 kt. to 165 kt.
        5) Category E: 166 kt. or more
    b. If it is necessary, while you are circling to land, to maneuver at an airspeed in excess of the upper limit of the speed range for each category, you should use the circling minimum for the next higher approach category.
        1) This minimum is necessary because the higher airspeed may extend the circling maneuver beyond the area for which obstruction clearance is provided.
            a) See Task VI.D., Circling Approach, on page 317 for a discussion on the extent of the circling area.
        2) EXAMPLE: If your airplane falls into Category A, but you are circling to land at an indicated airspeed between 91 kt. and 120 kt., you must use Category B minimums on the circling approach.
2. When you are operating on an unpublished route or being radar vectored and you receive an approach clearance, you must maintain the last assigned altitude unless a different altitude is assigned by ATC, or until you are established on a segment of a published route or IAP.
    a. After you are established on a segment of a published route or IAP, the altitudes shown on the chart apply to descent within each succeeding route or approach segment unless a different altitude is assigned by ATC.
        1) Thus, once you are established on a segment of IAP, you can descend to the charted altitude without a clearance by ATC.
        2) If ATC wants you to fly at a higher altitude, the controller will inform you.
        3) If you are uncertain, immediately request clarification from ATC.
    b. When you are operating on unpublished routes or while you are being radar vectored, ATC will (except when conducting a radar approach) issue an approach clearance only after you are established on a segment of a published route or IAP, or it will assign an altitude to maintain until you are established on a published route or IAP.
        1) EXAMPLES of ATC clearances:
            a) "Cross Redding VOR at or above 5,000, cleared VOR runway 34 approach."
            b) "Five miles from outer marker, turn right heading 330, maintain 2,000 until established on the localizer, cleared ILS runway 36 approach."

3. Since several IAPs, using various navigation and approach aids, may be authorized for an airport, ATC may advise that a particular IAP is being used, primarily to expedite traffic.
   a. If ATC issues you a clearance that specifies a particular IAP, notify ATC immediately if you desire a different approach.
      1) In this event, it may be necessary for ATC to withhold clearance for the different approach until such time as traffic conditions permit.
4. At times, ATC may not specify a particular approach in the clearance but will state "cleared approach."
   a. This clearance means that you may execute any IAP authorized for that airport.
   b. This clearance DOES NOT mean you may execute a visual approach or a contact approach.
5. When you receive approach clearance, you must execute the entire approach starting at an initial approach fix (IAF) or an associated feeder route shown on the IAP chart, unless
   a. You are being radar vectored to the final approach course.
   b. You receive an appropriate new or revised ATC clearance.
   c. You cancel your IFR flight plan with ATC.
6. You should not rely on radar to identify a fix unless the fix is indicated as RADAR on the IAP chart.
   a. You may request radar identification of an outer marker (OM), but ATC may not be able to provide the service due either to workload or not having the fix on the video map.

H. **Procedure Turn (PT)**

1. A procedure turn (PT) is the maneuver prescribed when it is necessary to perform a course reversal to establish yourself inbound on an intermediate or final approach course.
   a. The procedure turn, or hold in lieu of a procedure turn, is a required maneuver.
2. A procedure turn is a required maneuver except when
   a. The symbol NoPT is shown.
   b. Radar vectoring to the final approach course is provided.
   c. You are conducting a timed approach.
   d. A procedure turn is not authorized.
3. The hold in lieu of a procedure turn is not required when radar vectoring to the final approach course is provided or when NoPT is shown.
4. The altitude prescribed for the procedure turn on the IAP chart is a minimum altitude until you are established on the inbound course.
5. The procedure turn must be completed within the distance specified in the profile view of the IAP chart.
   a. The normal procedure turn distance is 10 NM.
      1) This distance may be reduced to a minimum of 5 NM where only Category A aircraft or helicopters are to be operated or may be increased as much as 15 NM to accommodate high-performance aircraft.
6. The direction or side of the outbound course on which the procedure turn is made is depicted on IAP charts. The point at which the turn is started, the type of turn, and the rate of turn are left to your discretion. Some procedure turns are specified by procedural track, which must be flown exactly as depicted. There are four types of procedure turns:

a. The **45° procedure turn** is the most common type depicted on IAP charts.

1) From the IAF, proceed outbound for 2 or 3 min. (to remain within the specified distance) and then turn 45° to the heading (or in the direction) depicted on the chart.

a) The objective is to remain within the specified distance and to be in a position at the completion of the procedure turn to execute the IAP using normal maneuvers.

2) Maintain this heading for 1 min. and then turn 180° in the opposite direction to intercept the inbound course.

b. The **80°/260° procedure turn** consists of flying outbound from the IAF for 2 or 3 min. (to remain within the specified distance) and then turning 80° in the depicted direction. Once you complete the 80° turn, you should turn 260° in the opposite direction to intercept the inbound course.

c. The **teardrop procedure or penetration turn** consists of flying outbound from the IAF on a specific course (for a specific amount of time or to a fix), followed by a turn to intercept the inbound course, which is different from the outbound course.

1) The purpose of a teardrop procedure turn is to allow an aircraft to reverse direction and to lose considerable altitude within reasonably limited airspace.

d. The **racetrack (or holding) pattern procedure turn** requires the appropriate entry into the holding pattern and then an appropriate turn to intercept the inbound course.

7. The maximum indicated airspeed for a procedure turn is 200 kt.

8. In some IAPs, a holding pattern that is used instead of a procedure turn may be specified for course reversal. In such cases, the holding pattern is established over an intermediate fix (IF) or a final approach fix (FAF).

a. The holding pattern distance or time specified in the profile view must be observed.

b. Maximum holding airspeed limitations as prescribed for holding patterns apply.

c. The holding pattern maneuver is completed when you are established on the inbound course after executing the appropriate entry procedure.

d. If you are cleared for the approach prior to returning to the holding fix and are at the prescribed altitude, additional turns in the holding pattern are not necessary nor expected by ATC.

1) If you desire to make additional turns in the holding pattern to lose excessive altitude or to become better established on course, it is your responsibility to advise ATC upon receipt of your approach clearance.

9. A procedure turn is not required when an approach can be made directly from a specified IF to the FAF.

a. In such cases, the term "NoPT" is used with the appropriate course and altitude to denote that the procedure turn is not required.

b. If you desire to execute a procedure turn and are cleared to do so by ATC, you should not descend below the procedure turn altitude until you are established on the inbound course since some NoPT altitudes may be lower than the procedure turn altitudes.

10. **Limitations on procedure turns**

a. In the case of radar vectors to a FAF or final approach course, a timed approach, or an approach in which the procedure specifies NoPT, you cannot execute a procedure turn unless you obtain a clearance to do so from ATC.

b. A teardrop procedure turn must be executed when it is depicted on the IAP chart and when a course reversal is required.

c. When a holding pattern replaces a procedure turn, the holding pattern must be followed, except when radar vectoring is provided or when NoPT is shown for the IAP.

1) The recommended entry procedure will ensure that your airplane remains within the holding pattern's protected airspace.

2) As in the procedure turn, the descent from the minimum holding pattern altitude to the FAF altitude (when lower) may not be started until you are established on the inbound course.

d. The absence of a procedure turn depicted on the plan view of the IAP chart indicates that a procedure turn is not authorized for that IAP.

I. **Radar Monitoring of Instrument Approaches**

1. Precision approach radar (PAR) facilities, operated by the FAA and the military services at some joint-use (civil and military) and military installations, monitor aircraft on instrument approaches and issue radar advisories to the pilot in weather that is below VFR minimums (1,000-ft. ceiling and/or 3-SM visibility), at night, or at the request of a pilot.

a. This service is provided only when the PAR final approach course coincides with the final approach of the navigational aid and only during the operational hours of the PAR facility.

b. The radar advisories serve only as a secondary aid since the pilot has selected the navigational aid as the primary aid for the approach.

2. Prior to starting final approach, you will be advised of the frequency on which the advisories will be transmitted.

a. If, for any reason, radar advisories cannot be furnished, you will be so advised.

3. Advisory information, derived from radar observations, includes the following information:

a. You will be advised when you are passing the final approach fix inbound (nonprecision approach) or passing the outer marker, or fix used in lieu of the outer marker, inbound (precision approach).

1) At this point, you may be requested to report sighting the approach lights or the runway.

b. Trend advisories with respect to elevator and/or azimuth radar position and movement will be provided.

1) Whenever your airplane nears the PAR safety limit, you will be advised that your airplane is well above or below the glide path or well right or left of course.

2) Glide path information is given only to those executing a precision ILS approach.

a) Altitude information is not provided to aircraft executing a nonprecision approach because the descent portion of these approaches generally does not coincide with the depicted PAR glide path.

c. If, after repeated advisories, you proceed outside the PAR safety limit or if a radical deviation is observed, you will be advised to execute a missed approach unless the prescribed visual reference with the surface is established.

4. Radar service is automatically terminated upon completion of the approach.

J. **Timed Approaches from a Holding Fix**

1. Timed approaches may be conducted only when the following conditions are met:
   a. A control tower is in operation at the airport where the approach is to be conducted.
   b. Communication is maintained between you and the center or approach controller until you are instructed to contact the tower.
   c. If more than one missed approach procedure is available, none may require a course reversal.
   d. If only one missed approach procedure is available, the following conditions are met:
      1) Course reversal is not required.
      2) Reported ceiling and visibility are equal to or greater than the highest prescribed circling minimums for the IAP.
   e. When cleared for the approach, you will not execute a procedure turn.
2. The holding fix will be at the final approach fix (nonprecision approach) or the outer marker or fix used instead of the outer marker (precision approach).
3. Although the controller will not specifically state that "timed approaches are in progress," the assigning of a time to depart the FAF inbound (nonprecision approach) or the outer marker, or fix used in lieu of the outer marker, inbound (precision approach) is an indication that time approaches are being conducted.
   a. Alternatively, the controller may use radar vectors to the final approach course (instead of holding) to establish a mileage interval between aircraft that will ensure the appropriate time sequence between the final approach fix/outer marker, or fix used instead of the outer marker, and the airport.
4. You will be given advance notice as to the time to leave the holding fix on approach to the airport.
   a. You should adjust your holding pattern as necessary to leave the holding fix as closely as possible to the designated time.
   b. EXAMPLE: At 12:03, you receive instructions to leave the holding fix inbound at 12:07, just as you have completed the turn at the outbound end of the holding pattern and are proceeding towards the holding fix. Arriving over the holding fix, you note that the time is 12:04 and that 3 min. remain until your assigned time to leave the holding fix. Since the time remaining is greater than 2 min., you decide to fly a holding pattern rather than a 360° turn, which would take 2 min. The turns at the end of the holding pattern will take 1 min. each, which will use up 2 min. Thus, 3 min. to go minus 2 min. for the turns leaves 1 min. for level flight. Since there are two portions of level flight, you must decide how long to fly outbound.
      1) If the winds are negligible, you can fly the outbound leg for 30 sec. before starting the turn back to the fix on final approach.
      2) If expecting a headwind on final approach, you should shorten the 30-sec. outbound leg somewhat, knowing that the wind will move you away from the fix faster while outbound and decrease your groundspeed while returning to the fix.
      3) If expecting a tailwind, you should lengthen the outbound leg somewhat, knowing that the wind will tend to hold you closer to the fix while outbound and increase the groundspeed while returning to the fix.

K. Simultaneous Converging Instrument Approaches

1. ATC may conduct instrument approaches simultaneously to converging runways at airports where such a program has been specifically approved.
   a. Converging runways are runways having an angle of 15° to 100° to each other.
2. The basic concept requires that dedicated, separate standard instrument approach procedures be developed for each converging runway included.
   a. Missed approach points must be at least 3 NM apart, and missed approach procedures ensure that missed approach airspace does not overlap.
3. Other requirements for simultaneous converging instrument approaches include
   a. Radar availability
   b. Nonintersecting final approach courses
   c. Precision (ILS) approach systems on each runway
4. If runways intersect, controllers must be able to apply visual separation as well as intersecting runway separation criteria.
   a. Intersecting runways also require minimums of at least 700-ft. ceilings and 2-SM visibility.
   b. Straight-in approaches and landings must be made.
5. Whenever simultaneous converging approaches are in progress, you will be informed by the controller as soon as feasible after initial contact or via ATIS.
   a. Additionally, the radar controller will have direct communications capability with the tower controller where separation responsibility has not been delegated to the tower.

L. Radar Approaches

1. A radar approach is an instrument approach procedure that uses **precision approach radar (PAR)** or **airport surveillance radar (ASR)**.
   a. During a radar approach, the radar controller vectors your aircraft to align it with the runway centerline. A PAR approach is considered to be a precision approach.
      1) The controller continues the vectors to keep your aircraft on course until you can complete the approach and landing by visual reference to the surface.
   b. You may request a radar approach, if one is available at the airport.
   c. Radar approach (PAR and ASR) minimums are listed in table form in the front of the NACO *Terminal Procedures Publication* (TPP) and in chart form by airport in the Jeppesen IAP charts.
2. A **precision approach** (PAR) is one in which a controller provides you with highly accurate navigational guidance in azimuth (runway centerline) and elevation (glide slope).
   a. You are given headings to fly to direct you to, and to keep your airplane aligned with, the extended centerline of the runway.
   b. Prior to intercepting the glide path, you will be advised of communication failure/missed approach procedures and may be told not to acknowledge further transmissions (so only the controller will be talking).
      1) You will be instructed to anticipate glide path interception approximately 10 to 30 seconds before it occurs and advised when to start your descent.
      2) The published decision height (DH) will be given only on your request.

c. During the final approach, the controller will supply guidance information, as "slightly/well above" or "slightly/well below" glide path and "slightly/well right" or "slightly/well left" of course.

   1) Trend information is also issued and may be modified by the terms "rapidly" and "slowly."

      a) EXAMPLE: "Well above glide path, coming down rapidly."

d. Accuracy in maintaining and correcting headings and rate of descent is essential.

e. Range from touchdown is given at least once each mile.

f. If your airplane is observed by the controller to proceed outside of specified safety zone limits in azimuth and/or elevation and to continue to operate outside the prescribed limits, you will be directed to execute a missed approach or to fly a specified course unless you have the runway environment in sight.

g. You will be provided navigational guidance in azimuth and elevation to the DH, at which point you must have the runway environment in sight to continue.

   1) Advisory course and glide path information will be provided until you cross the landing threshold, at which point you will be advised of any deviation from the runway centerline.

h. Radar service is automatically terminated upon completion of the approach.

3. A **surveillance approach** (ASR) is one in which a controller provides navigational guidance in azimuth only; i.e., you are given headings to fly to align your airplane with the extended centerline of the landing runway. An ASR approach is considered to be a nonprecision approach.

   a. During the initial part of the approach, you will be given communication failure/missed approach procedures, and, during final approach, you will be told not to acknowledge transmissions.

   b. Since the radar information used for an ASR approach is considerably less precise than that used for a PAR approach, accuracy for the approach will not be as great and higher minimums will apply.

      1) The controller will give you the published straight-in MDA.

      2) You will not be given the circling MDA unless you request it and provide the controller your aircraft approach category.

   c. You will be advised when to start your descent to the MDA or, if appropriate, to an intermediate step-down fix minimum crossing altitude (MCA) and subsequently to the published MDA.

   d. The controller will advise you of the location of the missed approach point (MAP) and your position each mile on final from the runway, airport, or MAP, as appropriate.

   e. If you so request, the controller will issue recommended altitudes each mile, based on the descent gradient established for the procedure, down to the last mile that is at or above the MDA.

   f. Normally, navigational guidance will be provided until you reach the MAP, and course correction instructions are similar to those used in a PAR approach.

      1) The controller will terminate guidance and instruct you to execute a missed approach unless at the MAP you have the runway environment in sight.

g. If at any time during the approach the controller considers that safe guidance for the remainder of the approach cannot be provided, you will be instructed to execute a missed approach.

1) Similarly, guidance termination and missed approach will be effected upon your request, and the controller may terminate guidance when you report the runway environment in sight.

h. Radar service is automatically terminated at the completion of the approach.

4. A **no-gyro approach** is available when you are operating in a radar environment and your airplane's heading indicator is inoperative or inaccurate.

a. When you experience a loss of your heading indicator, you should advise ATC and request a no-gyro vector or approach while operating in instrument meteorological conditions (IMC).

b. You should make all turns at standard rate and should execute a turn immediately upon receipt of instructions.

1) EXAMPLES: "Turn right." "Stop turn."

c. When an ASR or a PAR approach is made, you will be advised to make turns at half standard rate after you have been turned onto the final approach course.

M. Side-Step Maneuver

1. ATC may authorize a nonprecision approach procedure that serves either one of parallel runways separated by 1,200 ft. or less, followed by a straight-in landing on the adjacent runway.

2. If you are to execute a side-step maneuver, you will be cleared for a specific nonprecision approach and landing on the adjacent parallel runway.

a. EXAMPLE: "Cleared ILS runway 7 left approach, side-step to runway 7 right."

b. You are expected to commence the side-step maneuver as soon as possible after the runway or runway environment is in sight.

3. Landing minimums to the adjacent runway will be based on nonprecision criteria and thus will be higher than the precision minimums to the primary runway.

a. The landing minimums will normally be lower than the published circling minimums.

N. Visual Approach

1. A visual approach is conducted on an IFR flight plan and authorizes you to proceed visually and clear of clouds to the airport.

a. Visual approaches are an IFR procedure conducted under IFR in visual meteorological conditions (VMC).

1) Thus, cloud clearance requirements of FAR 91.155, Basic VFR Weather Minimums, are not applicable.

b. Visual approaches reduce pilot and ATC workload and expedite traffic by shortening flight paths to the airport.

1) It is your responsibility to advise ATC as soon as possible if a visual approach is not desired.

2. ATC may issue a visual approach clearance when

a. You have either the airport or the preceding identified aircraft in sight, and

b. The reported weather at the airport must have a ceiling at or above 1,000 ft. and visibility of 3 SM or greater.

1) When operating to an airport without a weather reporting service, ATC may initiate a visual approach provided there is reasonable assurance (e.g., area weather reports, PIREPs, etc.) that weather at the airport has a ceiling at or above 1,000 ft. and visibility of 3 SM or greater.

3. When operating to an airport with an operating control tower, you may be cleared to conduct a visual approach to one runway while other aircraft are conducting IFR or VFR approaches to another parallel, intersecting, or converging runway.

a. When operating to an airport with parallel runways separated by less than 2,500 ft., you must report sighting the aircraft you are to follow, unless ATC provides standard separation.

b. When operating to an airport with parallel runways separated by at least 2,500 ft. but less than 4,300 ft., you will be vectored by ATC to the final approach course at an angle of not more than 30° unless radar, vertical, or visual separation is provided during the turn-on.

1) The purpose of the 30° intercept angle is to reduce the potential for overshoots of the final approach course and to preclude side-by-side operations with one or both aircraft in a belly-up configuration during the turn.

c. When the parallel runways are separated by 4,300 ft. or more, or intersecting or converging runways are in use, ATC may authorize a visual approach after advising all aircraft involved that other aircraft are conducting operations to the other runway.

1) ATC may advise pilots of these operations through the use of ATIS.

4. If you have the airport in sight but cannot see the aircraft you are to follow, ATC may clear you for the visual approach.

a. However, ATC retains both separation and wake turbulence separation responsibility.

5. When you accept a visual approach clearance to follow an identified preceding airplane, you are required to establish a safe landing interval behind that airplane.

a. Additionally, you are responsible for wake turbulence separation.

1) ATC is responsible for informing you when the airplane you are following is a heavy airplane.

6. During the visual approach, you must notify ATC immediately if you are unable to continue following the preceding aircraft, cannot remain clear of clouds, or lose sight of the airport.

7. A visual approach is not an IAP and thus has no missed approach segment.

a. At an airport with an operating control tower, if you must go around, you will be issued an appropriate advisory/clearance/instruction by the tower.

b. At an airport without an operating control tower, you are expected to remain clear of clouds and complete a landing as soon as possible.

1) If a landing cannot be accomplished, you are expected to remain clear of clouds and contact ATC as soon as possible for further clearance.

8. Authorization to conduct a visual approach is an IFR authorization and does not alter IFR flight plan cancellation responsibility.

9. Radar service is automatically terminated, without your being advised, when you are instructed to change to advisory frequency.

O. Charted Visual Flight Procedures (CVFP)

1. CVFPs are charted visual approaches established for environmental/noise considerations and/or, when necessary, for the safety and efficiency of air traffic operations.
   a. Approach charts depict prominent landmarks, courses, and recommended altitudes to specific runways.
   b. While CVFPs are designed to be used primarily for turbojet aircraft, these procedures may be used by all aircraft.
2. CVFPs will be used only at airports with an operating control tower.
3. Most approach charts will depict some navigation aid information for supplemental navigational guidance only.
4. Unless indicating a Class B airspace floor, all depicted altitudes are for noise abatement purposes and are recommended only.
   a. You are not prohibited from flying at other-than-recommended altitudes if operational requirements dictate.
5. When landmarks used for navigation are not visible at night, the approach will be annotated "Procedure Not Authorized at Night."
6. CVFPs usually begin within 20 NM of the airport.
7. Published weather minimums for CVFPs are based on minimum vectoring altitudes rather than the minimum altitudes depicted on the charts.
   a. ATC will not issue a clearance for a CVFP when the weather is less than the published minimums.
8. ATC will issue a clearance for a CVFP after you report sighting a charted landmark or a preceding aircraft.
   a. If instructed to follow a preceding aircraft, you will be responsible for maintaining a safe approach interval and wake turbulence separation.
9. Remember that CVFPs are NOT instrument approaches and do not have missed approach segments.
   a. You should advise ATC if at any point you are unable to continue an approach or lose sight of a preceding aircraft.
   b. Missed approaches will be handled as a go-around.

P. Contact Approach

1. A contact approach is an approach procedure that you may use (with prior authorization from ATC) instead of conducting a standard (or special) IAP to an airport.
   a. A contact approach is not intended for use by a pilot on an IFR clearance to operate to an airport not having a published and functioning IAP.
      1) Nor is it intended for a pilot to conduct an instrument approach to one airport and then, when "in the clear," to discontinue that approach and proceed to another airport.
2. You may request ATC authorization for a contact approach provided you are clear of clouds, have at least 1-SM flight visibility, and reasonably expect to continue to your destination airport in those conditions.
   a. During the execution of a contact approach, you assume the responsibility for obstruction clearance.

b. You must advise ATC immediately if you are unable to continue the contact approach (cannot remain clear of clouds) or if you encounter less than 1-SM flight visibility.

c. If radar service is being received, it will automatically terminate when you are instructed to change to advisory frequency.

3. ATC may authorize a contact approach provided

a. You specifically request the approach. ATC cannot initiate this approach.

b. The reported ground visibility at your destination airport is at least 1 SM.

c. The approach will be made to an airport that has a standard or special IAP.

d. Approved separation is applied between aircraft so cleared and between these aircraft and other IFR or special VFR aircraft.

Q. **Landing Priority**

1. A clearance for a specific type of approach (ILS, VOR, GPS, or straight-in approach) given to an aircraft operating on an IFR flight plan does not mean that landing priority will be given to that aircraft over other traffic.

2. Air traffic control towers handle all aircraft, regardless of type of flight plan, on a "first-come, first-served" basis.

3. Because of local traffic or a runway in use, the controller may find it necessary, in the interest of safety, to provide a different landing sequence.

a. In any case, you will be issued a landing sequence as soon as possible to allow you to adjust your flight path.

NACO INSTRUMENT APPROACH CHART CHANGES

A. The NACO is currently reformatting the IAP charts. This will be a gradual process to convert all of the charts, so you may use the old format at one airport and the new format at another airport.

B. **Pilot Briefing (EZ Brief) Section**

1. The pilot briefing, or EZ brief, section is located at the top of the chart and condenses, into one location, pertinent information you need to conduct the approach, and/or to brief to other flight crewmembers.

2. There are three rows in the pilot briefing section which contains information such as the final approach course, runway/airport data, procedure restrictions, approach light data, missed approach text, and the various ATC and NAVAID frequencies to use.

C. **Missed Approach Symbology**

1. In order to make the missed approach guidance more readily understood, a method has been developed to display missed approach guidance in the profile view section through the use of icons.

a. Due to limited space in the profile section, a maximum of four icons will be shown.

2. While these icons are easily and rapidly understandable, they may or may not provide representation of the entire missed approach procedure.

a. The entire set of textual missed approach instructions are provided at the top of the approach chart in the pilot briefing section.

D. **New GPS Instrument Charting Format**

1. Reliance on area navigation (RNAV) systems for instrument approach operations is becoming more commonplace as systems such as GPS, Wide Area Augmentation System (WAAS), and Local Area Augmentation System (LAAS) are developed and placed into operation.
    a. The new format will avoid unnecessary duplication, by publishing on the same approach chart the approach minimums for unaugmented GPS (the present GPS approaches) and augmented GPS (WAAS and LAAS when they become operational).
        1) Until WAAS and LAAS become operational, the minima lines associated with augmented GPS will be marked "NA."
    b. The approach chart will be titled "RNAV RWY XX."
        1) During the transition period when GPS procedures are undergoing revision to the new title, both "RNAV" and "GPS" approach charts and formats will be published.
2. The RNAV chart may contain as many as four lines of approach minimums: GLS (Global Navigation Satellite System [GNSS] Landing System); LNAV/VNAV (lateral navigation/ vertical navigation); LNAV; and CIRCLING.
    a. GLS is based on WAAS and LAAS. We will add more to this discussion when WAAS and LAAS become operational.
    b. LNAV/VNAV is a new type of instrument approach with lateral and vertical navigation.
        1) The use of this approach is based on the aircraft's navigation system meeting required navigation performance (RNP), which would be in the aircraft's Airplane Flight Manual (AFM)/*Pilot's Operating Handbook* (*POH*) or its supplement.
            a) Additionally, pilots must have special training.
        2) Since single-engine general aviation airplanes do not have the navigation equipment to meet RNP criteria, we will not discuss these minimums.
    c. LNAV is the minima for lateral navigation only, and the approach minimum altitude will be published as an MDA.
        1) LNAV provides the same level of service as the present stand-alone GPS approaches.
        2) The LNAV line on the RNAV chart will allow the present certified GPS receivers to fly the new approaches.
3. Chart terminology will change slightly to support the new procedure types.
    a. **Decision altitude (DA)** replaces the currently used term decision height (DH).
        1) DA conforms to the international convention where altitudes relate to MSL and heights relate to AGL.
            a) DA will eventually be published for other types of IAPs with vertical guidance, such as the ILS.
    b. Minimum descent altitude (MDA) will continue to be used for the LNAV only and circling procedures.

4. The minima format will also change slightly.
   a. Each line of minima on the RNAV IAP will be titled to reflect the RNAV system applicable; e.g., GLS, LNAV/VNAV, LNAV, and CIRCLING.
   b. The minima title box will also indicate the nature of the minimum altitude for the IAP, e.g., DA or MDA.
   c. Where two or more systems (e.g., GLS and LNAV/VNAV), share the same minima, each line will be displayed separately.
5. Chart symbology will change slightly.
   a. The published descent profile and a graphical depiction of the vertical path to the runway will be shown.
   b. A visual descent point (VDP) will be published on most RNAV IAPs, but the VDP will only apply to aircraft using the LNAV minima.
   c. The new RNAV chart will depict the Terminal Arrival Areas (TAA) through the use of icons representing each TAA area (i.e., straight-in, right base, or left base) associated with the RNAV procedure.
      1) These icons will be depicted in the plan view of the chart.
      2) The waypoint (WP), to which navigation is appropriate and expected within each specific TAA area, will be named and depicted on the associated TAA icon.
         a) Each named and depicted WP is the IAF for arrivals in that area.
      3) TAAs may not be depicted on all RNAV charts because of the inability for ATC to accommodate the TAA due to airspace congestion.

NONPRECISION INSTRUMENT APPROACH

VI.A. TASK: NONPRECISION INSTRUMENT APPROACH

REFERENCES: 14 CFR Parts 61, 91; AC 61-27; IAP; AIM.

Note: Any reference to DME arcs, ADF, or GPS shall be disregarded if the aircraft is not equipped with the above specified navigational systems. If the aircraft is equipped with any of the above navigational systems, the examiner may ask the applicant to demonstrate those types of approaches. The examiner shall select two nonprecision approaches utilizing different approach systems.

Objective. To determine that the applicant:

1. Exhibits adequate knowledge of the elements related to an instrument approach procedure.
2. Selects and complies with the appropriate instrument approach procedure to be performed.
3. Establishes two-way communications with ATC, as appropriate, to the phase of flight or approach segment, and uses proper radio communications phraseology and technique.
4. Selects, tunes, identifies, and confirms the operational status of navigation equipment to be used for the approach procedure.
5. Complies with all clearances issued by ATC or the examiner.
6. Recognizes if heading indicator and/or attitude indicator is inaccurate or inoperative, advises controller, and proceeds with approach.
7. Advises ATC or examiner any time the aircraft is unable to comply with a clearance.
8. Establishes the appropriate aircraft configuration and airspeed considering turbulence and wind shear, and completes the aircraft checklist items appropriate to the phase of the flight.
9. Maintains, prior to beginning the final approach segment, altitude within 100 ft. (30 meters), heading within 10° and allows less than a full-scale deflection of the CDI or within 10° in the case of an RMI, and maintains airspeed within 10 kt.
10. Applies the necessary adjustments to the published MDA and visibility criteria for the aircraft approach category when required, such as --
    a. FDC and Class II NOTAMs.
    b. Inoperative aircraft and ground navigation equipment.
    c. Inoperative visual aids associated with the landing environment.
    d. National Weather Service (NWS) reporting factors and criteria.
11. Establishes a rate of descent and track that will ensure arrival at the MDA prior to reaching the MAP with the aircraft continuously in a position from which descent to a landing on the intended runway can be made at a normal rate using normal maneuvers.
12. Allows, while on the final approach segment, no more than a three-quarter-scale deflection of the CDI or within 10° in case of an RMI, and maintains airspeed within 10 kt.
13. Maintains the MDA, when reached, within +100 ft. (30 meters), –0 ft. to the MAP.
14. Executes the missed approach procedure when the required visual references for the intended runway are not distinctly visible and identifiable at the MAP.
15. Executes a normal landing from a straight-in or circling approach when instructed by the examiner.

A. General Information

1. The objective of this task is to determine your knowledge and ability to perform a nonprecision instrument approach.
2. During your practical test, you will be required to perform two of the following nonprecision approaches: VOR, NDB, GPS, localizer (LOC), localizer-type directional aid (LDA), or simplified directional facility (SDF).
    a. During your practical test, you can perform a VOR and a LOC approach, but you cannot perform a VOR/DME and a VOR approach.
    b. Remember that the types of approaches will be selected by your examiner, not you.
    c. One of your nonprecision approaches will be conducted using partial panel, i.e., without the use of your attitude and heading indicators.

3. During your flight training, you should become proficient in all of the various types of approaches.
4. Since this task is general to any nonprecision approach to be flown on your practical test, we have included discussion on the various nonprecision approach systems. Each type of approach is listed below with the page on which its discussion begins.
   a. VOR instrument approach -- See below.
   b. NDB instrument approach -- See page 242.
   c. GPS instrument approach -- See page 251.
   d. LOC instrument approach -- See page 265.
   e. LDA instrument approach -- See page 275.
   f. SDF instrument approach -- See page 285.

B. VOR/VORTAC Instrument Approach Task Objectives

1. **Exhibit adequate knowledge of the elements related to a VOR/VORTAC instrument approach procedure.**
   a. For the purpose of this discussion, the term VOR will include VOR, VORTAC, and VOR/DME ground stations.
   b. The VOR instrument approach procedure is a nonprecision approach with the course guidance based on a specific radial.
   c. The VOR is normally the initial approach fix (IAF), but the approach may have others (e.g., the start of a DME arc).
      1) If the VOR IAP chart has a note stating (usually in large letters), "RADAR REQUIRED," an IAF may not be available.
         a) ATC will provide radar vectors to join the final approach course before arrival at the final approach fix (FAF).
   d. When the VOR is located off the airport, the VOR will normally be the final approach fix (FAF).
      1) On the profile view of the NACO IAP chart, the FAF is identified by a Maltese cross symbol.
   e. When the VOR is located on the airport, there is no depicted FAF. In this case, the final approach segment begins at the final approach point.
      1) The final approach point is the point where you are established inbound on the final approach course from the procedure turn (or radar vector) and where you may start the final approach descent.
   f. If the IAP indicates a procedure turn, you will fly outbound from an off-airport VOR for 1 or 2 min. and then execute the procedure turn.
      1) EXAMPLE: An airplane using an indicated approach airspeed of 90 kt. would travel 3 NM under no-wind conditions. This would keep the airplane within the 10-NM limits of the procedure turn and allow enough time for it to be established on the final approach course before passing over the FAF.
   g. If the VOR is located on the airport, you will fly outbound for 3 or 4 min. and then execute the procedure turn.
      1) This is to allow you enough distance to become established on the final approach course and then begin your descent to the MDA.
      2) You will need to take into account the wind conditions and make adjustments as necessary to remain within the specified limits of a procedure turn.

h. With an off-airport VOR, the missed approach point (MAP) may be determined by a specific time after crossing the FAF, by a fix defined by another NAVAID, or by a DME distance.

  1) When the VOR is located on the airport, the MAP is the VOR station.

i. VOR approaches can also be used in conjunction with DME distance indications; these are called VOR/DME approaches.

  1) The VOR/DME approach is flown exactly like a standard VOR approach except that DME distances are used for one, all, or combinations of the following:

    a) Transition procedures
    b) IAFs
    c) Procedure turn limits
    d) FAF
    e) Step-down altitudes
    f) MAP

2. Select and comply with the appropriate VOR/VORTAC instrument approach procedure to be performed.

a. Once you know which VOR approach you will be conducting, you should select the appropriate IAP and study it to determine that you can comply with the procedure.

  1) If the inbound course of the approach is within 30° of the centerline of a runway, the approach will be designated to that runway (e.g., VOR RWY 31).

    a) If the course deviation is more than 30°, the approach will be designated with an alphabetic character (e.g., VOR-A).

  2) To accept a VOR/DME approach, you must have an operable DME.

b. We will use the VOR-A IAP to Gainesville Regional Airport (KGNV) on the following page as an example throughout our discussion.

c. Comprising the top half of the IAP chart, the **plan view** shows a bird's-eye view of the approach.

d. Assuming you are cleared by ATC to "proceed direct to Gainesville VORTAC (IAF), cleared for the VOR-A approach," you will use the appropriate entry to the published holding pattern. (See Task III.C., Holding Procedures, on page 122.)

  1) The holding pattern is used in lieu of a procedure turn to line you up on the final approach course.

  2) You are now on the initial approach segment.

  3) Since you have been cleared for the approach, and you are established in the published holding pattern, you may descend to 2,000 ft. MSL, the published altitude for the hold, shown in the **profile view** just below the plan view.

  4) You might receive an alternate clearance to execute a GNV 8 DME arc instead of the procedure described above to intercept the inbound course.

    a) You must be DME equipped to accept this clearance.

e. When you have turned inbound and are tracking the R-213 inbound, you are on the intermediate approach segment.

f. After you have passed the FAF, shown by the Maltese cross in the profile view (i.e., the GNV VORTAC), you are on the final approach segment.

  1) You should begin timing your approach to the MAP. Times are based on groundspeed and are shown below the airport diagram.

    a) In the example, using a groundspeed of 90 kt., the MAP is 5 min. 36 sec. from the FAF.

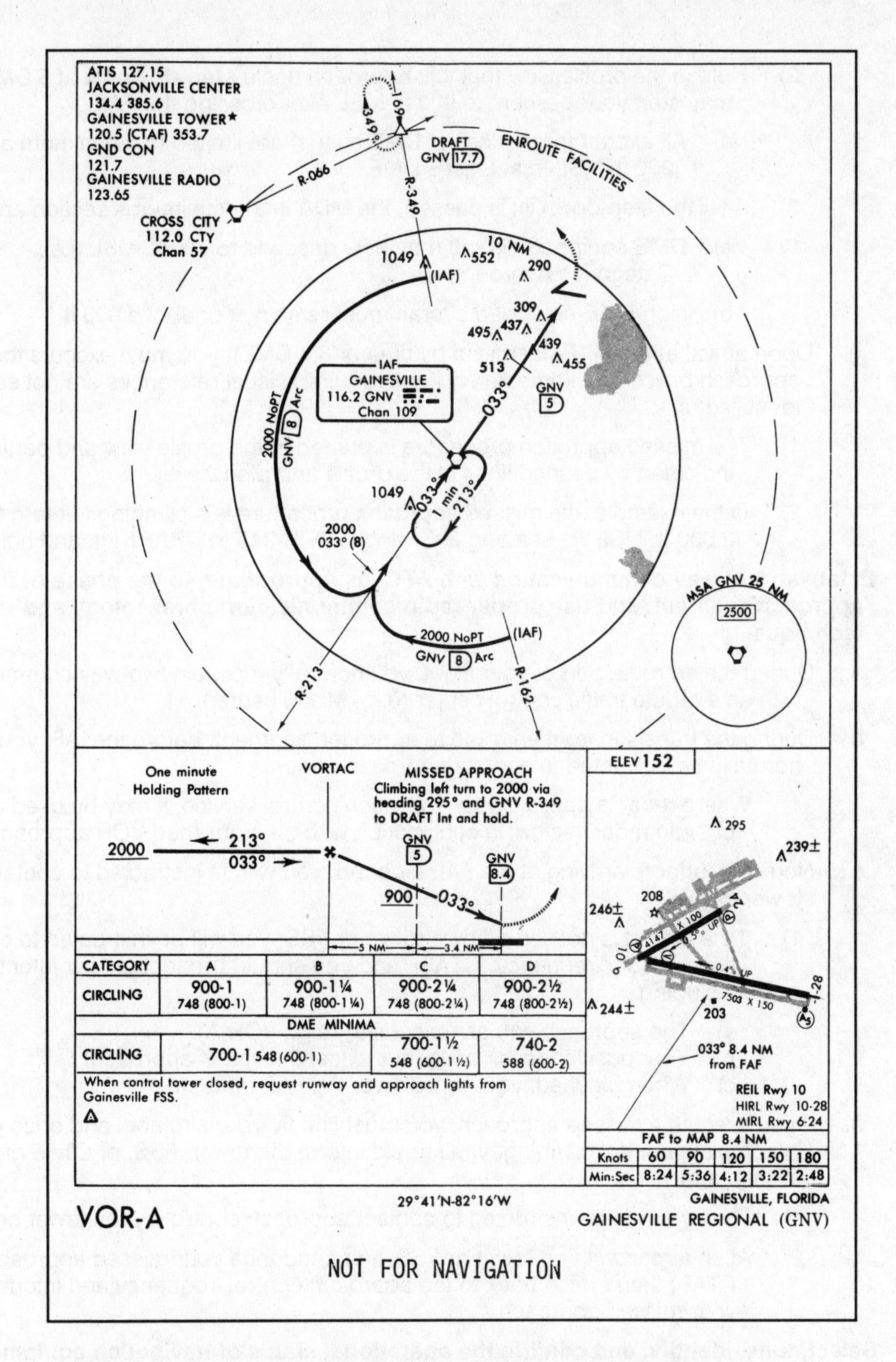

NOTE: This IAP chart is shown in the old format, which is still in use. Charts using the new format are shown on pages 255 and 298.

2) Note in the profile view that this approach has a **step-down fix** at 5 DME. You may start your descent to 900 ft. MSL after crossing the FAF.

a) All aircraft (regardless of DME status) are limited to a minimum altitude of 900 ft. MSL to at least 5 DME.

3) After the step-down fix is passed, the MDA in the minimums section applies.

a) DME-equipped aircraft may now descend to 700 ft. MSL (i.e., Categories A through C).

b) Non-DME-equipped aircraft must remain at or above 900 ft.

g. Upon arrival at the MAP (identified by time or 8.4 DME), you must execute the missed approach procedure immediately if the required visual references are not seen and identified.

1) The missed approach procedure is printed in the profile view and partially indicated by dashed lines in the profile and plan views.

2) In the example, the missed approach procedure is a climbing left turn to 2,000 ft. MSL via heading 295° and GNV R-349 to DRAFT Int. and hold.

3. Establish two-way communication with ATC, as appropriate, to the phase of flight or approach segment, and use proper radio communication phraseology and technique.

a. During the en route part of your flight, you normally maintain two-way communication with an air route traffic control center (e.g., Miami center).

b. During the transition from en route to approach segment, before the IAF, you will normally be instructed to contact approach control.

1) Where radar is approved for approach control service, it may be used to provide radar vectors in conjunction with the published VOR approach.

c. Normally, prior to arriving at the FAF inbound, you will be instructed to contact the tower or FSS.

1) If the airport is not served by a tower or FSS, you will be instructed to change to the advisory frequency (CTAF), and you should broadcast your intentions, including

a) The approach you are executing (e.g., VOR-A)
b) Your position (every mile for the last 5 miles of approach)
c) When reached, your arrival over the FAF inbound

d. If you execute a missed approach, you must first fly your airplane, and once you are in complete control climbing, you should inform the tower, FSS, or CTAF of your missed approach.

1) Then you will be instructed to contact approach control by the tower or FSS.

2) At an airport with no tower or FSS, first announce your missed approach on CTAF; then switch back to the approach control frequency and inform ATC of your missed approach.

4. Select, tune, identify, and confirm the operational status of navigation equipment to be used for the approach procedure.

a. After you have selected the correct IAP chart and have studied it, you should tune in the VOR frequency on your navigation equipment.

1) The frequency and the Morse code identifier are presented on the chart.

2) In the example VOR-A IAP on page 235, you will tune and identify GNV VORTAC.

b. Next, correctly identify the VOR facility by its Morse code identification or by a recorded voice identification that states the name of the facility followed by the letters "V-O-R."

   1) If the VOR is out of service, the coded identification will not be transmitted.

   2) The DME-coded identification is transmitted one time for each three or four times that the VOR-coded identification is transmitted. When either the VOR or the DME is inoperative, it is important to recognize which identifier is retained for the operative facility.

      a) A single coded identification with a repetition interval of approximately 30 sec. indicates that the DME is operative.

c. Operational checks that you can perform in your airplane include seeing that

   1) The VOR alarm flag is not on.
   2) The CDI has no erratic movements, unless at station passage.

d. Check each NAVAID that is to be used during approach and missed approach.

5. Comply with all clearances issued by ATC or your examiner.

a. When ATC (or your examiner) issues a clearance or an instruction, you are expected to execute its provisions upon receipt.

   1) You must not deviate from the provisions of any clearance or instructions unless an amended clearance is obtained or an emergency arises.

b. At times ATC may not specify a particular approach procedure in the approach clearance but will state "cleared for approach." Such a clearance indicates that you may execute any one of the authorized IAPs for that airport.

   1) This clearance does not constitute approval for you to execute a contact approach or a visual approach.

c. When cleared for a specifically prescribed IAP (e.g., "cleared VOR-A approach") or when "cleared for approach," you are required to execute the entire approach as described on the IAP chart unless an appropriate new or revised ATC clearance is received or you cancel your IFR flight plan.

6. Recognize if heading indicator (HI) and/or attitude indicator (AI) is inaccurate or inoperative, advise controller, and proceed with approach.

a. For information on recognizing whether the HI and/or AI is inaccurate or inoperative, see Task VII.B., Loss of Gyro Attitude and/or Heading Indicators, beginning on page 332.

b. You should report the malfunction of the instruments to ATC and advise that you will proceed with the approach.

7. Advise ATC (or your examiner) anytime your airplane is unable to comply with a clearance.

a. You must inform ATC (or your examiner) anytime your airplane's operating limitations forbid compliance with the clearance issued.

b. Before you accept a clearance, you must determine whether you can comply with it. If not, inform ATC of the reason you cannot accept the clearance, and request an amended clearance.

8. Establish the appropriate airplane configuration and airspeed considering turbulence and wind shear, and complete your airplane's checklist items appropriate to the phase of the flight.

a. During the initial segment of the IAP, you should slow your airplane to your desired approach speed. This is normally an airspeed within, or just above, flap operating range, from which your airplane can readily transition to a landing configuration.

1) In your airplane, approach speed is ________.
2) Based on known weather conditions, you may want to increase your approach airspeed due to turbulence, gusty winds, and possible wind shear.
3) If appropriate for your airplane, you should lower flaps to the approach setting.

b. During the initial approach segment, you should also complete your before-landing checklist as described in your *POH*.
1) All fuel-related items, such as fuel selectors, fuel pumps, and mixture, should be set for landing.
2) Most pilots will lower the landing gear (if applicable) at the beginning of the final approach segment (i.e., over the FAF).
3) At the FAF, you should have completed your before-landing checklist.

c. You should commit to memory certain important items prior to reaching the FAF:
1) MDA (or step-down minimums)
2) Time (or DME) from FAF to MAP
3) Visibility minimums
4) Missed approach procedure (at least initial part, e.g., "Climbing left turn to 2,000 ft. . . .")

9. *Maintain, prior to beginning the final approach segment, altitude within 100 ft. and heading within 10°; allow less than a full-scale deflection of the CDI or within 10° in the case of an RMI; and maintain airspeed within 10 kt.*

a. Remember that the final approach segment begins at the FAF or at a point where you begin your descent to the MDA.
1) By this time, you should be established on the final approach course having determined the necessary wind-drift corrections to maintain the desired radial.
2) You should have your airspeed stabilized for the approach airspeed.

b. The VOR-A example on page 235 shows that before the FAF you must be no lower than 2,000 ft. MSL.
1) The FAF is the GNV VORTAC.

10. Apply the necessary adjustments to the published MDA and visibility criteria for your airplane approach category when required.

a. FDC and Class II NOTAMs
1) Flight Data Center (FDC) NOTAMs are regulatory in nature and inform you of amendments to IAPs or aeronautical charts prior to their normal publication.
   a) FDC NOTAMs are available from FSS and are published in the *Notices to Airmen Publication* (*NTAP*).
2) The *NTAP* (formerly Class II NOTAMs) contains all FDC NOTAMs that are current at the time of publication and those NOTAMs (D) that are expected to remain in effect for 7 days after the issuance of the publication. The *NTAP* is issued every 28 days.
   a) During your preflight briefing, you must ask your FSS specialist for any published NOTAMs that are pertinent to your flight.
   b) Alternatively, you may view the *NTAP* web version at:

www.faa.gov/NTAP/default.htm

b. Inoperative airplane and ground navigation equipment

1) You cannot perform a VOR IAP unless the VOR equipment in your airplane and the desired VOR station are operating properly.

2) Without an operational DME, you cannot perform a VOR/DME IAP.

c. Inoperative visual aids associated with the landing environment

1) Higher minimums are required with inoperative visual aids.

a) NACO charts will have this information listed in the Inoperative Components Table located on the inside front cover.

b) JEPP charts will have this information listed within the minimums section of the IAP chart.

2) For VOR IAPs, an inoperative visual aid will increase the visibility requirement but not the MDA.

d. National Weather Service (NWS) reporting factors and criteria

1) On NACO IAP charts, this information is included below the minimums section.

a) Some IAP charts will instruct you to use an alternate altimeter setting when the local altimeter setting is not available and to increase all MDAs by a certain amount.

11. Establish a rate of descent and track that will ensure arrival at the MDA prior to reaching the MAP with your airplane continuously in a position from which descent to a landing on the intended runway can be made at a normal rate using normal maneuvers.

a. At the FAF, you can start your descent to the MDA while tracking on the desired radial to the MAP.

1) Some recommend that you should establish an expeditious, but safe, descent (i.e., a constant airspeed descent) to allow you more time at the MDA and increase your chances of seeing the runway environment.

a) For approach airspeeds of 90 kt., a descent rate of no more than 700 fpm is usually adequate.

b) The sooner you can see the runway, the sooner you can establish a normal landing approach using normal maneuvers.

2) The FAA recommends that, when the IAP chart provides a vertical descent angle (VDA), you should calculate a rate of descent at the given VDA and your actual or estimated groundspeed. This will provide a stabilized approach descent.

a) EXAMPLE: For a groundspeed of 90 kt. and a VDA of 3.00°, a rate of descent of approximately 480 fpm should be established for a stabilized approach.

b. You will want to maintain a constant airspeed, whether in level flight, descending to the MDA, or leveling off at the MDA.

1) To accomplish this, all you will need to do initially is to have the power and trim set for your desired approach airspeed, e.g., 90 kt.

2) To descend, reduce power to a predetermined setting to establish the rate of descent desired.

a) As the power is reduced, the nose of the airplane will lower. All you may need to do is ensure that the nose does not pitch down too much.

b) Since the airplane is trimmed for your approach speed, no change of the trim or elevator control is required.

c) You will be able to adjust the rate of descent with minor power adjustments.

c. See Task IV.C., Constant Airspeed Climbs and Descents, beginning on page 154.

12. *Allow, while on the final approach segment, no more than a three-quarter-scale deflection of the CDI or within 10° in case of an RMI, and maintain airspeed within 10 kt.*

a. The final approach segment is from the FAF to the MAP.

b. Maintain the final approach course (i.e., R-033 in our example) so that the CDI is no greater than the third dot (i.e., three-quarter-scale deflection).

1) If you are using an RMI, it should indicate that you are on the desired radial, ±10°.

c. Throughout the final approach segment, you must maintain your approach airspeed at all times, ±10 kt.

13. *Maintain the MDA, when reached, within +100 ft., –0 ft. to the MAP.*

a. Since you must not descend below the MDA, your author suggests flying at 50 ft. above the MDA to allow for turbulence (e.g., for an MDA of 700 ft. MSL, use 750 ft. MSL).

1) If it is a calm day, you should have no problem maintaining the MDA.

b. Using predetermined pitch and power settings, lead your level-off by approximately 100 ft. to allow you to maintain a constant airspeed and smooth control of your airplane.

14. Execute the missed approach procedure when the required visual references for the intended runway are not distinctly visible and identifiable at the MAP.

a. The MAP can be identified by a DME fix or the VOR station itself, or it may be determined by time from the FAF.

1) When the VOR station is the MAP, a reversal of the TO/FROM indicator marks station passage, thus the MAP.

2) In the example VOR-A IAP on page 235, the MAP is 8.4 DME from GNV VORTAC or may be determined by time based on groundspeed.

3) When using time to determine the MAP, you must estimate your groundspeed and use the chart below the airport diagram (NACO chart) to determine the time.

a) You can estimate your groundspeed by using the indicated airspeed (or true airspeed for high density altitude airports) and the known wind.

i) For a headwind or tailwind, estimate the groundspeed by subtracting or adding the speed of the wind to the indicated airspeed.

ii) For a quartering wind, use one-half the wind speed.

b) You can also use the groundspeed function on the DME.

c) When a time is provided for an approach, it is a good procedure to use it to back up other means of identifying the MAP.

i) This will allow you to continue the approach if another method fails (e.g., DME).

ii) Your examiner will expect you to follow this procedure.

b. To descend below the MDA (FAR 91.175), you are required to

1) Have the runway environment in sight
2) Have visibility at or above the minimums for your approach category
3) Be in a position to make a normal descent to the intended runway

c. **Runway environment** is defined as any one of the following visual references required for descent below the MDA:

1) Approach light system

a) However, you are not allowed to descend below 100 ft. above the touchdown zone elevation (TDZE) using the approach lights as a reference unless the red terminating bars or the red side row bars are also distinctly visible and identifiable.

2) The threshold
3) The threshold markings
4) The threshold lights
5) The runway end identifier lights
6) The visual approach slope indicator
7) The touchdown zone or touchdown zone markings
8) The touchdown zone lights
9) The runway or runway markings
10) The runway lights

d. You are required to execute the missed approach procedure when either of the following happens:

1) You cannot identify one of the required visual references at either of the following times:

a) Upon arrival at the MAP
b) At any time that you are below the MDA until touchdown

2) An identifiable part of the airport is not distinctly visible to you during a circling maneuver, at or above MDA, unless this is caused only from a bank of your airplane during the circling approach.

e. The missed approach procedure is written in the profile view of NACO charts.

1) In the example VOR-A IAP on page 235, the missed approach procedure is to execute a climbing left turn to 2,000 ft. via heading 295° and GNV R-349 to DRAFT Int. and hold.

f. Protected obstacle clearance areas for missed approach procedures are predicated on the assumption that the missed approach procedure is initiated at the MAP not lower than the MDA. Reasonable buffers are provided for normal maneuvers.

1) When an early missed approach is executed (i.e., before the MAP), you should, unless otherwise cleared by ATC, fly the IAP as specified on the chart to the MAP at or above the MDA before executing a turning maneuver.

2) If you lose visual reference while circling to land, you must follow the prescribed missed approach procedure.

a) To become established on the missed approach course, you should make an initial climbing turn toward the landing runway and continue the turn until you are established on the missed approach course.

15. Execute a normal landing from a straight-in or circling approach when instructed by your examiner.

a. See Task VI.E., Landing from a Straight-in or Circling Approach, beginning on page 322.

C. Common Errors during a VOR/VORTAC Instrument Approach Procedure

1. **Failure to have essential knowledge of the information on VOR/VORTAC instrument approach procedure chart.**
   a. Know your IAF and the way to arrive at the final approach course.
   b. Know how to identify the FAF.
   c. Know your minimum altitudes during each segment of the approach, including the MDA.
   d. Be able to identify the MAP, and know the missed approach procedure.
2. **Incorrect communication procedures or noncompliance with ATC clearances.**
   a. Always follow correct communication procedures.
   b. You are required to follow all ATC clearances. If you cannot comply with a clearance, you must request an amended clearance from ATC.
3. **Failure to accomplish checklist items.**
   a. You must complete the before-landing checklist for your airplane. The checklist is normally begun during the initial segment and completed by the start of the final approach segment (i.e., FAF).
   b. An approach is a busy time in the cockpit. Attempt to complete your checklist as much as possible before the FAF.
4. **Faulty basic instrument flying technique.**
   a. It is a busy time, but you must continue your cross-check and instrument interpretation throughout the approach.
   b. Flying an approach is nothing more than basic attitude instrument flying.
   c. Remember to fly your airplane first, then track your course, and then talk.
5. **Inappropriate descent below the MDA.**
   a. You can descend below the MDA only when you have the required visual references and the visibility is equal to or better than published on the IAP chart.
   b. Give yourself a cushion (e.g., add 50 ft.) above the MDA to allow for altitude variations.

D. NDB Instrument Approach Task Objectives

1. **Exhibit adequate knowledge of the elements related to an NDB instrument approach procedure.**
   a. The NDB instrument approach procedure is a nonprecision approach with the course guidance based on a specific bearing.
   b. The NDB is normally the initial approach fix (IAF), but the approach may have others (e.g., a VOR station).
      1) If the NDB IAP chart has a note stating (usually in large letters), "RADAR REQUIRED," an IAF may not be available.
         a) ATC will provide radar vectors to join the final approach course before arrival at the final approach fix (FAF).
   c. When the NDB is located off the airport, the NDB will normally be the final approach fix (FAF).
      1) On the profile view of the NACO IAP chart, the FAF is identified by a Maltese cross symbol.

d. When the NDB is located on the airport, there is no depicted FAF. In this case, the final approach segment begins at the final approach point.

  1) The final approach point is the point where you are established inbound on the final approach course from the procedure turn (or radar vector) and where you may start the final approach descent.

e. If the IAP indicates a procedure turn, you will fly outbound from an off-airport NDB for 1 or 2 min. and then execute the procedure turn.

  1) EXAMPLE: An airplane using an indicated approach airspeed of 90 kt. would travel 3 NM under no-wind conditions. This would keep the airplane within the 10-NM limits of the procedure turn and allow enough time for it to be established on the final approach course before passing over the FAF.

f. If the VOR is located on the airport, you will fly outbound for 3 or 4 min. and then execute the procedure turn.

  1) This is to allow you enough distance to become established on the final approach course and then begin your descent to the MDA.

  2) You will need to take into account the wind conditions and make adjustments as necessary to remain within the specified limits of a procedure turn.

g. The MAP for an NDB approach occurs at the NDB when it is located on the airport.

  1) The MAP occurs a specified time after crossing the FAF for an off-airport NDB.

2. Select and comply with the appropriate NDB instrument approach procedure to be performed.

a. Once you know which IAP you will be conducting, you should select the appropriate IAP chart and study it to determine that you can comply with the procedure.

  1) If the inbound course of the approach is within 30° of the centerline of a runway, the approach will be designated to that runway (e.g., NDB RWY 10).

    a) If the course deviation is more than 30°, the approach will be designated with an alphabetic character (e.g., NDB-A).

b. We will use the NDB RWY 10 IAP to New Orleans International (KMSY) on page 244 as an example throughout our discussion.

c. Comprising the top half of the chart, the **plan view** shows a bird's-eye view of the approach.

d. Assuming you are cleared by ATC to "proceed direct to KINTE LOM (IAF), cleared for the NDB RWY 10 approach," you will turn outbound upon reaching KINTE, on the 282° bearing from the LOM.

  1) This is the procedure turn outbound.

  2) You are now on the initial approach segment.

  3) Since you have been cleared for the approach and you are established in the published procedure turn, you may descend to 2,100 ft. MSL, the published altitude for the procedure turn outbound, shown in the **profile view** just below the plan view.

  4) After flying outbound for approximately 1 to 2 min., you will turn left 45° to a heading of 237° and fly that heading for 1 min. before turning right 180° to a heading of 057° to intercept the 102° bearing inbound to the LOM.

e. When you have intercepted and are tracking the inbound course, you are on the intermediate approach segment.

f. After passing the FAF, shown by the Maltese cross in the profile view (i.e., KINTE LOM), you will be on the final approach segment.

1) You should begin timing your approach to the MAP. Times are based on groundspeed and are shown below the airport diagram.

a) In the example, using a groundspeed of 90 kt., the MAP is 4 min. 24 sec. from the FAF.

2) You may start your descent to the published MDA (since there are no step-down fixes) for your approach category, as shown in the **minimums section** just below the profile view.

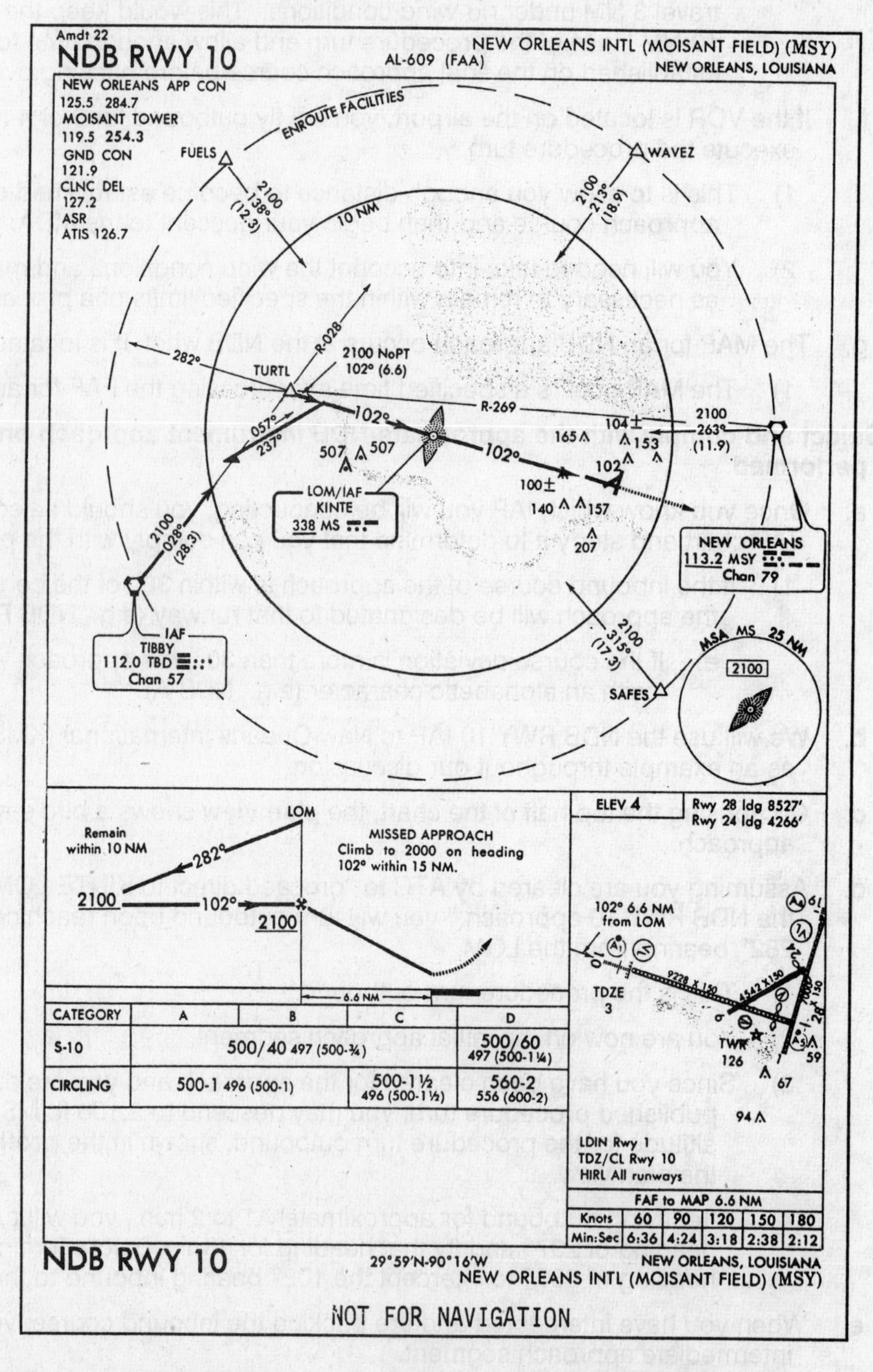

CATEGORY	A	B	C	D
S-10	500/40 497 (500-¾)			500/60 497 (500-1¼)
CIRCLING	500-1 496 (500-1)		500-1½ 496 (500-1½)	560-2 556 (600-2)

FAF to MAP 6.6 NM					
Knots	60	90	120	150	180
Min:Sec	6:36	4:24	3:18	2:38	2:12

NOTE: This IAP chart is shown in the old format, which is still in use. Charts using the new format are shown on pages 255 and 298.

g. Upon arrival at the MAP (identified by time), you must execute the missed approach procedure immediately if the required visual references are not seen and identified.

1) The missed approach procedure is printed in the profile view and partially indicated by dashed lines in the profile and plan views.

2) In the example, the missed approach procedure is a climb to 2,000 ft. MSL on heading 102° within 15 NM.

3. Establish two-way communication with ATC, as appropriate, to the phase of flight or approach segment, and use proper radio communication phraseology and technique.

a. During the en route part of your flight, you normally maintain two-way communication with an air route traffic control center (e.g., Miami center).

b. During the transition from en route to approach segment, before the IAF, you will normally be instructed to contact approach control.

1) Where radar is approved for approach control service, it may be used to provide radar vectors in conjunction with the published NDB approach.

c. Normally, prior to arriving at the FAF inbound, you will be instructed to contact the tower or FSS.

1) If the airport is not served by a tower or FSS, you will be instructed to change to the advisory frequency (CTAF), and you should broadcast your intentions, including

a) The approach you are executing (e.g., NDB RWY 10)
b) Your position (every mile for the last 5 miles of approach)
c) When reached, your arrival over the FAF inbound

d. If you execute a missed approach, you must first fly your airplane, and once you are in complete control, you should inform the tower, FSS, or CTAF of your missed approach.

1) Then you will be instructed to contact approach control by the tower or FSS.

2) At an airport with no tower or FSS, first announce your missed approach on CTAF; then switch back to the approach control frequency and inform ATC of your missed approach.

4. Select, tune, identify, confirm, and monitor the operational status of navigation equipment to be used for the approach procedure.

a. After you have selected the correct IAP chart and have studied it, you should tune the NDB frequency into your ADF navigation equipment.

1) The frequency and the Morse code identifier are presented on the chart.

2) In the example NDB RWY 10 IAP on page 244, you will tune and identify KINTE LOM. A LOM is a collocated NDB and outer marker (used for ILS approaches).

b. Next, correctly identify the NDB facility by its Morse code identification, which consists of three letters, unless the facility is a LOM, which is identified by two letters.

1) If the NDB is out of service, the coded identification will not be transmitted.

2) Once the NDB has been identified, the volume on your ADF should be set so that you can hear the Morse code identifier.

a) Since the ADF is not equipped with an alarm flag, monitoring the Morse code identifier is the only way to ensure continuous reception.

c. Use the appropriate test procedure for the equipment in your airplane to test the operational status of your ADF.

1) In some airplanes, the ADF's operational status is tested by pressing a test button, which moves the ADF needle. Once released, the needle should return to the original position.

d. Check each NAVAID that is to be used during approach and missed approach.

5. Comply with all clearances issued by ATC or your examiner.

a. When ATC (or your examiner) issues a clearance or instruction, you are expected to execute its provisions upon receipt.

1) You must not deviate from the provisions of any clearance or instructions unless an amended clearance is obtained or an emergency arises.

b. At times ATC may not specify a particular approach procedure in the approach clearance but will state "cleared for approach." Such a clearance indicates that you may execute any one of the authorized IAPs for that airport.

1) This clearance does not constitute approval for you to execute a contact approach or a visual approach.

c. When cleared for a specifically prescribed IAP (i.e., "cleared NDB RWY 10 approach") or when "cleared for approach," you are required to execute the entire approach as described on the IAP chart unless an appropriate new or revised ATC clearance is received or you cancel your IFR flight plan.

6. Recognize when heading indicator (HI) and/or attitude indicator (AI) is inaccurate or inoperative, advise controller, and proceed with approach.

a. For information on recognizing whether the HI and/or AI is inaccurate or inoperative, see Task VII.B., Loss of Gyro Attitude and/or Heading Indicators, beginning on page 332.

b. You should report the malfunction of the instruments to ATC and advise that you will proceed with the approach.

7. Advise ATC (or your examiner) anytime your airplane is unable to comply with a clearance.

a. You must inform ATC (or your examiner) anytime your airplane's operating limitations forbid compliance with the clearance issued.

b. Before you accept a clearance, you must determine whether you can comply with it. If not, inform ATC of the reason you cannot accept the clearance and request an amended clearance.

8. Establish the appropriate airplane configuration and airspeed considering turbulence and wind shear, and complete your airplane's checklist items appropriate to the phase of the flight.

a. During the initial segment of the IAP, you should slow your airplane to your desired approach speed. This is normally an airspeed within, or just above, flap operating range, from which your airplane can readily transition to a landing configuration.

1) In your airplane, approach speed is ________.

2) Based on known weather conditions, you may want to increase your approach airspeed due to turbulence, gusty winds, and possible wind shear.

3) If appropriate for your airplane, you should lower flaps to the approach setting.

b. During the initial approach segment, you should also complete your before-landing checklist as described in your *POH*.

   1) All fuel-related items, such as fuel selectors, fuel pumps, and mixture, should be set for landing.

   2) Most pilots will lower the landing gear (if applicable) at the beginning of the final approach segment (i.e., over the FAF).

   3) At the FAF, you should have completed your before-landing checklist.

c. You should commit to memory certain important items prior to reaching the FAF:

   1) MDA (or step-down minimums)
   2) Time from FAF to MAP
   3) Visibility minimums
   4) Missed approach procedure (at least initial part, e.g., "Climb to 2,000 ft. . . .")

9. *Maintain, prior to beginning the final approach segment, altitude within 100 ft., heading and bearing within 10°, and airspeed within 10 kt.*

a. Remember that the final approach segment begins at the FAF or at a point where you begin your descent to the MDA.

   1) By this time, you should be established on the final approach course having determined the necessary wind-drift corrections to maintain the desired bearing.

   2) You should have your airspeed stabilized for the approach airspeed.

b. The NDB RWY 10 example on page 244 shows that before the FAF you must be no lower than 2,100 ft. MSL.

   1) The FAF is KINTE LOM (NDB).

10. Apply the necessary adjustments to the published MDA and visibility criteria for your airplane approach category when required.

a. FDC and Class II NOTAMs

   1) Flight Data Center (FDC) NOTAMs are regulatory in nature and inform you of amendments to IAPs or aeronautical charts prior to their normal publication.

      a) FDC NOTAMs are available from FSS and are published in the *Notices to Airmen Publication* (*NTAP*).

   2) The *NTAP* (formerly Class II NOTAMs) contains all FDC NOTAMs that are current at the time of publication and those NOTAMs (D) that are expected to remain in effect for 7 days after the issuance of the publication. The *NTAP* is issued every 28 days.

      a) During your preflight briefing, you must ask your FSS specialist for any published NOTAMs that are pertinent to your flight.

      b) Alternatively, you may view the *NTAP* web version at:

      www.faa.gov/NTAP/default.htm

b. Inoperative airplane and ground navigation equipment

   1) You cannot perform an NDB IAP unless the ADF equipment in your airplane and the desired NDB facility are operating properly.

   2) You must monitor the NDB identifier throughout the approach to ensure that it remains operational.

c. Inoperative visual aids associated with the landing environment

   1) Higher minimums are required with inoperative visual aids.

      a) NACO charts will have this information listed in the Inoperative Components Table located on the inside front cover.

b) JEPP charts will have this information listed within the minimums section of the IAP chart.

2) For NDB IAPs, an inoperative visual aid will increase the visibility requirement but not the MDA.

d. National Weather Service (NWS) reporting factors and criteria

1) On NACO IAP charts, this information is included below the minimums section.

a) Some IAP charts will instruct you to use an alternate altimeter setting when the local altimeter setting is not available and to increase all MDAs by a certain amount.

11. Establish a rate of descent and track that will ensure arrival at the MDA prior to reaching the MAP with your airplane continuously in a position from which descent to a landing on the intended runway can be made at a normal rate using normal maneuvers.

a. At the FAF, you can start your descent to the MDA while tracking the desired bearing to the MAP.

1) Some recommend that you should establish an expeditious, but safe, descent (i.e., constant airspeed descent) to allow you more time at the MDA and increase your chances of seeing the runway environment.

a) For approach airspeeds of 90 kt., a descent rate of no more than 700 fpm is usually adequate.

b) The sooner you can see the runway, the sooner you can establish a normal landing approach using normal maneuvers.

2) The FAA recommends that, when the IAP chart provides a vertical descent angle (VDA), you should calculate a rate of descent at the given VDA and your actual or estimated groundspeed. This will provide a stabilized approach descent.

a) EXAMPLE: For a groundspeed of 90 kt. and a VDA of 3.00°, a rate of descent of approximately 480 fpm should be established for a stabilized approach.

b. You will want to maintain a constant airspeed, whether in level flight, descending to the MDA, or leveling off at the MDA.

1) To accomplish this, all you will need to do initially is to have the power and trim set for your desired approach airspeed, e.g., 90 kt.

2) To descend, reduce power to a predetermined setting to establish the rate of descent desired.

a) As the power is reduced, the nose of the airplane will lower. All you may need to do is ensure that the nose does not pitch down too much.

b) Since the airplane is trimmed for your approach speed, no change of the trim or elevator control is required.

c) You will be able to adjust the rate of descent with minor power adjustments.

3) See Task IV.C., Constant Airspeed Climbs and Descents, beginning on page 154.

12. Maintain, while on the final approach segment, a deviation of not more than 10° from the desired bearing, and maintain airspeed within 10 kt.

a. The final approach segment is from the FAF to the MAP.

b. Maintain the final approach course (i.e., 102° bearing from the NDB in our example) so that the ADF needle is displaying the correct wind-drift correction to maintain the bearing, ±10°.

c. Throughout the final approach segment, you must maintain your approach airspeed at all times, ±10 kt.

13. *Maintain the MDA, when reached, within +100 ft., –0 ft. to the MAP.*

a. Since you must not descend below the MDA, your author suggests flying at 50 ft. above the MDA to allow for turbulence (e.g., for an MDA of 700 ft. MSL, use 750 ft. MSL).

1) If it is a calm day, you should have no problem maintaining the MDA.

b. Using predetermined pitch and power settings, lead your level-off by approximately 100 ft. to allow you to maintain a constant airspeed and smooth control of your airplane.

c. In the example NDB RWY 10 IAP on page 244, the straight-in MDA for all approach categories is 500 ft. MSL, and the MDA for a circling approach is the same, except Category D has an MDA of 560 ft. MSL.

14. Execute the missed approach procedure when the required visual references for the intended runway are not distinctly visible and identifiable at the MAP.

a. The MAP can be identified by the NDB facility itself, or it may be determined by time from the FAF.

1) When the NDB facility is the MAP, a 180° reversal of the ADF needle marks station passage, thus the MAP.

2) In the example NDB RWY 10 IAP on page 244, the MAP is timed. When using time to determine the MAP, you must estimate your groundspeed and use the chart below the airport diagram (NACO chart) to determine the time.

a) You can estimate your groundspeed by using the indicated airspeed (or true airspeed for high density altitude airports) and the known wind.

i) For a headwind or tailwind, estimate the groundspeed by subtracting or adding the speed of the wind to the indicated airspeed.

ii) For a quartering wind, use one-half the wind speed.

b. To descend below the MDA (FAR 91.175), you are required to

1) Have the runway environment in sight
2) Have visibility at or above the minimums for your approach category
3) Be in a position to make a normal descent to the intended runway

c. Runway environment is defined as any one of the following visual references required for descent below the MDA:

1) Approach light system

a) However, you are not allowed to descend below 100 ft. above the touchdown zone elevation (TDZE) using the approach lights as a reference unless the red terminating bars or the red side row bars are also distinctly visible and identifiable.

2) The threshold
3) The threshold markings
4) The threshold lights
5) The runway end identifier lights
6) The visual approach slope indicator
7) The touchdown zone or touchdown zone markings
8) The touchdown zone lights
9) The runway or runway markings
10) The runway lights

d. You are required to execute the missed approach procedure when either of the following happens:

1) You cannot identify one of the required visual references at either of the following times:

a) Upon arrival at the MAP
b) At any time that you are below the MDA until touchdown

2) An identifiable part of the airport is not distinctly visible to you during a circling maneuver, at or above MDA, unless this is caused only from a bank of your airplane during the circling approach.

e. The missed approach procedure is written in the profile view of NACO charts.

1) In the example NDB RWY 10 IAP on page 244, the missed approach procedure is a climb to 2,000 ft. on heading 102° within 15 NM.

f. Protected obstacle clearance areas for missed approach procedures are predicated on the assumption that the missed approach procedure is initiated at the MAP not lower than the MDA. Reasonable buffers are provided for normal maneuvers.

1) When an early missed approach is executed (i.e., before the MAP), you should, unless otherwise cleared by ATC, fly the IAP as specified on the chart to the MAP at or above the MDA before executing a turning maneuver.

2) If you lose visual reference while circling to land, you must follow the prescribed missed approach procedure.

a) To become established on the missed approach course, you should make an initial climbing turn toward the landing runway and continue the turn until you are established on the missed approach course.

15. Execute a normal landing from a straight-in or circling approach when instructed by your examiner.

a. See Task VI.E., Landing from a Straight-in or Circling Approach, beginning on page 322.

E. Common Errors during an NDB Instrument Approach

1. Failure to have essential knowledge of the information on the NDB instrument approach procedure chart.

a. Know your IAF and the way to arrive at the final approach course.

b. Know how to identify the FAF.

c. Know your minimum altitudes during each segment of the approach, including the MDA.

d. Be able to identify the MAP, and know the missed approach procedure.

2. Incorrect communication procedures or noncompliance with ATC clearances.

a. Always follow correct communication procedures.

b. You are required to follow all ATC clearances. If you cannot comply with a clearance, you must request an amended clearance from ATC.

3. **Failure to accomplish checklist items.**
   a. You must complete the before-landing checklist for your airplane. The checklist is normally begun during the initial segment and completed by the start of the final approach segment.
   b. An approach is a busy time in the cockpit. Attempt to complete your checklist as much as possible before the FAF.
   c. Set and check your HI. Failure to do so could result in a dangerous track on an NDB approach.
4. **Faulty basic instrument flying technique.**
   a. It is a busy time, but you must continue your cross-check and instrument interpretation throughout the approach.
   b. Flying an approach is nothing more than basic attitude instrument flying.
   c. Remember to fly your airplane first, then track your course, and then talk.
5. **Inappropriate descent below the MDA.**
   a. You can descend below the MDA only when you have the required visual references and the visibility is equal to or better than that published on the IAP chart.
   b. Give yourself a cushion (e.g., add 50 ft.) above the MDA to allow for altitude variations.

F. GPS (RNAV) Task Objectives

1. **Exhibit adequate knowledge of the elements related to a GPS instrument approach procedure.**
   a. GPS standard instrument approach procedure design concepts
      1) Knowledge of how GPS IAPs are designed will provide you with the knowledge of what to expect during a GPS approach.
         a) The GPS approaches are designed to be flown as depicted, and your being vectored onto the final approach segment would be very unlikely.
      2) The objective of the Terminal Arrival Area (TAA) procedure design is to provide a new transition method for an arriving aircraft equipped with a flight management system (FMS) and/or GPS navigational equipment.
         a) The TAA contains within it a "T" structure that normally provides an approach without a procedure turn (NoPT).
         b) The TAA provides you and ATC with a very efficient method for routing traffic from en route to terminal structure.

3) The basic "T" that is contained in the TAA normally aligns the IAP on the runway centerline. Each of the waypoints has a five-character pronounceable name. The basic "T" design is shown below.

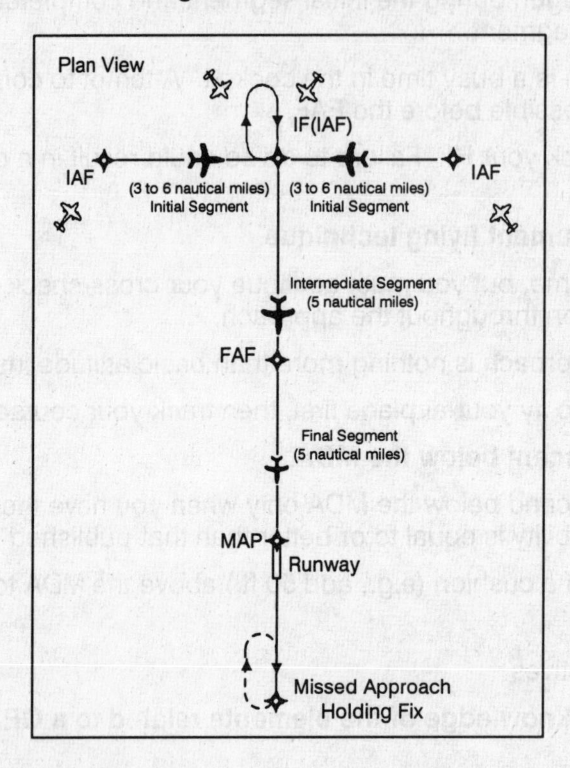

a) The missed approach point (MAP) will be located at the runway threshold, the final approach fix (FAF) 5 NM from the MAP, and the intermediate fix (IF) 5 NM from the FAF.

b) Two initial approach fixes (IAF) are located 3 to 6 NM from the center IF (IAF).

i) The length of the initial segment (from the IAF to the IF) varies with the category of aircraft using the procedure or descent gradient requirements.

ii) These initial segments are constructed perpendicular (90°) to the intermediate segment.

- The intermediate segment is from the IF to the FAF.

c) A holding pattern will be located at the IF (IAF) for course reversal requirements.

i) Some pilots may desire to execute a procedure turn holding pattern to meet a descent gradient requirement.

d) The missed approach segment is ideally aligned with the final approach course and terminates in a direct entry into a holding pattern.

i) Conditions may require a different routing.

4) The standard TAA may be modified because of operational requirements.

a) The left and/or right IAFs may be modified (other than at a 90° angle to the IF) or eliminated.

b) By looking at the IAP chart, you will easily see how the procedure is designed.

b. A waypoint is a geographical position (fix) used for route/instrument approach definition that is defined by latitude/longitude coordinates. GPS approaches make use of both flyover and flyby waypoints.

1) **Flyby waypoints** are used when an airplane should begin a turn to the next course prior to reaching the waypoint separating the two route segments.

a) This is known as **turn anticipation** and is compensated for in the airspace and terrain clearances.

b) A flyby waypoint is depicted by the symbol ✧.

c) Approach waypoints, except for the missed approach waypoint (MAWP) and the missed approach holding waypoint (MAHWP), are normally flyby waypoints.

2) **Flyover waypoints** are used when the airplane must fly over the point prior to starting a turn.

a) New approach charts depict flyover waypoints by the symbol ⟐.

i) Overlay approach charts (titled "or GPS") and some early stand-alone approach charts may not use this symbol.

c. An overlay approach is a GPS approach that is based on an existing nonprecision approach (except LOC, LDA, or SDF approaches), i.e., VOR and NDB.

1) When the FAA began authorizing GPS approaches, they used existing VOR and NDB approaches as a basis for the GPS approach.

2) Now the FAA designs stand-alone GPS approaches; i.e., they are not based on an existing approach.

d. On overlay approaches, if no pronounceable five-character name is published for an approach waypoint or fix, it was given a database identifier consisting of letters and numbers.

1) These waypoints will appear in the list of waypoints in the IAP database but may not appear on the IAP chart.

2) Procedures without a FAF (approaches with the VOR or NDB located on the airport) will have a sensor **final approach waypoint (FAWP)** added to the database at least 4 NM prior to the MAWP to allow the receiver to transition to the approach mode.

a) Some approaches also contain an additional waypoint in the holding pattern when the MAWP and MAHWP are colocated.

3) Arc and radial approaches have an additional waypoint that is used for turn anticipation computation where the arc joins the final approach course.

a) These coded names will not be used by ATC.

e. The **runway threshold waypoint**, which is normally the MAWP, may have a five-letter identifier (e.g., SNEEZ) or be coded by the runway identifier (e.g., RW36).

1) Those thresholds coded as five-letter identifiers are being changed to the runway designation.

a) As a result, the IAP chart and the database may differ until all changes are complete.

2) MAWPs not located at the runway threshold will have a five-letter identifier.

f. You should pay particular attention to the exact operation of your GPS for performing holding patterns and, in the case of overlay approaches, operations such as procedure turns.

1) These procedures may require manual intervention by the pilot to stop the sequencing of waypoints by the receiver and to resume automatic GPS navigation sequencing once the maneuver is complete.

g. A fix on an overlay approach identified by a DME fix will not be in the waypoint sequence on the GPS receiver unless a published name is assigned to it.

1) When a name is assigned, the along-track distance (ATD) to the waypoint may be zero rather than the DME stated on the approach chart.

2) You should be alert to this discrepancy on any overlay procedure when the original approach used DME.

2. Select and comply with the appropriate GPS instrument approach procedure to be performed.

a. Once you know what IAP you will be conducting, you should select the appropriate IAP chart and study it to determine that you can comply with the procedure.

1) The RNAV approach will be designated to a specific runway (e.g., RNAV RWY 9).

b. We will use the RNAV RWY 9 IAP to Lawrenceville-Vincennes International (KLWV) on page 255 as an example throughout our discussion.

c. At the top of the chart, the three rows of information is the **pilot briefing**.

1) This area contains information on the final approach course, runway/airport data, missed approach text, and various frequencies.

d. Below the pilot briefing, the **plan view** shows a bird's-eye view of the approach.

e. In the plan view for the RNAV RWY 9 IAP, you will see the three Terminal Arrival Areas (TAA) icons, with the appropriate waypoint IAF (MALRY, VIHKE, or PODAQ) for the approach area.

1) When crossing the 30-NM boundary of each of these areas or when released by ATC, within the area, you are expected to proceed to the appropriate IAF.

a) EXAMPLE: If your course to MALRY is from 180° clockwise to 270°, then you are in the left base area and VIHKE will be your IAF (see the TAA icon at the upper right corner of the plan view).

2) You will specify the approach and the IAF to begin the approach.

a) Once the approach is loaded and activated, the GPS will provide you guidance from your present position to the selected IAF.

3) Once you arrive at the appropriate IAF, the GPS will sequence to the next appropriate waypoint.

a) If you used VIHKE or PODAQ, then the next waypoint would be MALRY, then a turn toward YUKUB.

b) If the IAF is MALRY, then YUKUB would be the next waypoint.

c) The minimum altitude for this segment is 2,600 ft. MSL.

d) Note these are fly-by waypoints. Some GPS units will provide guidance on leading the turn. If not, then you will need to do it manually.

f. After passing MALRY, you will be tracking a course of 090° toward YUKUB, which is the FAF.

1) You are now on the intermediate approach segment.
2) You can now descend to 2,100 ft. MSL.

g. After passing the FAF, shown by the Maltese cross in the profile view (i.e., YUKUB), you will be on the final approach segment.

1) Your GPS will now show the next waypoint as RW09, which is the MAP.

2) You may start your descent to the published MDA (since there are no step-down fixes) for your approach category, as shown in the **minimums section** just below the profile view.

   a) You will use the line labeled LNAV for a straight-in approach or CIRCLING for a circling approach.

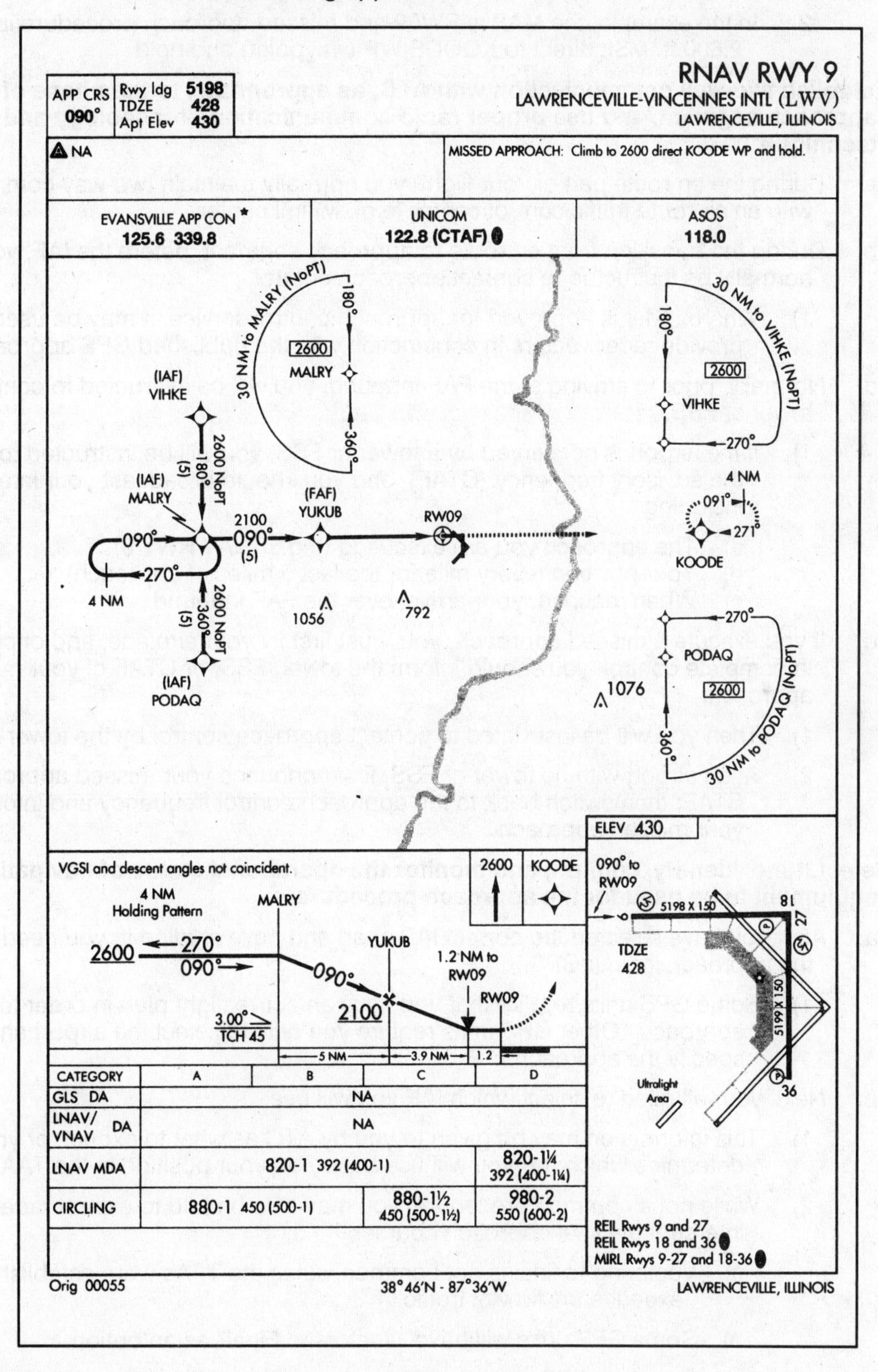

h. Upon arrival at the MAP (identified by a waypoint), you must execute the missed approach procedure immediately if the required visual references are not seen and identified.

    1) The missed approach procedure is printed in the pilot briefing and also by the use of icons in the upper right portion of the profile view.

    2) In the example, the MAP is RW09 and missed approach procedure is a climb to 2,600 ft. MSL direct to KOODE WP (waypoint) and hold.

3. Establish two-way communication with ATC, as appropriate, to the phase of flight or approach segment, and use proper radio communication phraseology and technique.

a. During the en route part of your flight, you normally maintain two-way communication with an air route traffic control center (e.g., Miami center).

b. During the transition from en route to approach segment, before the IAF, you will normally be instructed to contact approach control.

    1) Where radar is approved for approach control service, it may be used to provide radar vectors in conjunction with the published GPS approach.

c. Normally, prior to arriving at the FAF inbound, you will be instructed to contact the tower or FSS.

    1) If the airport is not served by a tower or FSS, you will be instructed to change to the advisory frequency (CTAF), and you should broadcast your intentions, including

        a) The approach you are executing (e.g., RNAV RWY 9)
        b) Your position (every mile for the last 5 miles of approach)
        c) When reached, your arrival over the FAF inbound

d. If you execute a missed approach, you must first fly your airplane, and once you are in complete control, you should inform the tower, FSS, or CTAF of your missed approach.

    1) Then you will be instructed to contact approach control by the tower or FSS.

    2) At an airport with no tower or FSS, first announce your missed approach on CTAF; then switch back to the approach control frequency and inform ATC of your missed approach.

4. Select, tune, identify, confirm, and monitor the operational status of navigation equipment to be used for the approach procedure.

a. After you have selected the correct IAP chart and have studied it, you need to select the approach in your GPS.

    1) Some GPS units require that you have an active flight plan in order to select an approach. Other GPS units require you only to select the airport and then specify the approach.

b. Next, you will need to select which IAF you will use.

    1) This information may be given to you by ATC as what to expect, or you can determine which IAF you will use based on your position in the TAA.

    2) While not a common procedure, you may also be told to expect radar vectors to intercept the final approach course.

        a) Vectoring to final is not common, since the TAAs were established to expedite the flow of traffic.

        b) Some GPS units will have "Vectors to Final" as an option.

        c) For other GPS units, you will not select an IAF, but you may have to follow some other procedures so the GPS sequences the approach waypoints properly.

c. Following are some items to consider when receiving vectors to final on a GPS IAP:

1) This procedure provides an extended final approach course in cases in which you are vectored onto the final approach course outside of any existing segment aligned with the runway.

2) Assigned altitudes must be maintained until you are established on a published segment of the approach.

3) Required altitudes at waypoints outside the FAWP or stepdown fixes must be considered.

4) Calculating the distance to the FAWP may be required in order to descend at the proper location.

d. Some GPS units will require you to set the current altimeter setting in the GPS.

1) This is done to improve the receiver autonomous integrity monitoring (RAIM) capability.

2) Some GPS units have a default altimeter setting of 29.92 that does not need to be changed for the approach. However, you may change the altimeter setting if you wish, and the change will improve the RAIM.

e. When an approach has been loaded into the GPS receiver, an "arm" or "apch" annunciation will be activated at a straight-line distance of 30 NM from the airport reference point.

1) You should arm the approach mode at this time if it has not already been armed. Some GPS receivers arm automatically.

a) Without arming, the GPS receiver will not change from en route CDI and RAIM sensitivity of ±5 NM on either side of the centerline to ±1 NM terminal sensitivity.

i) The CDI will smoothly and gradually make this change in sensitivity. In some units, the change may take up to 5 min.

f. Do not attempt to fly an approach unless the procedure is contained in the current GPS database.

1) Flying point to point on the approach does not assure compliance with the published IAP.

2) The proper RAIM sensitivity will not be available, and the CDI sensitivity will not automatically change to ±0.3 NM.

a) Manually setting CDI sensitivity does not automatically change the RAIM sensitivity on some receivers.

3) Some existing nonprecision approach procedures cannot be coded for use with GPS and will not be available as overlays.

5. Comply with all clearances issued by ATC or your examiner.

a. When ATC (or your examiner) issues a clearance or an instruction, you are expected to execute its provisions upon receipt.

1) You must not deviate from the provisions of any clearance or instructions unless an amended clearance is obtained or an emergency arises.

b. At times ATC may not specify a particular approach procedure in the approach clearance but will state "cleared for approach." Such a clearance indicates that you may execute any one of the authorized IAPs for that airport.

1) This clearance does not constitute approval for you to execute a contact approach or a visual approach.

c. When cleared for a specifically prescribed IAP (i.e., "cleared RNAV RWY 9 approach") or when "cleared for approach," you are required to execute the entire approach as described on the IAP chart unless an appropriate new or revised ATC clearance is received or you cancel your IFR flight plan.

6. Recognize when heading indicator (HI) and/or attitude indicator (AI) is inaccurate or inoperative, advise controller, and proceed with approach.

a. For information on recognizing whether the HI and/or AI is inaccurate or inoperative, see Task VII.B., Loss of Gyro Attitude and/or Heading Indicators, beginning on page 332.

b. You should report the malfunction of the instruments to ATC and advise that you will proceed with the approach.

7. Advise ATC (or your examiner) anytime your airplane is unable to comply with a clearance.

a. You must inform ATC (or your examiner) anytime your airplane's operating limitations forbid compliance with the clearance issued.

b. Before you accept a clearance, you must determine whether you can comply with it. If not, inform ATC of the reason you cannot accept the clearance and request an amended clearance.

8. Establish the appropriate airplane configuration and airspeed considering turbulence and wind shear, and complete your airplane's checklist items appropriate to the phase of the flight.

a. During the initial segment of the IAP, you should slow your airplane to your desired approach speed. This is normally an airspeed within, or just above, flap operating range, from which your airplane can readily transition to a landing configuration.

1) In your airplane, approach speed is ________.

2) Based on known weather conditions, you may want to increase your approach airspeed due to turbulence, gusty winds, and possible wind shear.

3) If appropriate for your airplane, you should lower flaps to the approach setting.

b. During the initial approach segment, you should also complete your before-landing checklist as described in your *POH*.

1) All fuel-related items, such as fuel selectors, fuel pumps, and mixture, should be set for landing.

2) Most pilots will lower the landing gear (if applicable) at the beginning of the final approach segment (i.e., over the FAF).

3) At the FAF, you should have completed your before-landing checklist.

c. You should commit to memory certain important items prior to reaching the FAF:

1) MDA (or step-down minimums)

2) MAP

3) Visibility minimums

4) Missed approach procedure (at least initial part, e.g., "Climb to 2,600 ft. . . .")

9. *Maintain, prior to beginning the final approach segment, altitude within 100 ft. and heading within 10°, allow less than a full-scale deflection of the CDI, and maintain airspeed within 10 kt.*

a. In the "T" design of a GPS approach, you must make a 90° turn at the IF.

1) The IF is a flyby waypoint, and your GPS unit will provide you with guidance on when to make the turn to establish yourself on the intermediate segment of the approach.

2) You must know what bank angle/turn rate the GPS receiver uses to compute turn anticipation and whether wind and airspeed are included in the receiver's calculations. This information should be in the GPS operating manual.

a) Over- or underbanking the turn onto the final approach course may significantly delay getting on course and may result in high descent rates to achieve the next segment altitude.

b. When you are within 2 NM of the FAF with the approach mode armed, the approach mode will switch to active, resulting in a RAIM change to approach sensitivity and a change in CDI sensitivity.

1) Beginning 2 NM prior to the FAF, the full-scale CDI sensitivity will change from ±1 NM to ±0.3 NM at the FAF.

2) As sensitivity changes from ±1 NM to ±0.3 NM approaching the FAF, with the CDI not centered, the corresponding increase in CDI displacement may give the impression that the airplane is moving farther away from the intended course even though it is on an acceptable intercept heading.

a) Referencing the digital track displacement (or cross track error), if it is available in the approach mode, may help you to remain position-oriented in this situation.

3) Being established on the final approach course prior to the beginning of the sensitivity change at 2 NM will help prevent problems in interpreting the CDI.

a) Thus, requesting or accepting vectors that will cause you to intercept the final approach course within 2 NM is not recommended.

c. If a RAIM failure/status annunciation occurs prior to the FAF, the approach should not be completed since the GPS may no longer provide the required accuracy.

1) The GPS receiver performs a RAIM prediction by 2 NM from the FAF to ensure RAIM is available at the FAF as a condition for entering the approach mode.

2) You should ensure that the GPS receiver has sequenced from "armed" to "approach" prior to the FAF (normally 2 NM prior to the FAF).

a) Failure to sequence may indicate the detection of a satellite anomaly, a failure to arm the receiver (if required), or other problems that will prevent completing the approach.

d. If the GPS receiver does not sequence into the approach mode or a RAIM failure/status annunciation occurs prior to the FAF, you should not descend to the MDA but should proceed to the FAF, execute the missed approach, proceed to the MAP, and contact ATC.

1) Refer to your GPS operating manual for specific indications and instructions associated with the loss of RAIM prior to the FAF.

e. The RNAV RWY 9 example on page 255 indicates that you will maintain 2,600 ft. MSL to MALRY.

1) Once established on the final approach course of 090° to YUKUB (FAF), you can descend to 2,100 ft. MSL.

f. The final approach segment begins at the FAF.

1) You should have your airplane stabilized (speed, tracking, etc.) prior to the FAF.

10. Apply the necessary adjustments to the published MDA and visibility criteria for your airplane approach category when required.

a. FDC and Class II NOTAMs

1) Flight Data Center (FDC) NOTAMs are regulatory in nature and inform you of amendments to IAPs or aeronautical charts prior to their normal publication.

a) FDC NOTAMs are available from FSS and are published in the *Notices to Airmen Publication* (*NTAP*).

2) The *NTAP* (formerly Class II NOTAMs) contains all FDC NOTAMs that are current at the time of publication and those NOTAMs (D) that are expected to remain in effect for 7 days after the issuance of the publication. The *NTAP* is issued every 28 days.

a) During your preflight briefing, you must ask your FSS specialist for any published NOTAMs that are pertinent to your flight.

b) Alternatively, you may view the NTAP web version at:

www.faa.gov/NTAP/default.htm

b. Inoperative airplane and ground navigation equipment

1) You cannot perform an RNAV (GPS) IAP unless the GPS equipment in your airplane is working properly.

2) If RAIM is not available, you cannot conduct the RNAV (GPS) approach.

a) Monitor your GPS units for RAIM messages.

c. Inoperative visual aids associated with the landing environment

1) Higher minimums are required with inoperative visual aids.

a) NACO charts will have this information listed in the Inoperative Components Table located on the inside front cover.

b) JEPP charts will have this information listed within the minimums section of the IAP chart.

2) For RNAV (GPS) IAPs, an inoperative visual aid will increase the visibility requirement but not the MDA.

d. National Weather Service (NWS) reporting factors and criteria

1) On NACO IAP charts, this information is included below the minimums section.

a) Some IAP charts will instruct you to use an alternate altimeter setting when the local altimeter setting is not available and to increase all MDAs by a certain amount.

11. Establish a rate of descent and track that will ensure arrival at the MDA prior to reaching the MAP with your airplane continuously in a position from which descent to a landing on the intended runway can be made at a normal rate using normal maneuvers.

a. At the FAF, you can start your descent to the MDA while tracking the desired course to the MAP.

1) In the profile view, the vertical descent angle (VDA) and the threshold crossing height (TCH) will be provided.

   a) In the RNAV RWY 9 IAP, the VDA is depicted as 3.00° and the TCH is 45 ft. AGL.

2) This information is provided so you can determine a rate of descent in order to establish a stabilized approach descent from the FAF (or stepdown fix) to the TCH.

   a) NACO has a rate of descent table on the inside back cover of the IAP chart book.

   b) JEPP has a rate of descent table on the IAP chart itself.

3) EXAMPLE: For an approach speed of 90 kt. and a VDA of 3.00°, a target rate of descent of approximately 478 fpm should be established for a stabilized approach.

4) In the profile view of the RNAV RWY 9 IAP on page 255, there is a note which states, VGSI [vertical glide slope indicator] and descent angles not coincident.

   a) This means that the glide path projected by the VGSI is not the same as the VDA.

b. You will want to maintain a constant airspeed, whether in level flight, descending to the MDA, or leveling off at the MDA.

1) To accomplish this, all you will need to do initially is to have the power and trim set for your desired approach airspeed, e.g., 90 kt.

2) To descend, reduce power to a predetermined setting to establish the rate of descent desired.

   a) As the power is reduced, the nose of the airplane will lower. All you may need to do is ensure that the nose does not pitch down too much.

   b) Since the airplane is trimmed for your approach speed, no change of the trim or elevator control is required.

   c) You will be able to adjust the rate of descent with minor power adjustments.

c. See Task IV.C., Constant Airspeed Climbs and Descents, beginning on page 154.

12. *Allow, while on the final approach segment, no more than a three-quarter-scale deflection of the CDI, and maintain airspeed within 10 kt.*

a. The final approach segment is from the FAF to the MAP.

b. Maintain the final approach course (i.e., 090° in our example) so that the CDI needle is centered.

1) Use all available information from the GPS (i.e., track, track angle error, etc.) to maintain the course.

2) You should have no problem keeping the CDI within a three-quarter-scale deflection.

   a) If you do have problems, the error is in your basic instrument flying skills, and your instructor will help you correct the problem.

c. Throughout the final approach segment, you must maintain your approach airspeed at all times, ±10 kt.

d. If a RAIM failure occurs after the FAF, the GPS receiver is allowed to continue operating without displaying an annunciation for up to 5 min. to allow completion of the approach (refer to your GPS operating manual).

1) If the RAIM flag/status annunciation DOES appear after the FAF, you must execute the missed approach immediately, unless you have the required runway environment in sight to continue for a landing.

13. *Maintain the MDA, when reached, within +100 ft., –0 ft. to the MAP.*

a. Since you must not descend below the MDA, your author suggests flying at approximately 50 ft. above the MDA to allow for turbulence (e.g., for an MDA of 3,680 ft. MSL, use 3,700 ft. MSL).

   1) On a calm day, you should have no problem holding the MDA.

b. Using predetermined pitch and power settings, lead your level-off by approximately 100 ft. to allow you to maintain a constant airspeed and smooth control of your airplane.

c. In the example RNAV RWY 9 IAP on page 255, the straight-in LNAV MDA for all approach categories is 820 ft. MSL, and the MDA for a circling approach is 880 ft. MSL (Cat. A and B), 880 ft. MSL (Cat. C), and 980 ft. MSL (Cat. D).

d. If a visual descent point (VDP) is published, it will not be included in the sequence of waypoints.

   1) You are expected to use normal piloting techniques for beginning the visual descent.
   2) Additionally, any unnamed step-down fixes in the final approach segment will not be coded in the waypoint sequence.
      a) You must calculate the ATD (along-track distance) to identify these fixes.
   3) In the example RNAV RWY 9 IAP on page 255, the VDP is 1.2 NM to RW09 (MAP).

14. Execute the missed approach procedure when the required visual references for the intended runway are not distinctly visible and identifiable at the MAP.

a. The MAP is identified by MAP waypoint.

   1) When the distance to the MAP is 0.0 NM, you are at the MAP.
   2) In the example RNAV RWY 9 IAP on page 255, the MAP is RW09.

b. To descend below the MDA (FAR 91.175), you are required to

   1) Have the runway environment in sight
   2) Have visibility at or above the minimums for your approach category
   3) Be in a position to make a normal descent to the intended runway

c. Runway environment is defined as any one of the following visual references required for descent below the MDA:

   1) Approach light system
      a) However, you are not allowed to descend below 100 ft. above the touchdown zone elevation (TDZE) using the approach lights as a reference unless the red terminating bars or the red side row bars are also distinctly visible and identifiable.
   2) The threshold
   3) The threshold markings
   4) The threshold lights
   5) The runway end identifier lights
   6) The visual approach slope indicator
   7) The touchdown zone or touchdown zone markings
   8) The touchdown zone lights
   9) The runway or runway markings
   10) The runway lights

d. You are required to execute the missed approach procedure when either of the following happens:
   1) You cannot identify one of the required visual references at either of the following times:
      a) Upon arrival at the MAP
      b) At any time that you are below the MDA until touchdown
   2) An identifiable part of the airport is not distinctly visible to you during a circling maneuver, at or above MDA, unless this is caused only from a bank of your airplane during the circling approach.

e. The missed approach procedure is written in the pilot briefing of NACO charts. Additionally, the procedure is also depicted by icons in the profile view.
   1) In the example RNAV RWY 9 IAP on page 255, the missed approach procedure is a climb to 2,600 ft. direct to KOODE and hold.

f. Protected obstacle clearance areas for missed approach procedures are predicated on the assumption that the missed approach procedure is initiated at the MAP not lower than the MDA. Reasonable buffers are provided for normal maneuvers.
   1) When an early missed approach is executed (i.e., before the MAP), you should, unless otherwise cleared by ATC, fly the IAP as specified on the chart to the MAP at or above the MDA before executing a turning maneuver.
   2) If you lose visual reference while circling to land, you must follow the prescribed missed approach procedure.
      a) To become established on the missed approach course, you should make an initial climbing turn toward the landing runway and continue the turn until you are established on the missed approach course.

g. A GPS missed approach requires you to sequence the GPS past the MAP to the missed approach portion of the procedure.
   1) You must be thoroughly familiar with the activation procedure.
   2) Activating the missed approach prior to the MAP will cause CDI sensitivity to change immediately to terminal (±1 NM) sensitivity, and the GPS will continue to navigate to the MAP.
      a) The GPS will not sequence past the MAP.
   3) Turns should not begin prior to the MAP.
   4) If the missed approach is not activated, the GPS will display an extension of the inbound final approach course, and the along-track distance (ATD) will increase from the MAP until it is manually sequenced after crossing the MAP.

h. Missed approach routings in which the first track is via a course rather than direct to the next waypoint requires you to set the course.
   1) Being familiar with all of the inputs required is especially critical during this phase of flight.

15. Execute a normal landing from a straight-in or circling approach when instructed by your examiner.

a. See Task VI.E., Landing from a Straight-in or Circling Approach, beginning on page 322.

G. Common Errors during an GPS Instrument Approach

1. **Failure to have essential knowledge of the information on the GPS instrument approach procedure chart.**
   a. Know your IAF and the way to arrive at the final approach course.
   b. Know how to identify the FAF.
   c. Know your minimum altitudes during each segment of the approach, including the MDA.
   d. Be able to identify the MAP, and know the missed approach procedure.
   e. Ensure that the sequence of waypoints in the GPS matches the IAP chart.
2. **Incorrect communication procedures or noncompliance with ATC clearances.**
   a. Always follow correct communication procedures.
   b. You are required to follow all ATC clearances. If you cannot comply with a clearance, you must request an amended clearance from ATC.
3. **Failure to accomplish checklist items.**
   a. You must complete the before-landing checklist for your airplane. This is normally begun during the initial segment and completed by the start of the final approach segment.
   b. An approach is a busy time in the cockpit. Attempt to complete your checklist as much as possible before the FAF.
   c. Set and check your HI. Failure to accomplish this could result in tracking errors on a GPS approach.
4. **Faulty basic instrument flying technique.**
   a. It is a busy time, but you must continue your cross-check and instrument interpretation throughout the approach.
   b. Remember to fly your airplane first, then track your course, and then talk.
5. **Inappropriate descent below the MDA.**
   a. You can descend below the MDA only when you have the required visual references and the visibility is equal to or better than that published on the IAP chart.
   b. Give yourself a cushion (e.g., add 50 ft.) above the MDA to allow for altitude variations.
6. **Failure to operate your GPS unit properly.**
   a. Since there is no established standard on information presentation and operation of GPS receivers (such as with a VOR), you must become knowledgeable about the make and model of GPS in your airplane.
   b. Differences exist between manufacturers and even within various models of a manufacturer.
   c. Practice all the operations using the demo mode of the GPS and under VFR while in the airplane before flying into IMC.
   d. Incorrect inputs into the GPS receiver are especially critical during approaches. In some instances, an incorrect entry can cause the receiver to exit the approach mode.

H. Localizer (LOC) Instrument Approach Task Objectives

1. **Exhibit adequate knowledge of the elements related to a LOC instrument approach procedure.**

   a. The localizer (LOC) or localizer back course (LOC BC) instrument approach procedure is a nonprecision approach with the course guidance provided by a localizer.

      1) The localizer signals provide you with course guidance to the runway centerline.

      2) The approach course of the localizer is called the front course.

         a) The course line along the extended centerline of a runway, in the opposite direction to the front course, is called the back course (BC).

      3) Distance-measuring equipment (DME) may be colocated with the localizer transmitter to provide distance information.

   b. The localizer course width is defined as the angular displacement at any point along the course between a full-scale fly-left and a full-scale fly-right CDI indication.

      1) The localizer signal is adjusted to produce an angular width between 3° and 6°, as necessary, beginning at the antenna site to provide a linear width of 700 ft. at the runway approach threshold (i.e., the front course).

      2) Since a full-scale right (or left) deflection of the CDI indicates 1.5° to 3.0° off course, the localizer is approximately four times as sensitive as a VOR signal.

   c. The initial approach fix (IAF) may be identified by an NDB, a VOR, an intersection, or a DME fix.

      1) In some instances, an IAF may not be listed on the IAP charts. These approaches normally require ATC radar vectors to the approach course.

   d. The final approach fix (FAF) may be identified by an NDB, a VOR, an intersection, or a DME fix.

      1) The FAF will be identified by a Maltese cross on the profile view of the NACO IAP chart.

   e. The missed approach point (MAP) is normally identified by timing from the FAF.

      1) On some LOC (or LOC BC) approaches, the MAP may be identified by a DME fix.

2. **Select and comply with the appropriate LOC instrument approach procedure to be performed.**

   a. Once you know what IAP you will be conducting, you should select the appropriate IAP chart and study it to determine that you can comply with the procedure.

      1) The title of the approach may list the required equipment.

         a) EXAMPLE: LOC/DME RWY 36 requires you to receive the localizer and DME.

      2) The approach lists any required navigation equipment as a note on the IAP chart (e.g., ADF REQUIRED).

         a) Your airplane must have that equipment operational to conduct the approach.

      3) If you do not have the proper navigation equipment, notify ATC immediately for another approach.

   b. A LOC (or LOC BC) approach will be designated to a specific runway (e.g., LOC RWY 35).

c. We will use the LOC RWY 35 IAP to Calhoun/Tom B. David Field (KCZL) on page 267 as an example throughout our discussion.

d. Comprising the top half of the chart, the **plan view** shows a bird's-eye view of the approach.

   1) The note in the upper-left portion of the plan view informs you that an ADF is required for this approach.

e. Assuming you are cleared by ATC to "proceed direct to CALHOUN NDB (i.e., the IAF), cleared for the LOC RWY 35 approach," you will turn outbound upon reaching the CALHOUN NDB, on the localizer outbound course of 170°.

   1) This is the procedure turn outbound.
   2) You are now on the initial approach segment.
   3) Since you have been cleared for the approach and you are established in the published procedure turn, you may descend to 2,700 ft. MSL, the published altitude for the procedure turn outbound, shown in the **profile view** just below the plan view.
   4) After flying outbound for approximately 1 to 2 min., you will turn left 45° to a heading of 125° and fly that heading for 1 min. before turning right 180° to a heading of 305° to intercept the localizer front course of 350° to the FAF.

f. When you have intercepted and are tracking the inbound course, you are on the intermediate approach segment.

g. After passing the FAF, shown by the Maltese cross in the profile view (i.e., CALHOUN NDB), you will be on the final approach segment.

   1) You should begin timing your approach to the MAP. Times are based on groundspeed and are shown below the airport diagram.

      a) In the example, using a groundspeed of 90 kt., the MAP is 2 min. from the FAF.

   2) You may start your descent to the published MDA (since there are no step-down fixes) for your approach category, as shown in the **minimums section** just below the profile view.

h. Upon arrival at the MAP (identified by time), you must execute the missed approach procedure immediately if the required visual references are not seen and identified.

   1) The missed approach procedure is printed in the profile view and partially indicated by dashed lines in the profile and plan views.
   2) In the example, the missed approach procedure is a climbing right turn to 2,700 ft. MSL direct to OUK (CALHOUN) NDB and hold.

3. Establish two-way communication with ATC, as appropriate, to the phase of flight or approach segment, and use proper radio communication phraseology and technique.

a. During the en route part of your flight, you normally maintain two-way communication with an air route traffic control center (e.g., Atlanta center).

b. During the transition from en route to approach segment, before the IAF, you will normally be instructed to contact approach control, if available.

   1) Where radar is approved for approach control service, it may be used to provide radar vectors in conjunction with the published LOC approach.

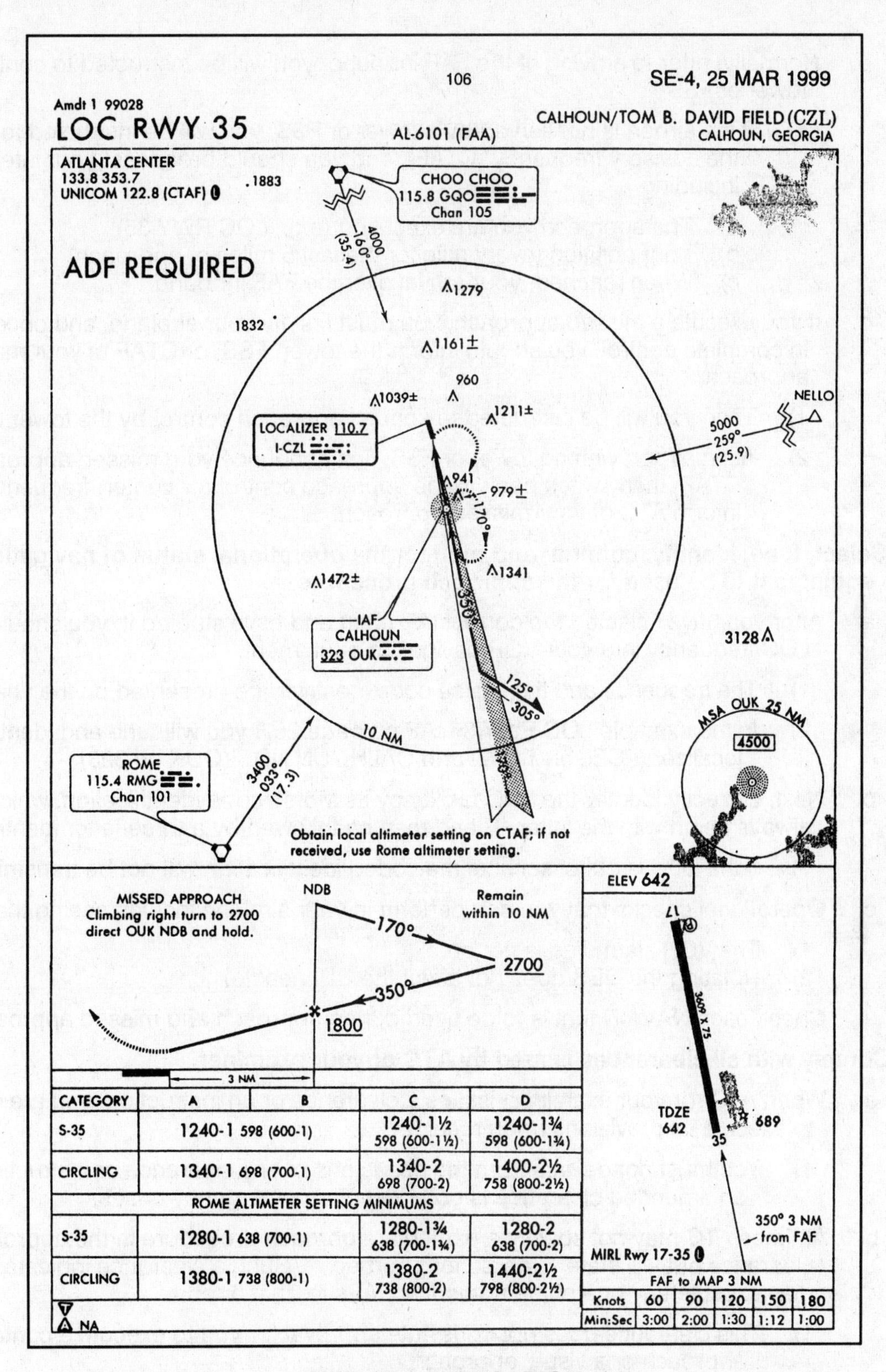

CATEGORY	A	B	C	D
S-35	1240-1 598 (600-1)		1240-1½ 598 (600-1½)	1240-1¾ 598 (600-1¾)
CIRCLING	1340-1 698 (700-1)		1340-2 698 (700-2)	1400-2½ 758 (800-2½)
ROME ALTIMETER SETTING MINIMUMS				
S-35	1280-1 638 (700-1)		1280-1¾ 638 (700-1¾)	1280-2 638 (700-2)
CIRCLING	1380-1 738 (800-1)		1380-2 738 (800-2)	1440-2½ 798 (800-2½)

FAF to MAP 3 NM					
Knots	60	90	120	150	180
Min:Sec	3:00	2:00	1:30	1:12	1:00

NOT FOR NAVIGATION

NOTE: This IAP chart is shown in the old format, which is still in use. Charts using the new format are shown on pages 255 and 298.

c. Normally, prior to arriving at the FAF inbound, you will be instructed to contact the tower or FSS.

   1) If the airport is not served by a tower or FSS, you will be instructed to change to the advisory frequency (CTAF), and you should broadcast your intentions, including

      a) The approach you are executing (e.g., LOC RWY 35)
      b) Your position (every mile for the last 5 miles of approach)
      c) When reached, your arrival over the FAF inbound

d. If you execute a missed approach, you must first fly your airplane, and once you are in complete control, you should inform the tower, FSS, or CTAF of your missed approach.

   1) Then you will be instructed to contact approach control by the tower or FSS.

   2) At an airport with no tower or FSS, first announce your missed approach on CTAF; then switch back to the approach control (or center) frequency and inform ATC of your missed approach.

4. Select, tune, identify, confirm, and monitor the operational status of navigation equipment to be used for the approach procedure.

a. After you have selected the correct IAP chart and have studied it, you should tune the LOC frequency into your VOR navigation equipment.

   1) The frequency and the Morse code identifier are presented on the chart.

   2) In the example LOC RWY 35 IAP on page 267, you will tune and identify the localizer (I-CZL on 110.7) and CALHOUN NDB (OUK on 323).

b. Next, correctly identify the LOC facility by its Morse code identification, which will always begin with the letter "I" and then be followed by a three-letter identifier.

   1) If the LOC is out of service, the coded identification will not be transmitted.

c. Operational checks that you can perform in your airplane include seeing that

   1) The VOR alarm flag is not on.
   2) Rotating the OBS does not affect the CDI needle.

d. Check each NAVAID that is to be used during approach and missed approach.

5. Comply with all clearances issued by ATC or your examiner.

a. When ATC (or your examiner) issues a clearance or an instruction, you are expected to execute its provisions upon receipt.

   1) You must not deviate from the provisions of any clearance or instructions unless an amended clearance is obtained or an emergency arises.

b. At times ATC may not specify a particular approach procedure in the approach clearance but will state "cleared for approach." Such a clearance indicates that you may execute any one of the authorized IAPs for that airport.

   1) This clearance does not constitute approval for you to execute a contact approach or a visual approach.

c. When cleared for a specifically prescribed IAP (i.e., "cleared LOC RWY 35 approach") or when "cleared for approach," you are required to execute the entire approach as described on the IAP chart unless an appropriate new or revised ATC clearance is received or you cancel your IFR flight plan.

6. Recognize when heading indicator (HI) and/or attitude indicator (AI) is inaccurate or inoperative, advise controller, and proceed with approach.

a. For information on recognizing whether the HI and/or AI is inaccurate or inoperative, see Task VII.B., Loss of Gyro Attitude and/or Heading Indicators, beginning on page 332.

b. You should report the malfunction of the instruments to ATC and advise that you will proceed with the approach.

7. Advise ATC (or your examiner) anytime your airplane is unable to comply with a clearance.

a. You must inform ATC (or your examiner) anytime your airplane's operating limitations forbid compliance with the clearance issued.

b. Before you accept a clearance, you must determine whether you can comply with it. If not, inform ATC of the reason you cannot accept the clearance and request an amended clearance.

8. Establish the appropriate airplane configuration and airspeed considering turbulence and wind shear, and complete your airplane's checklist items appropriate to the phase of the flight.

a. During the initial segment of the IAP, you should slow your airplane to your desired approach speed. This is normally an airspeed within, or just above, flap operating range, from which your airplane can readily transition to a landing configuration.

1) In your airplane, approach speed is _______.

2) Based on known weather conditions, you may want to increase your approach airspeed due to turbulence, gusty winds, and possible wind shear.

3) If appropriate for your airplane, you should lower flaps to the approach setting.

b. During the initial approach segment, you should also complete your before-landing checklist as described in your *POH*.

1) All fuel-related items, such as fuel selectors, fuel pumps, and mixture, should be set for landing.

2) Most pilots will lower the landing gear (if applicable) at the beginning of the final approach segment (i.e., over the FAF).

3) At the FAF, you should have completed your before-landing checklist.

c. You should commit to memory certain important items prior to reaching the FAF:

1) MDA (or step-down minimums)

2) Time from FAF to MAP

3) Visibility minimums

4) Missed approach procedure (at least initial part, e.g., "Climbing right turn to 2,700 ft. . . .")

9. Maintain, prior to beginning the final approach segment, altitude within 100 ft. and heading within 10°; allow less than a full-scale deflection of the CDI; and maintain airspeed within 10 kt.

a. During the LOC, or LOC BC, instrument approach, the CDI needle indicates, by deflection, whether you are to the right or left of the localizer centerline, regardless of the position or heading of your airplane.

1) The following applies when you are using a standard VOR indicator:

a) When you are inbound on the front course or outbound on the back course (e.g., to execute a procedure turn), the needle is deflected toward the localizer course, and you turn toward the needle to correct your track.

b) When you are tracking outbound on the front course (e.g., to execute a procedure turn) or inbound on the back course, you will need to turn away from the direction of needle deflection (reverse sensing) to return to the localizer course.

2) If your airplane is equipped with a horizontal situation indicator (HSI), reverse sens-ing does not occur as long as you always set the localizer front course on the HSI.

a) Once this is done, always turn toward the direction of the needle deflection to return to the course.

b. Because of the narrow localizer course width, overcontrolling of heading corrections is a common error.

1) Unless the CDI shows a full deflection, heading corrections in 5° increments (or less) should keep you close to the centerline.

2) The sooner you establish the correct wind correction angle, the easier it will be to track the localizer without chasing the CDI needle.

3) Remember to fly a heading on your HI and then check the result on the CDI. DO NOT fly by the CDI needle.

c. When reintercepting the localizer after a procedure turn, you will need to cross-check the CDI frequently to prevent overshooting the course centerline.

1) If your interception angle is 45°, start your turn to the inbound course as soon as the CDI moves from the full deflection position.

d. When you are being radar vectored to the localizer course, ATC will normally try to provide you with a 30° intercept angle.

e. During the interception of the localizer course, use other navigation aids, such as the ADF (if available) or GPS, to monitor the position of your airplane relative to the localizer course.

f. Once you have reached the localizer centerline, maintain the inbound heading until the CDI moves off center.

1) Drift corrections should be small and reduced proportionally as the course narrows.

2) By the time you reach the FAF, your wind correction angle should be established.

g. The final approach segment begins at the FAF.

1) At this point, you should also have your airspeed stabilized at the approach airspeed.

2) In the LOC RWY 35 example on page 267, the FAF is CALHOUN NDB, and the minimum altitude at the FAF is 1,800 ft. MSL.

10. Apply the necessary adjustments to the published MDA and visibility criteria for your airplane approach category when required.

a. FDC and Class II NOTAMs

1) Flight Data Center (FDC) NOTAMs are regulatory in nature and inform you of amendments to IAPs or aeronautical charts prior to their normal publication.

a) FDC NOTAMs are available from FSS and are published in the *Notices to Airmen Publication* (*NTAP*).

2) The *NTAP* (formerly Class II NOTAMs) contains all FDC NOTAMs that are current at the time of publication and those NOTAMs (D) that are expected to remain in effect for 7 days after the issuance of the publication. The *NTAP* is issued every 28 days.

a) During your preflight briefing, you must ask your FSS specialist for any published NOTAMs that are pertinent to your flight.

b) Alternatively, you may view the *NTAP* web version at:

www.faa.gov/NTAP/default.htm

b. Inoperative airplane and ground navigation equipment

1) You cannot perform a LOC IAP unless the VOR equipment in your airplane and the desired LOC facility are operating properly.

2) If the IAP requires a DME (i.e., LOC/DME) or a note is on the IAP chart requiring certain navigation system(s), all of these systems must be operational.

c. Inoperative visual aids associated with the landing environment

1) Higher minimums are required with inoperative visual aids.

a) NACO charts will have this information listed in the Inoperative Components Table located on the inside front cover.

b) JEPP charts will have this information listed within the minimums section of the IAP chart.

2) For LOC IAPs, an inoperative visual aid will increase the visibility requirement but not the MDA.

d. National Weather Service (NWS) reporting factors and criteria

1) On NACO IAP charts, this information is included below the minimums section.

a) Some IAP charts will instruct you to use an alternate altimeter setting when the local altimeter setting is not available and to increase all MDAs by a certain amount.

11. Establish a rate of descent and track that will ensure arrival at the MDA prior to reaching the MAP with your airplane continuously in a position from which descent to a landing on the intended runway can be made at a normal rate using normal maneuvers.

a. At the FAF, you can start your descent to the MDA while tracking the localizer course to the MAP.

1) Some recommend that you should establish an expeditious, but safe, descent (i.e., constant airspeed descent) to allow you more time at the MDA and increase your chances of seeing the runway environment.

a) For approach airspeeds of 90 kt., a descent rate of no more than 700 fpm is usually adequate.

b) The sooner you can see the runway, the sooner you can establish a normal landing approach using normal maneuvers.

2) The FAA recommends that when the IAP chart provides a vertical descent angle (VDA), you should calculate a rate of descent at the given VDA and your actual or estimated groundspeed. This will provide a stabilized approach descent.

a) EXAMPLE: For a groundspeed of 90 kt. and a VDA of 3.00°, a rate of descent of approximately 480 fpm should be established for a stabilized approach.

b. You will want to maintain a constant airspeed, whether in level flight, descending to the MDA, or leveling off at the MDA.

   1) To accomplish this, all you will need to do initially is to have the power and trim set for your desired approach airspeed, e.g., 90 kt.

   2) To descend, reduce power to a predetermined setting to establish the rate of descent desired.

      a) As the power is reduced, the nose of the airplane will lower. All you may need to do is ensure that the nose does not pitch down too much.

      b) Since the airplane is trimmed for your approach speed, no change of the trim or elevator control is required.

      c) You will be able to adjust the rate of descent with minor power adjustments.

c. See Task IV.C., Constant Airspeed Climbs and Descents, beginning on page 154.

12. *Allow, while on the final approach segment, no more than a three-quarter-scale deflection of the CDI, and maintain airspeed within 10 kt.*

a. The final approach segment is from the FAF to the MAP.

b. If you established the necessary wind-drift correction to track the localizer course prior to the FAF, you should be able to track the localizer accurately to the MAP with heading corrections no greater than 2°. Remember that

   1) You are to fly a heading; DO NOT chase the CDI needle.

   2) The localizer course width becomes narrower as you get closer to the antenna site. At the runway threshold, the width is 700 ft. (or 350 ft. either side of the center).

c. Maintain the final approach course (i.e., 350° in our example) so that the CDI is no greater than a three-quarter-scale deflection.

d. Throughout the final approach segment, you must maintain your approach airspeed at all times, ±10 kt.

13. *Maintain the MDA, when reached, within +100 ft., –0 ft. to the MAP.*

a. Since you must not descend below the MDA, your author suggests flying at 50 ft. above the MDA to allow for turbulence (e.g., for an MDA of 1,240 ft. MSL, use approximately 1,300 ft. MSL).

   1) If it is a calm day, you should have no problem maintaining the MDA.

b. Using predetermined pitch and power settings, lead your level-off by approximately 100 ft. to allow you to maintain a constant airspeed and smooth control of your airplane.

c. In the example LOC RWY 35 IAP on page 267, the straight-in MDA for all approach categories is 1,240 ft. MSL, and the MDA for a circling approach is 1,340 ft. MSL, except Category D has an MDA of 1,400 ft. MSL.

14. Execute the missed approach procedure when the required visual references for the intended runway are not distinctly visible and identifiable at the MAP.

a. The MAP can be identified by time from the FAF or in some approaches by a DME fix.

1) In the example LOC RWY 35 IAP on page 267, the MAP is timed. When using time to determine the MAP, you must estimate your groundspeed and use the chart below the airport diagram (NACO chart) to determine the time.

a) You can estimate your groundspeed by using the indicated airspeed (or true airspeed for high density altitude airports) and the known wind.

i) For a headwind or tailwind, estimate the groundspeed by subtracting or adding the speed of the wind to the indicated airspeed.

ii) For a quartering wind, use one-half the wind speed.

b. To descend below the MDA (FAR 91.175), you are required to

1) Have the runway environment in sight
2) Have visibility at or above the minimums for your approach category
3) Be in a position to make a normal descent to the intended runway

c. Runway environment is defined as any one of the following visual references required for descent below the MDA:

1) Approach light system

a) However, you are not allowed to descend below 100 ft. above the touchdown zone elevation (TDZE) using the approach lights as a reference unless the red terminating bars or the red side row bars are also distinctly visible and identifiable.

2) The threshold
3) The threshold markings
4) The threshold lights
5) The runway end identifier lights
6) The visual approach slope indicator
7) The touchdown zone or touchdown zone markings
8) The touchdown zone lights
9) The runway or runway markings
10) The runway lights

d. You are required to execute the missed approach procedure when either of the following happens:

1) You cannot identify one of the required visual references at either of the following times:

a) Upon arrival at the MAP
b) At any time that you are below the MDA until touchdown

2) An identifiable part of the airport is not distinctly visible to you during a circling maneuver, at or above MDA, unless this is caused only from a bank of your airplane during the circling approach.

e. The missed approach procedure is written in the profile view of NACO charts.

1) In the example LOC RWY 35 IAP on page 267, the missed approach procedure is a climbing right turn to 2,700 ft., direct to OUK (Calhoun) NDB and hold.

f. Protected obstacle clearance areas for missed approach procedures are predicated on the assumption that the missed approach procedure is initiated at the MAP not lower than the MDA. Reasonable buffers are provided for normal maneuvers.

1) When an early missed approach is executed (i.e., before the MAP), you should, unless otherwise cleared by ATC, fly the IAP as specified on the chart to the MAP at or above the MDA before executing a turning maneuver.

2) If you lose visual reference while circling to land, you must follow the prescribed missed approach procedure.

a) To become established on the missed approach course, you should make an initial climbing turn toward the landing runway and continue the turn until you are established on the missed approach course.

15. Execute a normal landing from a straight-in or circling approach when instructed by your examiner.

a. See Task VI.E., Landing from a Straight-in or Circling Approach, beginning on page 322.

I. Common Errors during a LOC Instrument Approach

1. Failure to have essential knowledge of the information on the LOC instrument approach procedure chart.

a. Know your IAF and the way to arrive at the final approach course.

b. Know how to identify the FAF.

c. Know your minimum altitudes during each segment of the approach, including the MDA.

d. Be able to identify the MAP, and know the missed approach procedure.

2. Incorrect communication procedures or noncompliance with ATC clearances.

a. Always follow correct communication procedures.

b. You are required to follow all ATC clearances. If you cannot comply with a clearance, you must request an amended clearance from ATC.

3. Failure to accomplish checklist items.

a. You must complete the before-landing checklist for your airplane. The checklist is normally begun during the initial segment and completed by the start of the final approach segment.

b. An approach is a busy time in the cockpit. Attempt to complete your checklist as much as possible before the FAF.

c. Set and check your HI. Failure to accomplish this could result in a dangerous track on a LOC approach.

4. Faulty basic instrument flying technique.

a. It is a busy time, but you must continue your cross-check and instrument interpretation throughout the approach.

b. Remember to fly your airplane first, then track your course, and then talk.

5. Failure to understand the localizer course dimensions.

a. Since the VOR receiver is used on the localizer course, the assumption is that interception and tracking techniques are the same as for VOR radials.

b. Remember that the CDI is more sensitive and faster moving on the localizer course.

6. **Chasing the CDI needle.**

   a. Remember to make small corrections on the heading indicator and wait to see the effect on the CDI.

7. **Inappropriate descent below the MDA.**

   a. You can descend below the MDA only when you have the required visual references and the visibility is equal to or better than that published on the IAP chart.

   b. Give yourself a cushion (e.g., add 50 ft.) above the MDA to allow for altitude variations.

J. LDA Instrument Approach Task Objectives

1. **Exhibit adequate knowledge of the elements related to an LDA instrument approach procedure.**

   a. The localizer-type directional aid (LDA) is comparable in use and accuracy to an ILS localizer, but it is not part of a complete ILS.

      1) The major difference is that the LDA is NOT aligned with the runway centerline.

   b. The initial approach fix(es) (IAF) may be an NDB, LOM, DME fix, or intersection.

   c. The final approach fix (FAF) may also be an NDB, LOM, DME fix, or intersection.

      1) The FAF will be identified by a Maltese cross on the profile view of the NACO IAP chart.

   d. The missed approach point (MAP) is normally identified by timing from the FAF.

      1) On some LDA approaches, the MAP may also be identified by a DME fix.

2. **Select and comply with the appropriate LDA instrument approach procedure to be performed.**

   a. Once you know that you will be conducting an LDA approach, you should select the appropriate IAP chart and study it to determine that you can comply with the procedure.

      1) If the inbound course of the approach is within 30° of the centerline of a runway, the approach will be designated to that runway (e.g., LDA RWY 19R).

         a) If the course deviation is more than 30°, the approach will be designated with an alphabetic character (e.g., LDA-A).

   b. We will use the LDA RWY 19R IAP to Concord/Buchanan Field (KCCR) on page 277 as an example throughout our discussion.

   c. Comprising the top half of the chart, the **plan view** shows a bird's-eye view of the approach.

   d. You will notice that the LDA RWY 19R approach at KCCR has two IAFs. One IAF is at REJOY Intersection (Int.), and the second IAF is at KANAN LOM (NDB).

      1) If you are routed to REJOY Int. by ATC and then cleared for the LDA RWY 19R approach, you would proceed inbound on the R-074 of Scaggs Island VORTAC (SGD), or a heading of 254°.

         a) At REJOY Int., above the course is "2000 NoPT," which means that there is no procedure turn and the minimum altitude until joining the LDA is 2,000 ft. MSL.

            i) The ten (10) in parentheses means that REJOY Int. is 10 NM from the LDA course.

b) You are now on the initial approach segment.

c) Since you have been cleared for the approach and you are on a published route, you may descend to 2,000 ft. MSL, unless ATC has informed you to maintain a higher altitude.

d) Since you are approaching the LDA course at a 73° intercept angle, you will need to monitor the CDI movement to determine when to make your turn inbound on the LDA course of 181°.

i) During your training, your instructor should give you practice in intercepting courses using large intercept angles.

2) If you are cleared by ATC to "proceed direct to KANAN LOM (IAF), cleared for the LDA RWY 19R approach," you will turn outbound upon reaching KANAN, on the LDA course of 001°.

a) This is the procedure turn outbound.

b) You are now on the initial approach segment.

c) Since you have been cleared for the approach and you are established in the published procedure turn, you may descend to 2,500 ft. MSL, the published altitude for the procedure turn outbound, shown in the profile view just below the plan view.

d) After flying outbound for approximately 1 to 2 min., you will turn right 45° to a heading of 046° and fly that heading for 1 min. before turning left 180° to a heading of 226° to intercept the LDA course of 181° inbound to the LOM.

e. When you have intercepted and are tracking the inbound course, you are on the intermediate approach segment.

f. After passing the FAF, shown by the Maltese cross in the profile view (i.e., KANAN LOM), you will be on the final approach segment.

1) You should begin timing your approach to the MAP. Times are based on groundspeed and are shown below the airport diagram.

a) In the example, using a groundspeed of 90 kt., the MAP is 2 min. 16 sec. from the FAF.

2) You may start your descent to the published MDA for your approach category, as shown in the **minimums section** just below the profile view.

a) If you do not have DME, the MDA for a straight-in approach is 440 ft. MSL and for a circling approach is 580 ft. MSL.

b) If your airplane is DME equipped, you should tune the DME to Concord VOR/DME (CCR).

i) If you are conducting a straight-in approach, at 2.0 DME you can descend to the DME minimums of 380 ft. MSL.

- DME does not affect the circling MDA.

ii) The MAP can be identified at 3.1 DME.

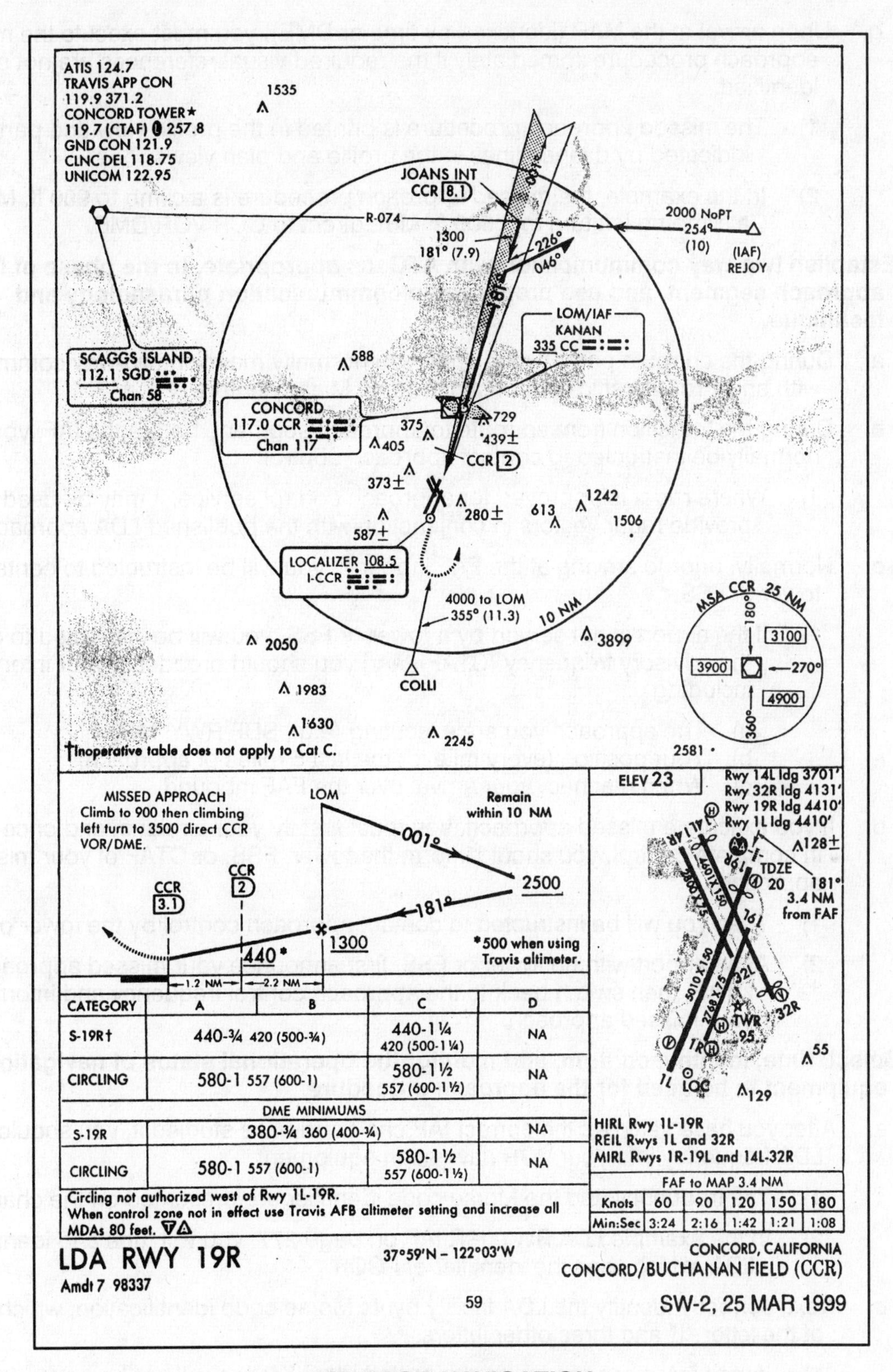

CATEGORY	A	B	C	D
S-19R†	440-¾ 420 (500-¾)		440-1¼ 420 (500-1¼)	NA
CIRCLING	580-1 557 (600-1)		580-1½ 557 (600-1½)	NA
DME MINIMUMS				
S-19R	380-¾ 360 (400-¾)			NA
CIRCLING	580-1 557 (600-1)		580-1½ 557 (600-1½)	NA

FAF to MAP 3.4 NM

Knots	60	90	120	150	180
Min:Sec	3:24	2:16	1:42	1:21	1:08

NOTE FOR NAVIGATION

NOTE: This IAP chart is shown in the old format, which is still in use. Charts using the new format are shown on pages 255 and 298.

g. Upon arrival at the MAP (identified by time or DME), you must execute the missed approach procedure immediately if the required visual references are not seen and identified.

   1) The missed approach procedure is printed in the profile view and partially indicated by dashed lines in the profile and plan views.

   2) In the example, the missed approach procedure is a climb to 900 ft. MSL, then a climbing left turn to 3,500 ft. MSL direct to CCR VOR/DME.

3. Establish two-way communication with ATC, as appropriate, to the phase of flight or approach segment, and use proper radio communication phraseology and technique.

a. During the en route part of your flight, you normally maintain two-way communication with an air route traffic control center (e.g., Miami center).

b. During the transition from en route to approach segment, before the IAF, you will normally be instructed to contact approach control.

   1) Where radar is approved for approach control service, it may be used to provide radar vectors in conjunction with the published LDA approach.

c. Normally, prior to arriving at the FAF inbound, you will be instructed to contact the tower or FSS.

   1) If the airport is not served by a tower or FSS, you will be instructed to change to the advisory frequency (CTAF), and you should broadcast your intentions, including

      a) The approach you are executing (e.g., SDF RWY 19R)
      b) Your position (every mile for the last 5 miles of approach)
      c) When reached, your arrival over the FAF inbound

d. If you execute a missed approach, you must first fly your airplane, and once you are in complete control, you should inform the tower, FSS, or CTAF of your missed approach.

   1) Then you will be instructed to contact approach control by the tower or FSS.

   2) At an airport with no tower or FSS, first announce your missed approach on CTAF; then switch back to the approach control frequency and inform ATC of your missed approach.

4. Select, tune, identify, confirm, and monitor the operational status of navigation equipment to be used for the approach procedure.

a. After you have selected the correct IAP chart and have studied it, you should tune the LDA frequency into your VOR navigation equipment.

   1) The frequency and the Morse code identifier are presented on the chart.

   2) In the example LDA RWY 19R IAP on page 277, you will tune and identify the LDA, which uses the identifier of I-CCR.

b. Next, correctly identify the LDA facility by its Morse code identification, which consists of the letter "I" and three other letters.

   1) If the LDA is out of service, the coded identification will not be transmitted.

c. Operational checks that you can perform in your airplane include seeing that

   1) The VOR alarm flag is not on.
   2) The CDI has no erratic movements, unless at station passage.

d. Check each NAVAID that is to be used during approach and missed approach.

5. **Comply with all clearances issued by ATC or your examiner.**
   a. When ATC (or your examiner) issues a clearance or an instruction, you are expected to execute its provisions upon receipt.
      1) You must not deviate from the provisions of any clearance or instructions unless an amended clearance is obtained or an emergency arises.
   b. At times ATC may not specify a particular approach procedure in the approach clearance but will state "cleared for approach." Such a clearance indicates that you may execute any one of the authorized IAPs for that airport.
      1) This clearance does not constitute approval for you to execute a contact approach or a visual approach.
   c. When cleared for a specifically prescribed IAP (i.e., "cleared LDA RWY 19R approach") or when "cleared for approach," you are required to execute the entire approach as described on the IAP chart unless an appropriate new or revised ATC clearance is received or you cancel your IFR flight plan.
6. **Recognize when heading indicator (HI) and/or attitude indicator (AI) is inaccurate or inoperative, advise controller, and proceed with approach.**
   a. For information on recognizing whether the HI and/or AI is inaccurate or inoperative, see Task VII.B., Loss of Gyro Attitude and/or Heading Indicators, beginning on page 332.
   b. You should report the malfunction of the instruments to ATC and advise that you will proceed with the approach.
7. **Advise ATC (or your examiner) anytime your airplane is unable to comply with a clearance.**
   a. You must inform ATC (or your examiner) anytime your airplane's operating limitations forbid compliance with the clearance issued.
   b. Before you accept a clearance, you must determine whether you can comply with it. If not, inform ATC of the reason you cannot accept the clearance and request an amended clearance.
8. **Establish the appropriate airplane configuration and airspeed considering turbulence and wind shear, and complete your airplane's checklist items appropriate to the phase of the flight.**
   a. During the initial segment of the IAP, you should slow your airplane to your desired approach speed. This is normally an airspeed within, or just above, flap operating range, from which your airplane can readily transition to a landing configuration.
      1) In your airplane, approach speed is ________.
      2) Based on known weather conditions, you may want to increase your approach airspeed due to turbulence, gusty winds, and possible wind shear.
      3) If appropriate for your airplane, you should lower flaps to the approach setting.
   b. During the initial approach segment, you should also complete your before-landing checklist as described in your *POH*.
      1) All fuel-related items, such as fuel selectors, fuel pumps, and mixture, should be set for landing.
      2) Most pilots will lower the landing gear (if applicable) at the beginning of the final approach segment (i.e., over the FAF).
      3) At the FAF, you should have completed your before-landing checklist.

c. You should commit to memory certain important items prior to reaching the FAF:

1) MDA (or step-down minimums)
2) Time from FAF to MAP (or DME indication at the MAP)
3) Visibility minimums
4) Missed approach procedure (at least initial part, e.g., "Climb to 900 ft. . . .")

9. *Maintain, prior to beginning the final approach segment, altitude within 100 ft. and heading within 10°; allow less than a full-scale deflection of the CDI; and maintain airspeed within 10 kt.*

a. Remember that the final approach segment begins at the FAF or at a point where you begin your descent to the MDA.

1) By this time, you should be established on the final approach course having determined the necessary wind-drift corrections to maintain the desired bearing.

2) You should have your airspeed stabilized for the approach airspeed.

b. The LDA RWY 19R example on page 277 shows that before the FAF you must be no lower than 1,300 ft. MSL.

1) The FAF is KANAN LOM (NDB).

10. Apply the necessary adjustments to the published MDA and visibility criteria for your airplane approach category when required.

a. FDC and Class II NOTAMs

1) Flight Data Center (FDC) NOTAMs are regulatory in nature and inform you of amendments to IAPs or aeronautical charts prior to their normal publication.

a) FDC NOTAMs are available from FSS and are published in the *Notices to Airmen Publication* (*NTAP*).

2) The *NTAP* (formerly Class II NOTAMs) contains all FDC NOTAMs that are current at the time of publication and those NOTAMs (D) that are expected to remain in effect for 7 days after the issuance of the publication. The *NTAP* is issued every 28 days.

a) During your preflight briefing, you must ask your FSS specialist for any published NOTAMs that are pertinent to your flight.

b) Alternatively, you may view the NTAP web version at

www.faa.gov/NTAP/default.htm

b. Inoperative airplane and ground navigation equipment

1) You cannot perform an LDA IAP unless the VOR equipment in your airplane and the desired LDA facility are operating properly.

c. Inoperative visual aids associated with the landing environment

1) Higher minimums are required with inoperative visual aids.

a) NACO charts will have this information listed in the Inoperative Components Table located on the inside front cover.

b) JEPP charts will have this information listed within the minimums section of the IAP chart.

2) For LDA IAPs, an inoperative visual aid will increase the visibility requirement but not the MDA.

d. National Weather Service (NWS) reporting factors and criteria

1) On NACO IAP charts, this information is included below the minimums section.

a) Some IAP charts will instruct you to use an alternate altimeter setting when the local altimeter setting is not available and to increase all MDAs by a certain amount.

2) On the LDA RWY 19R IAP on page 277, the note section (below the minimums section) informs you that, if the tower is closed, you must use the Travis AFB altimeter setting and increase all MDAs 80 ft.

11. Establish a rate of descent and track that will ensure arrival at the MDA prior to reaching the MAP with your airplane continuously in a position from which descent to a landing on the intended runway can be made at a normal rate using normal maneuvers.

a. At the FAF, you can start your descent to the MDA while tracking the desired bearing to the MAP.

1) Some recommend that you should establish an expeditious, but safe, descent (i.e., constant airspeed descent) to allow you more time at the MDA and increase your chances of seeing the runway environment.

a) For approach airspeeds of 90 kt., a descent rate of no more than 700 fpm is usually adequate.

b) The sooner you can see the runway, the sooner you can establish a normal landing approach using normal maneuvers.

2) The FAA recommends that when the IAP chart provides a vertical descent angle (VDA), you should calculate a rate of descent at the given VDA and your actual or estimated groundspeed. This will provide a stabilized approach descent.

a) EXAMPLE: For a groundspeed of 90 kt. and a VDA of 3.00°, a rate of descent of approximately 480 fpm should be established for a stabilized approach.

b. You will want to maintain a constant airspeed, whether in level flight, descending to the MDA, or leveling off at the MDA.

1) To accomplish this, all you will need to do initially is to have the power and trim set for your desired approach airspeed, e.g., 90 kt.

2) To descend, reduce power to a predetermined setting to establish the rate of descent desired.

a) As the power is reduced, the nose of the airplane will lower. All you may need to do is ensure that the nose does not pitch down too much.

b) Since the airplane is trimmed for your approach speed, no change of the trim or elevator control is required.

c) You will be able to adjust the rate of descent with minor power adjustments.

c. See Task IV.C., Constant Airspeed Climbs and Descents, beginning on page 154.

12. Allow, while on the final approach segment, no more than a three-quarter-scale deflection of the CDI, and maintain airspeed within 10 kt.

a. The final approach segment is from the FAF to the MAP.

b. Maintain the final approach course (i.e., 181 in our example) so that the CDI is no greater than the third dot (i.e., three-quarter-scale deflection).

1) If you are using an RMI, it should indicate that you are on the desired radial, ±10°.

c. Throughout the final approach segment, you must maintain your approach airspeed at all times, ±10 kt.

13. *Maintain the MDA, when reached, within +100 ft., –0 ft. to the MAP.*

a. Since you must not descend below the MDA, your author suggests flying at 50 ft. above the MDA to allow for turbulence (e.g., for an MDA of 700 ft. MSL, use 750 ft. MSL).

   1) If it is a calm day, you should have no problem maintaining the MDA.

b. Using predetermined pitch and power settings, lead your level-off by approximately 100 ft. to allow you to maintain a constant airspeed and smooth control of your airplane.

c. In the example LDA RWY 19R on page 277, MDAs are established for "without DME" and "with DME." Also note that Category D aircraft are not authorized (NA).

   1) The MDA without DME is

      a) 440 ft. MSL for a straight-in approach
      b) 580 ft. MSL for a circling approach

   2) The MDA with DME is

      a) 380 ft. MSL for a straight-in approach
      b) 580 ft. MSL for a circling approach

14. Execute the missed approach procedure when the required visual references for the intended runway are not distinctly visible and identifiable at the MAP.

a. The MAP can be identified by time from the FAF, or it may be determined by a DME fix.

   1) In the example LDA RWY 19R IAP on page 277, the MAP is determined by time or DME.

      a) When using time to determine the MAP, you must estimate your groundspeed and use the chart below the airport diagram (NACO chart) to determine the time.

         i) You can estimate your groundspeed by using the indicated airspeed (or true airspeed for high density altitude airports) and the known wind.

            - For a headwind or tailwind, estimate the groundspeed by subtracting or adding the speed of the wind to the indicated airspeed.
            - For a quartering wind, use one-half the wind speed.

      b) The MAP is also identified by 3.1 DME from CCR VOR/DME.

b. To descend below the MDA (FAR 91.175), you are required to

   1) Have the runway environment in sight
   2) Have visibility at or above the minimums for your approach category
   3) Be in a position to make a normal descent to the intended runway

c. Runway environment is defined as any one of the following visual references required for descent below the MDA:
   1) Approach light system
      a) However, you are not allowed to descend below 100 ft. above the touchdown zone elevation (TDZE) using the approach lights as a reference unless the red terminating bars or the red side row bars are also distinctly visible and identifiable.
   2) The threshold
   3) The threshold markings
   4) The threshold lights
   5) The runway end identifier lights
   6) The visual approach slope indicator
   7) The touchdown zone or touchdown zone markings
   8) The touchdown zone lights
   9) The runway or runway markings
   10) The runway lights

d. You are required to execute the missed approach procedure when either of the following happens:
   1) You cannot identify one of the required visual references at either of the following times:
      a) Upon arrival at the MAP
      b) At any time that you are below the MDA until touchdown
   2) An identifiable part of the airport is not distinctly visible to you during a circling maneuver, at or above MDA, unless this is caused only from a bank of your airplane during the circling approach.

e. The missed approach procedure is written in the profile view of NACO charts.
   1) In the example LDA RWY 19R IAP on page 277, the missed approach procedure is a climb to 900 ft., then a climbing left turn to 3,500 ft., proceeding direct to CCR VOR/DME.

f. Protected obstacle clearance areas for missed approach procedures are predicated on the assumption that the missed approach procedure is initiated at the MAP not lower than the MDA. Reasonable buffers are provided for normal maneuvers.
   1) When an early missed approach is executed (i.e., before the MAP), you should, unless otherwise cleared by ATC, fly the IAP as specified on the chart to the MAP at or above the MDA before executing a turning maneuver.
   2) If you lose visual reference while circling to land, you must follow the prescribed missed approach procedure.
      a) To become established on the missed approach course, you should make an initial climbing turn toward the landing runway and continue the turn until you are established on the missed approach course.

15. Execute a normal landing from a straight-in or circling approach when instructed by your examiner.

a. See Task VI.E., Landing from a Straight-in or Circling Approach, beginning on page 322.

K. Common Errors during an LDA Instrument Approach

1. **Failure to have essential knowledge of the information on the LDA instrument approach procedure chart.**
   a. Know your IAF and the way to arrive at the final approach course.
   b. Know how to identify the FAF.
   c. Know your minimum altitudes during each segment of the approach, including the MDA.
   d. Be able to identify the MAP, and know the missed approach procedure.
2. **Incorrect communication procedures or noncompliance with ATC clearances.**
   a. Always follow correct communication procedures.
   b. You are required to follow all ATC clearances. If you cannot comply with a clearance, you must request an amended clearance from ATC.
3. **Failure to accomplish checklist items.**
   a. You must complete the before-landing checklist for your airplane. The checklist is normally begun during the initial segment and completed by the start of the final approach segment.
   b. An approach is a busy time in the cockpit. Attempt to complete your checklist as much as possible before the FAF.
   c. Set and check your HI. Failure to accomplish this could result in a dangerous track on an LDA approach.
4. **Faulty basic instrument flying technique.**
   a. It is a busy time, but you must continue your cross-check and instrument interpretation throughout the approach.
   b. Flying an approach is nothing more than basic attitude instrument flying.
   c. Remember to fly your airplane first, then track your course, and then talk.
5. **Failure to understand the LDA course dimensions.**
   a. Since the VOR receiver is used on the LDA course, the assumption is that interception and tracking techniques are the same as for VOR radials.
   b. Remember that the CDI is more sensitive and faster moving on the LDA course.
6. **Chasing the CDI needle.**
   a. Remember to make small corrections on the heading indicator and wait to see the effect on the CDI.
7. **Inappropriate descent below the MDA.**
   a. You can descend below the MDA only when you have the required visual references and the visibility is equal to or better than that published on the IAP chart.
   b. Give yourself a cushion (e.g., add 50 ft.) above the MDA to allow for altitude variations.

L. SDF Instrument Approach Task Objectives

1. **Exhibit adequate knowledge of the elements related to an SDF instrument approach procedure.**
   a. The simplified directional facility (SDF) instrument approach procedure is a nonprecision approach that provides final approach course guidance similar to that of an ILS localizer.
      1) An SDF instrument approach does not provide glide slope information.
   b. The approach techniques and procedures used in an SDF instrument approach are essentially the same as those employed in executing a standard localizer approach except the SDF course may not be aligned with the runway and the course may be wider, resulting in less precision.
      1) The SDF antenna may be offset from the runway centerline. Because of this, the angle of convergence between the final approach course and the runway bearing should be determined by reference to the IAP chart.
         a) The angle is generally no more than 3°.
         b) Remember that the approach course originates at the antenna site and an approach continued beyond the runway threshold will lead you to the SDF antenna site rather than along the runway centerline.
      2) The SDF signal is fixed at either 6° or 12° width, as necessary to provide maximum flyability and optimum course quality.
      3) Usable off-course indications are limited to 35° either side of the course centerline.
         a) Instrument indications received beyond 35° should be disregarded.
   c. The initial approach fix(es) (IAF) may be an NDB, a VOR, or an intersection.
   d. The final approach fix (FAF) may be an NDB, a VOR, or an intersection.
      1) The FAF will be identified by a Maltese cross on the profile view of the NACO IAP chart.
   e. The missed approach point (MAP) is normally identified by timing from the FAF.
      1) On some SDF approaches, the MAP may also be identified by a DME fix.
2. **Select and comply with the appropriate SDF instrument approach procedure to be performed.**
   a. Once you know that you will be conducting an SDF approach, you should select the appropriate IAP chart and study it to determine that you can comply with the procedure.
      1) An SDF approach will be designated to a specific runway (e.g., SDF RWY 36).
   b. We will use the SDF RWY 36 IAP to Poplar Bluff Municipal (KPOF) on page 287 as an example throughout our discussion.
   c. Comprising the top half of the chart, the **plan view** shows a bird's-eye view of the approach.

d. You will notice that the SDF RWY 36 approach at KPOF has two IAFs. One IAF is at Malden (MAW) VORTAC, and the second IAF is at Earli NDB/Earli Intersection (Int.).

1) If you are cleared by ATC to "proceed direct to Malden VORTAC, cleared for the SDF RWY 36 approach," you will turn to a heading of 269° and track outbound on the MAW R-269.

a) The note below the Malden VORTAC states, "2100 NoPT to NDB/Int 269° (20.2) and 358° (6.1)."

i) This note means that no procedure turn should be performed and 2,100 ft. MSL should be maintained to Earli NDB/Int. You will track 269° for 20.2 NM to intercept the SDF course. From the SDF course intercept to the FAF is 6.1 NM.

b) You are now on the initial approach segment.

c) Since you have been cleared for the approach and you are on a published route, you may descend to 2,100 ft. MSL, unless ATC informed you to maintain a higher altitude.

d) Since you are approaching the SDF course at an 89° intercept angle, you will need to monitor the CDI movement to determine when to make your turn to track the SDF course.

i) During your training, your instructor should give you practice in intercepting courses with a 90° intercept angle.

2) If you are cleared by ATC to "proceed direct to Earli NDB/Int., cleared for the SDF RWY 36 approach," you will turn outbound upon reaching Earli, on the SDF course heading 178°.

a) This is the procedure turn outbound.

b) You are now on the initial approach segment.

c) Since you have been cleared for the approach and you are established in the published procedure turn, you may descend to 2,100 ft. MSL, the published altitude for the procedure turn outbound, shown in the **profile view** just below the plan view.

d) After flying outbound for approximately 1 to 2 min., you will turn left 45° to a heading of 133° and fly that heading for 1 min. before turning right 180° to a heading of 313° to intercept the SDF course of 358° inbound to the FAF.

e. When you have intercepted and are tracking the inbound course, you are on the intermediate approach segment.

f. After passing the FAF, shown by the Maltese cross in the profile view (i.e., EARLI NDB/INT), you will be on the final approach segment.

1) You should begin timing your approach to the MAP. Times are based on groundspeed and are shown below the airport diagram.

a) In the example, using a groundspeed of 90 kt., the MAP is 3 min. 56 sec. from the FAF.

2) You may start your descent to the published MDA (since there are no step-down fixes) for your approach category, as shown in the **minimums section** just below the profile view.

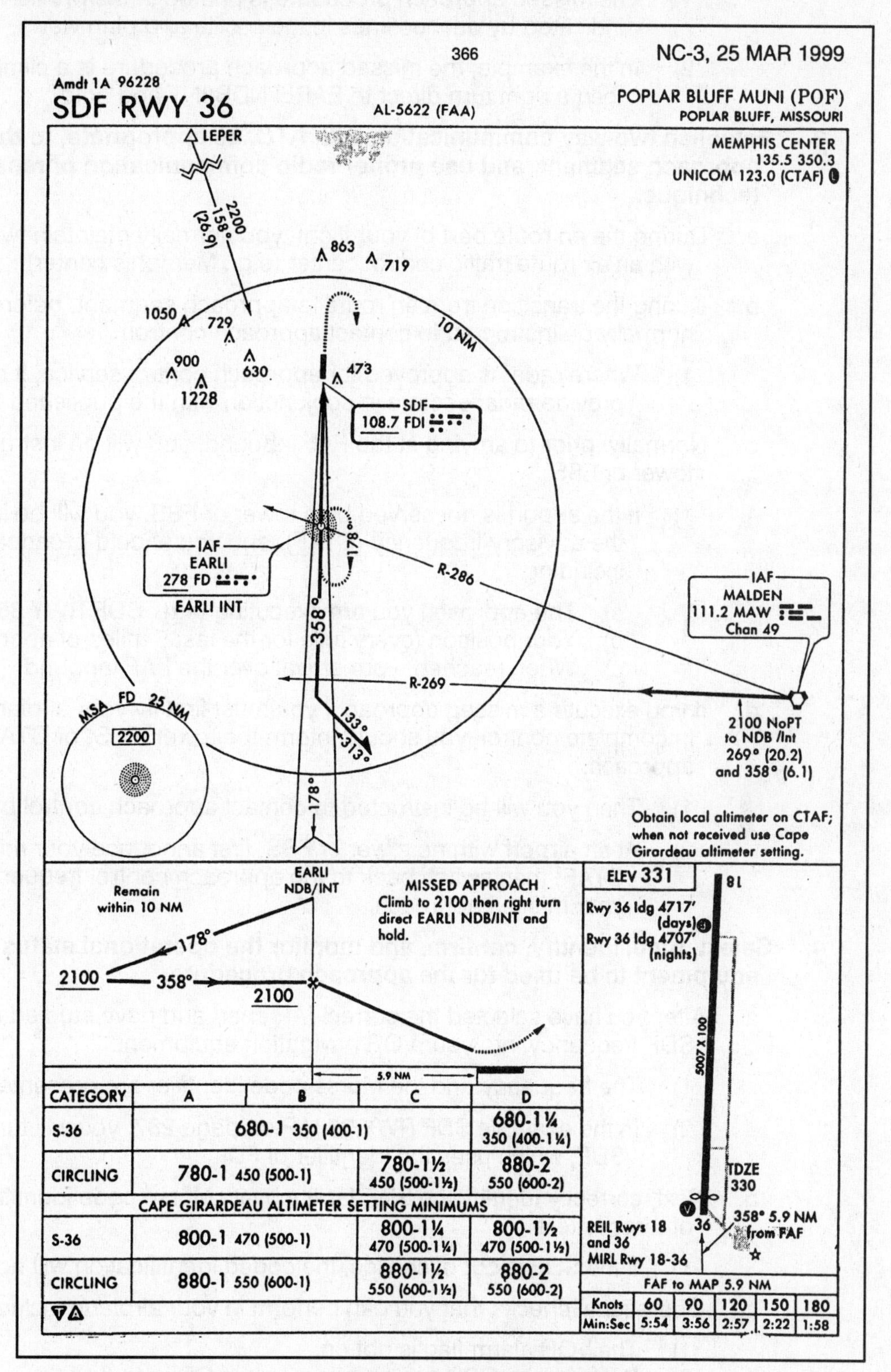

NOT FOR NAVIGATION

NOTE: This IAP chart is shown in the old format, which is still in use. Charts using the new format are shown on pages 255 and 298.

g. Upon arrival at the MAP (identified by time), you must execute the missed approach procedure immediately if the required visual references are not seen and identified.

  1) The missed approach procedure is printed in the profile view and partially indicated by dashed lines in the profile and plan views.

  2) In the example, the missed approach procedure is a climb to 2,100 ft. MSL, then a right turn direct to EARLI NDB/INT and hold.

3. Establish two-way communication with ATC, as appropriate, to the phase of flight or approach segment, and use proper radio communication phraseology and technique.

a. During the en route part of your flight, you normally maintain two-way communication with an air route traffic control center (e.g., Memphis center).

b. During the transition from en route to approach segment, before the IAF, you will normally be instructed to contact approach control.

  1) Where radar is approved for approach control service, it may be used to provide radar vectors in conjunction with the published SDF approach.

c. Normally, prior to arriving at the FAF inbound, you will be instructed to contact the tower or FSS.

  1) If the airport is not served by a tower or FSS, you will be instructed to change to the advisory frequency (CTAF), and you should broadcast your intentions, including

    a) The approach you are executing (e.g., SDF RWY 36)
    b) Your position (every mile for the last 5 miles of approach)
    c) When reached, your arrival over the FAF inbound

d. If you execute a missed approach, you must first fly your airplane, and once you are in complete control, you should inform the tower, FSS, or CTAF of your missed approach.

  1) Then you will be instructed to contact approach control by the tower or FSS.

  2) At an airport with no tower or FSS, first announce your missed approach on CTAF; then switch back to the approach control frequency and inform ATC of your missed approach.

4. Select, tune, identify, confirm, and monitor the operational status of navigation equipment to be used for the approach procedure.

a. After you have selected the correct IAP chart and have studied it, you should tune the SDF frequency into your VOR navigation equipment.

  1) The frequency and the Morse code identifier are presented on the chart.

  2) In the example SDF RWY 36 IAP on page 287, you will tune and identify the SDF, which uses the identifier of FDI.

b. Next, correctly identify the SDF facility by its Morse code identification, which consists of three letters.

  1) If the SDF is out of service, the coded identification will not be transmitted.

c. Operational checks that you can perform in your airplane include seeing that

  1) The VOR alarm flag is not on.
  2) Rotating the OBS has no effect on the CDI needle.

d. Check each NAVAID that is to be used during approach and missed approach.

5. **Comply with all clearances issued by ATC or your examiner.**
   a. When ATC (or your examiner) issues a clearance or an instruction, you are expected to execute its provisions upon receipt.
      1) You must not deviate from the provisions of any clearance or instructions unless an amended clearance is obtained or an emergency arises.
   b. At times ATC may not specify a particular approach procedure in the approach clearance but will state "cleared for approach." Such a clearance indicates that you may execute any one of the authorized IAPs for that airport.
      1) This clearance does not constitute approval for you to execute a contact approach or a visual approach.
   c. When cleared for a specifically prescribed IAP (i.e., "cleared SDF RWY 36 approach") or when "cleared for approach," you are required to execute the entire approach as described on the IAP chart unless an appropriate new or revised ATC clearance is received or you cancel your IFR flight plan.
6. **Recognize when heading indicator (HI) and/or attitude indicator (AI) is inaccurate or inoperative, advise controller, and proceed with approach.**
   a. For information on recognizing whether the HI and/or AI is inaccurate or inoperative, see Task VII.B., Loss of Gyro Attitude and/or Heading Indicators, beginning on page 332.
   b. You should report the malfunction of the instruments to ATC and advise that you will proceed with the approach.
7. **Advise ATC (or your examiner) anytime your airplane is unable to comply with a clearance.**
   a. You must inform ATC (or your examiner) anytime your airplane's operating limitations forbid compliance with the clearance issued.
   b. Before you accept a clearance, you must determine whether you can comply with it. If not, inform ATC of the reason you cannot accept the clearance and request an amended clearance.
8. **Establish the appropriate airplane configuration and airspeed considering turbulence and wind shear, and complete your airplane's checklist items appropriate to the phase of the flight.**
   a. During the initial segment of the IAP, you should slow your airplane to your desired approach speed. This is normally an airspeed within, or just above, flap operating range, from which your airplane can readily transition to a landing configuration.
      1) Most light, single-engine airplanes use 90 kt. indicated airspeed as the approach airspeed.
      2) In your airplane, approach speed is ______.
      3) Based on known weather conditions, you may want to increase your approach airspeed due to turbulence, gusty winds, and possible wind shear.
      4) If appropriate for your airplane, you should lower flaps to the approach setting.
   b. During the initial approach segment, you should also complete your before-landing checklist as described in your *POH*.
      1) All fuel-related items, such as fuel selectors, fuel pumps, and mixture, should be set for landing.

2) Most pilots will lower the landing gear (if applicable) at the beginning of the final approach segment (i.e., over the FAF).

3) At the FAF, you should have completed your before-landing checklist.

c. You should commit to memory certain important items prior to reaching the FAF:

1) MDA (or step-down minimums)
2) Time from FAF to MAP (or DME indication at the MAP)
3) Visibility minimums
4) Missed approach procedure (at least initial part, e.g., "Climb to 2,100 ft. . . .")

9. *Maintain, prior to beginning the final approach segment, altitude within 100 ft. and heading within 10°; allow less than a full-scale deflection of the CDI; and maintain airspeed within 10 kt.*

a. Remember that the final approach segment begins at the FAF or at a point where you begin your descent to the MDA.

1) By this time, you should be established on the final approach course having determined the necessary wind-drift corrections to maintain the desired bearing.

2) You should have your airspeed stabilized for the approach airspeed.

b. The SDF RWY 36 example on page 287 shows that before the FAF you must be no lower than 2,100 ft. MSL.

1) The FAF is EARLI NDB/INT.

10. Apply the necessary adjustments to the published MDA and visibility criteria for your airplane approach category when required.

a. FDC and Class II NOTAMs

1) Flight Data Center (FDC) NOTAMs are regulatory in nature and inform you of amendments to IAPs or aeronautical charts prior to their normal publication.

a) FDC NOTAMs are available from FSS and are published in the *Notices to Airmen Publication* (*NTAP*).

2) The *NTAP* (formerly Class II NOTAMs) contains all FDC NOTAMs that are current at the time of publication and those NOTAMs (D) that are expected to remain in effect for 7 days after the issuance of the publication. The *NTAP* is issued every 28 days.

a) During your preflight briefing, you must ask your FSS specialist for any published NOTAMs that are pertinent to your flight.

b) Alternatively, you may view the NTAP web version at:

www.faa.gov/NTAP/default.htm

b. Inoperative airplane and ground navigation equipment

1) You cannot perform an SDF IAP unless the VOR equipment in your airplane and the desired SDF facility are operating properly.

c. Inoperative visual aids associated with the landing environment

1) Higher minimums are required with inoperative visual aids.

a) NACO charts will have this information listed in the Inoperative Components Table located on the inside front cover.

b) JEPP charts will have this information listed within the minimums section of the IAP chart.

2) For SDF IAPs, an inoperative visual aid will increase the visibility requirement but not the MDA.

d. **National Weather Service (NWS) reporting factors and criteria**

1) On NACO IAP charts, this information is included below the minimums section.

a) Some IAP charts will instruct you to use an alternate altimeter setting when the local altimeter setting is not available and to increase all MDAs by a certain amount.

2) In the minimums section of the SDF RWY 36 IAP on page 287, you will notice minimums applicable if you cannot obtain the Poplar Bluff altimeter and must use the Cape Girardeau altimeter setting.

11. **Establish a rate of descent and track that will ensure arrival at the MDA prior to reaching the MAP with your airplane continuously in a position from which descent to a landing on the intended runway can be made at a normal rate using normal maneuvers.**

a. At the FAF, you can start your descent to the MDA while tracking the SDF course to the MAP.

1) Some recommend that you should establish an expeditious, but safe, descent (i.e., constant airspeed descent) to allow you more time at the MDA and increase your chances of seeing the runway environment.

a) For approach airspeeds of 90 kt., a descent rate of no more than 700 fpm is usually adequate.

b) The sooner you can see the runway, the sooner you can establish a normal landing approach using normal maneuvers.

2) The FAA recommends that when the IAP chart provides a vertical descent angle (VDA), you should calculate a rate of descent at the given VDA and your actual or estimated groundspeed. This will provide a stabilized approach descent.

a) EXAMPLE: For a groundspeed of 90 kt. and a VDA of 3.00°, a rate of descent of approximately 480 fpm should be established for a stabilized approach.

b. You will want to maintain a constant airspeed, whether in level flight, descending to the MDA, or leveling off at the MDA.

1) To accomplish this, all you will need to do initially is to have the power and trim set for your desired approach airspeed, e.g., 90 kt.

2) To descend, reduce power to a predetermined setting to establish the rate of descent desired.

a) As the power is reduced, the nose of the airplane will lower. All you may need to do is ensure that the nose does not pitch down too much.

b) Since the airplane is trimmed for your approach speed, no change of the trim or elevator control is required.

c) You will be able to adjust the rate of descent with minor power adjustments.

c. See Task IV.C., Constant Airspeed Climbs and Descents, beginning on page 154.

12. ***Allow, while on the final approach segment, no more than a three-quarter-scale deflection of the CDI, and maintain airspeed within 10 kt.***

a. The final approach segment is from the FAF to the MAP.

b. Maintain the final approach course (i.e., 358° in our example) so that the CDI is no greater than the third dot (i.e., three-quarter-scale deflection).

c. Throughout the final approach segment, you must maintain your approach airspeed at all times, ±10 kt.

13. *Maintain the MDA, when reached, within +100 ft., –0 ft. to the MAP.*

a. Since you must not descend below the MDA, your author suggests flying at 50 ft. above the MDA to allow for turbulence (e.g., for an MDA of 700 ft. MSL, use 750 ft. MSL).

   1) If it is a calm day, you should have no problem maintaining the MDA.

b. Using predetermined pitch and power settings, lead your level-off by approximately 100 ft. to allow you to maintain a constant airspeed and smooth control of your airplane.

c. In the example SDF RWY 36 IAP on page 287, the straight-in MDA (using local altimeter) for all approach categories is 680 ft. MSL, and the MDA for a circling approach is 780 ft. MSL, except Category D has an MDA of 880 ft. MSL.

14. Execute the missed approach procedure when the required visual references for the intended runway are not distinctly visible and identifiable at the MAP.

a. The MAP can be identified by time from the FAF or, in some approaches, by a DME fix.

   1) In the example SDF RWY 36 IAP on page 287, the MAP is timed. When using time to determine the MAP, you must estimate your groundspeed and use the chart below the airport diagram (NACO chart) to determine the time.

      a) You can estimate your groundspeed by using the indicated airspeed (or true airspeed for high density altitude airports) and the known wind.

         i) For a headwind or tailwind, estimate the groundspeed by subtracting or adding the speed of the wind to the indicated airspeed.

         ii) For a quartering wind, use one-half the wind speed.

b. To descend below the MDA (FAR 91.175), you are required to

   1) Have the runway environment in sight
   2) Have visibility at or above the minimums for your approach category
   3) Be in a position to make a normal descent to the intended runway

c. Runway environment is defined as any one of the following visual references required for descent below the MDA:

   1) Approach light system

      a) However, you are not allowed to descend below 100 ft. above the touchdown zone elevation (TDZE) using the approach lights as a reference unless the red terminating bars or the red side row bars are also distinctly visible and identifiable.

   2) The threshold
   3) The threshold markings
   4) The threshold lights
   5) The runway end identifier lights
   6) The visual approach slope indicator
   7) The touchdown zone or touchdown zone markings
   8) The touchdown zone lights
   9) The runway or runway markings
   10) The runway lights

d. You are required to execute the missed approach procedure when either of the following happens:

1) You cannot identify one of the required visual references at either of the following times:

a) Upon arrival at the MAP
b) At any time that you are below the MDA until touchdown

2) An identifiable part of the airport is not distinctly visible to you during a circling maneuver, at or above MDA, unless this is caused only from a bank of your airplane during the circling approach.

e. The missed approach procedure is written in the profile view of NACO charts.

1) In the example SDF RWY 36 IAP on page 287, the missed approach procedure is a climb to 2,100 ft., then a right turn direct to EARLI NDB/INT and hold.

f. Protected obstacle clearance areas for missed approach procedures are predicated on the assumption that the missed approach procedure is initiated at the MAP not lower than the MDA. Reasonable buffers are provided for normal maneuvers.

1) When an early missed approach is executed (i.e., before the MAP), you should, unless otherwise cleared by ATC, fly the IAP as specified on the chart to the MAP at or above the MDA before executing a turning maneuver.

2) If you lose visual reference while circling to land, you must follow the prescribed missed approach procedure.

a) To become established on the missed approach course, you should make an initial climbing turn toward the landing runway and continue the turn until you are established on the missed approach course.

15. Execute a normal landing from a straight-in or circling approach when instructed by your examiner.

a. See Task VI.E., Landing from a Straight-in or Circling Approach, beginning on page 322.

M. Common Errors during an SDF Instrument Approach

1. Failure to have essential knowledge of the information on the SDF instrument approach procedure chart.

a. Know your IAF and the way to arrive at the final approach course.

b. Know how to identify the FAF.

c. Know your minimum altitudes during each segment of the approach, including the MDA.

d. Be able to identify the MAP, and know the missed approach procedure.

2. Incorrect communication procedures or noncompliance with ATC clearances.

a. Always follow correct communication procedures.

b. You are required to follow all ATC clearances. If you cannot comply with a clearance, you must request an amended clearance from ATC.

3. **Failure to accomplish checklist items.**
   a. You must complete the before-landing checklist for your airplane. The checklist is normally begun during the initial segment and completed by the start of the final approach segment.
   b. An approach is a busy time in the cockpit. Attempt to complete your checklist as much as possible before the FAF.
   c. Set and check your HI. Failure to accomplish this could result in a dangerous track on an SDF approach.
4. **Faulty basic instrument flying technique.**
   a. It is a busy time, but you must continue your cross-check and instrument interpretation throughout the approach.
   b. Flying an approach is nothing more than basic attitude instrument flying.
   c. Remember to fly your airplane first, then track your course, and then talk.
5. **Failure to understand the SDF course dimensions.**
   a. Since the VOR receiver is used on the SDF course, the assumption is that interception and tracking techniques are the same as for VOR radials.
   b. Remember that the CDI is more sensitive and faster moving on the SDF course.
6. **Chasing the CDI needle.**
   a. Remember to make small corrections on the heading indicator and wait to see the effect on the CDI.
7. **Inappropriate descent below the MDA.**
   a. You can descend below the MDA only when you have the required visual references and the visibility is equal to or better than that published on the IAP chart.
   b. Give yourself a cushion (e.g., add 50 ft.) above the MDA to allow for altitude variations.

END OF TASK

PRECISION ILS INSTRUMENT APPROACH

VI.B. TASK: PRECISION ILS INSTRUMENT APPROACH

REFERENCES: 14 CFR Parts 61, 91; AC 61-27; IAP; AIM.

Objective. To determine that the applicant:

1. Exhibits adequate knowledge of the elements of an ILS instrument approach procedure.
2. Selects and complies with the appropriate ILS instrument approach procedure to be performed.
3. Establishes two-way communications with ATC, as appropriate to the phase of flight or approach segment, and uses proper radio communications phraseology and technique.
4. Selects, tunes, identifies, and confirms the operational status of ground and aircraft navigation equipment to be used for the approach procedure.
5. Complies with all clearances issued by ATC or the examiner.
6. Advises ATC or examiner any time the aircraft is unable to comply with a clearance.
7. Establishes the appropriate aircraft configuration and airspeed, considering turbulence and wind shear, and completes the aircraft checklist items appropriate to the phase of flight.
8. Maintains, prior to beginning the final approach segment, specified altitude within 100 ft. (30 meters), heading or course within 10°, and airspeed within 10 kt.
9. Applies the necessary adjustments to the published DH and visibility criteria for the aircraft approach category when required, such as --
   a. FDC and Class II NOTAMs.
   b. Inoperative aircraft and ground navigation equipment.
   c. Inoperative visual aids associated with the landing environment.
   d. National Weather Service (NWS) reporting factors and criteria.
10. Establishes an initial rate of descent at the point where the electronic glide slope is intercepted, which approximates that required for the aircraft to follow the glide slope.
11. Allows, while on the final approach segment, no more than three-quarter-scale deflection of either the localizer or glide slope indications, and maintains the specified airspeed within 10 kt.
12. Avoids descent below the DH before initiating a missed approach procedure or transitioning to a normal landing approach.
13. Initiates immediately the missed approach procedure when, at the DH, the required visual references for the intended runway are not distinctly visible and identifiable.
14. Transitions to a normal landing approach when the aircraft is continuously in a position from which a descent to a landing on the intended runway can be made at a normal rate of descent using normal maneuvers.

A. General Information

1. The objective of this task is to determine your knowledge and ability to perform an instrument landing system (ILS) instrument approach procedure.
2. For additional reading, see Task II.B., Aircraft Flight Instruments and Navigation Equipment, on page 72 for more information on the ILS components.
3. Following item C., Common Errors during an ILS Instrument Approach, various types of ILS approaches to parallel runways are discussed.

B. Task Objectives

1. **Exhibit adequate knowledge of the elements of an ILS instrument approach procedure.**

   a. The ILS instrument approach procedure is a precision approach, i.e., using descent path guidance.

      1) Given that the ILS localizer produces one fixed signal, the OBS has no effect on course interpretation.

      2) To fly the approach accurately, you must keep the localizer and glide slope (GS) needles centered in the fashion of a cross.

         a) Fly to the needles.

            i) If the GS needle is above center, you are below the GS and you should slow your descent.

               - Do not attempt to climb. You will go above the GS; i.e., climbing would be overcompensating.

            ii) If the localizer needle is to the left of center, you are to the right of course and need a slight (2° or 5°) left correction.

      3) Unlike a nonprecision approach, the missed approach point (MAP) is the point of arrival at a decision height (DH), not at a fix or after a certain amount of elapsed time.

         a) Follow the glide slope until the DH.

         b) At the DH, you must either have visual references or execute the missed approach procedure.

         c) The time from the final approach fix (FAF) to the MAP is noted only as a backup in case of glide slope failure or for use on the localizer (LOC) approach.

         NOTE: While this is a good operating procedure, you will not be required to time during the ILS approach during your practical test. Timing is not a requirement of this task.

      4) Some localizer sites have a calculated DME, which provides position backup and identifies step-down fixes. You still fly the glide slope for altitude references.

   b. The outer marker (OM) may be an initial approach fix (IAF), but the approach may have other IAFs (e.g., the start of a DME arc).

   c. The FAF on an ILS approach is the glide slope intercept point at the altitude shown below the lightning bolt arrow.

      1) The Maltese cross on the IAP chart shows the FAF for the LOC approach when the glide slope is not being used.

2. **Select and comply with the appropriate ILS instrument approach procedure to be performed.**

   a. Once you know which ILS IAP you will be conducting, you should select the appropriate instrument approach chart and study it to determine that you can comply with the procedure.

   b. We will use the ILS RWY 28 IAP to Pittsburgh/Allegheny County (AGC) on page 298 as an example throughout our discussion.

      1) Note that this IAP requires that the airplane is equipped with an ADF, or an appropriately certified GPS.

c. At the top of the chart, the three rows of information are the **pilot briefing**.

1) This section contains information on the LOC frequency, final approach course, runway/airport data, approach lighting, missed approach text, and various frequencies.

d. Below the pilot briefing, the **plan view** shows a bird's-eye view of the approach.

e. Assuming you are cleared by ATC to "proceed direct to MKP NDB (IAF), cleared for the ILS RWY 8 approach," you will enter the holding pattern upon reaching MKP NDB.

1) This is the procedure turn outbound.

2) You are now on the initial approach segment.

3) Since you have been cleared for the approach, and you are entering the published holding pattern, you may descend to 3,000 ft. MSL, the published altitude for the holding pattern, shown in the **profile view** just below the plan view.

4) Since the holding pattern is used for a course reversal, you should be established on the LOC course inbound MKP NDB or directly to MKP NDB, depending on the entry procedure.

f. When you have intercepted and are tracking inbound on the localizer, you are on the intermediate approach segment.

g. When you intercept the glide slope, you should begin a descent appropriate to your groundspeed so that you maintain the glide slope down to the DH.

1) You are now on the final approach segment.

2) You should begin timing as you pass MIFFY OM (shown by the Maltese cross in the profile view) in case the glide slope becomes inoperative during your approach.

a) This is the FAF for LOC only approach.

b) If the glide slope becomes inoperative, you must use LOC or circling minimums, as appropriate.

c) If the glide slope is working, but you are planning a circling approach, you must use circling minimums and use timing to determine the MAP.

h. Upon arrival at the DH (i.e., the MAP), you must execute the missed approach procedure immediately if the required visual references are not seen and identified.

1) The missed approach procedure is printed in the pilot briefing and depicted by icons in the profile view.

2) In the example, the missed approach procedure is to climb to 2,000 ft. MSL, then a climbing left turn to 3,000 ft. MSL direct to AGC VOR/DME and hold.

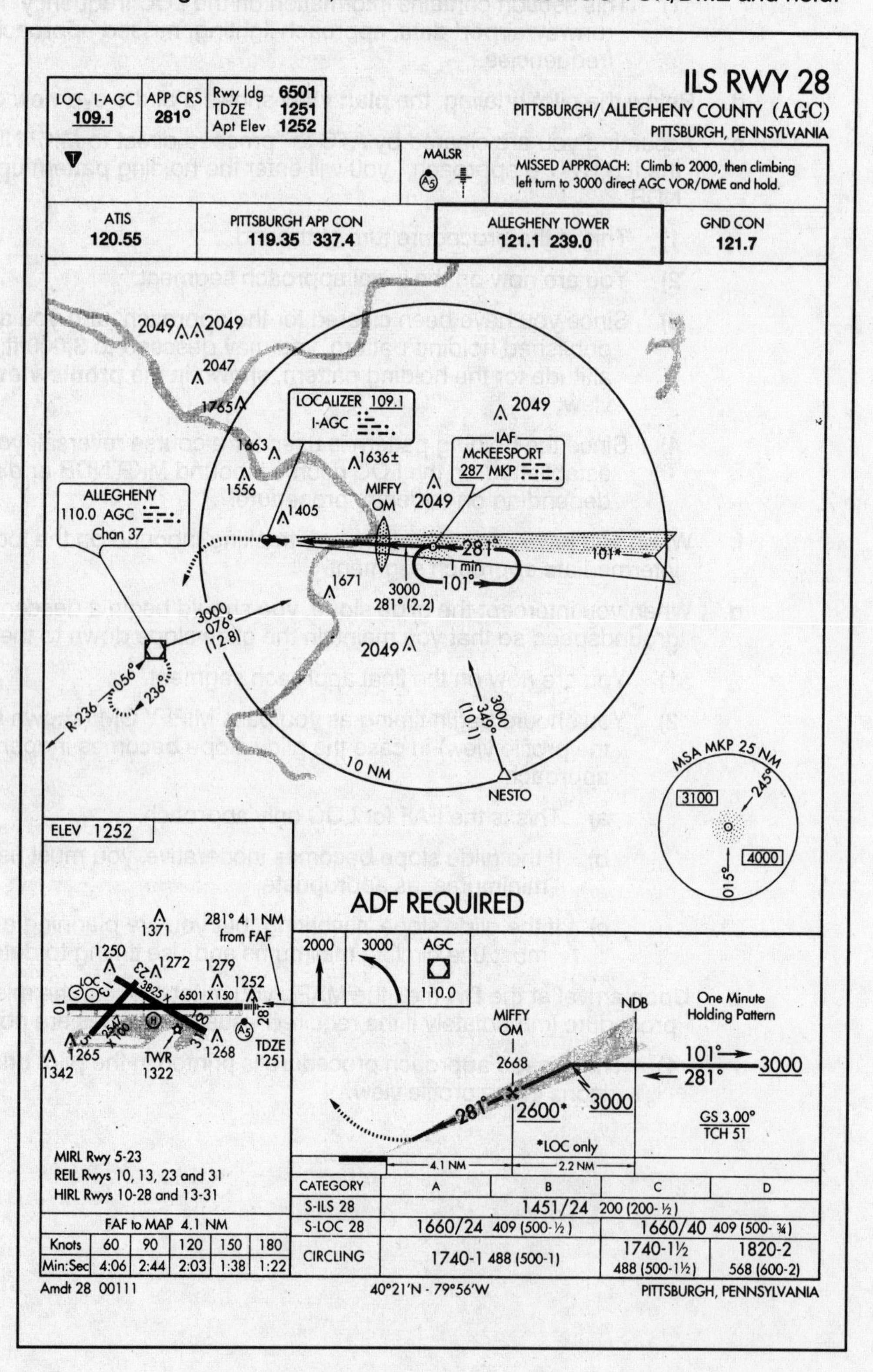

FAF to MAP 4.1 NM					
Knots	60	90	120	150	180
Min:Sec	4:06	2:44	2:03	1:38	1:22

CATEGORY	A	B	C	D
S-ILS 28	1451/24 200 (200-½)			
S-LOC 28	1660/24 409 (500-½)		1660/40 409 (500-¾)	
CIRCLING	1740-1 488 (500-1)		1740-1½ 488 (500-1½)	1820-2 568 (600-2)

3. **Establish two-way communication with ATC, as appropriate, to the phase of flight or approach segment, and use proper radio communication phraseology and technique.**
   a. During the en route part of your flight, you normally maintain two-way communication with an air route traffic control center (e.g., Miami center).
   b. During the transition from en route to approach segment, before the IAF, you will normally be instructed to contact approach control.
      1) Where radar is approved for approach control service, it may be used to provide radar vectors in conjunction with the published ILS approach.
   c. Normally, prior to intercepting the glide slope, you will be instructed to contact the tower or FSS.
      1) If the airport is not served by a tower or an FSS, you will be instructed to change to the advisory frequency (CTAF), and you should broadcast your intentions, including
         a) The approach you are executing (e.g., ILS RWY 28)
         b) Your position (every mile for the last 5 miles of approach)
         c) When reached, your arrival over the FAF inbound
   d. If you execute a missed approach, you must first fly your airplane, and once you are in complete control climbing, you should inform the tower, FSS, or CTAF of your missed approach.
      1) Then you will be instructed to contact approach control by the tower or FSS.
      2) At an airport with no tower or FSS, first announce your missed approach on CTAF; then switch back to the approach control frequency and inform ATC of your missed approach.

4. **Select, tune, identify, and confirm the operational status of ground and airplane navigation equipment to be used for the approach procedure.**
   a. After you have selected the correct instrument approach chart and have studied it, you should tune the localizer frequency into your appropriate VOR navigation equipment.
      1) The frequency and the Morse code identifier are presented on the chart.
      2) In the example ILS RWY 28 IAP on page 298, the localizer (LOC) is on frequency 109.1, and its identifier is I-AGC.
   b. Next, correctly identify the LOC by its Morse code identifier.
      1) Some airports may have a LOC antenna at each end of the runway and use the same frequency. The only way to determine which LOC you are using is by the identifier.
      2) An example of this is at the Spirit of St. Louis airport. The LOC for Rwy 8R is on frequency 111.9, and the identifier is I-SUS.
         a) The LOC for Rwy 26L (opposite direction of Rwy 8R) is also on frequency 111.9, but the identifier is I-FZU.

c. Operational checks that you can perform in your airplane include seeing that
   1) The VOR and glide slope alarm flags are not on.
   2) The correct CDI indication is given based on your location relative to the LOC course.
      a) There should be no erratic movements of the needle.

d. Check each NAVAID that is to be used during approach and missed approach.

5. Comply with all clearances issued by ATC or your examiner.

a. When ATC (or your examiner) issues a clearance or an instruction, you are expected to execute its provisions upon receipt.
   1) You must not deviate from the provisions of any clearance or instructions unless an amended clearance is obtained or an emergency arises.

b. At times, ATC may not specify a particular approach procedure in the approach clearance but will state "cleared for approach." Such a clearance indicates that you may execute any one of the authorized IAPs for that airport.
   1) This clearance does not constitute approval for you to execute a contact approach or a visual approach.

c. When cleared for a specifically prescribed IAP (i.e., "cleared for the ILS RWY 28 approach") or when "cleared for approach," you are required to execute the entire approach as described on the IAP chart unless an appropriate new or revised ATC clearance is received or you cancel your IFR flight plan.

6. Advise ATC (or your examiner) anytime your airplane is unable to comply with a clearance.

a. You must inform ATC (or your examiner) anytime your airplane's operating limitations forbid compliance with the clearance issued.

b. Before you accept a clearance, you must determine whether you can comply with it. If not, inform ATC of the reason you cannot accept the clearance and request an amended clearance.

7. Establish the appropriate airplane configuration and airspeed considering turbulence and wind shear, and complete your airplane's checklist items appropriate to the phase of the flight.

a. During the initial segment of the IAP, you should slow your airplane to your desired approach speed. This is normally an airspeed within, or just above, flap operating range, from which your airplane can readily transition to a landing configuration.
   1) In your airplane, approach speed is _______.
   2) Based on known weather conditions, you may want to increase your approach airspeed due to turbulence, gusty winds, and possible wind shear.
   3) If appropriate for your airplane, you should lower flaps to the approach setting.

b. During the initial approach segment, you should also complete your before-landing checklist as described in your *POH*.
   1) All fuel-related items, such as fuel selectors, fuel pumps, and mixture, should be set for landing.
   2) Most pilots will lower the landing gear (if applicable) at the beginning of the final approach segment (i.e., glide slope intercept).
   3) At the FAF, you should have completed your before-landing checklist.

c. You should commit to memory certain important items prior to reaching the FAF:
   1) DH
   2) Time (or DME) from FAF to MAP (in event of GS failure)
   3) Visibility minimums
   4) Missed approach procedure (at least initial part, e.g., "Climb to 2,000 ft. . . .")

8. *Maintain, prior to beginning the final approach segment, specified altitude within 100 ft., heading or course within 10°, and airspeed within 10 kt.*

a. During the ILS instrument approach, the CDI needle indicates, by deflection, whether you are to the right or left of the localizer centerline, regardless of the position or heading of your airplane.
   1) The following applies when you are using a standard VOR indicator:
      a) When you are inbound on the front course or outbound on the back course (e.g., to execute a procedure turn), the needle is deflected toward the localizer course, and you should turn toward the needle to correct your track.
      b) When you are tracking outbound on the front course (e.g., to execute a procedure turn) or inbound on the back course, you should turn away from the direction of needle deflection (reverse sensing) to return to the localizer course.
   2) If your airplane is equipped with a horizontal situation indicator (HSI), reverse sensing does not occur as long as you always set the localizer front course on the HSI.
      a) Once this is done, always turn toward the direction of the needle deflection to return on course.

b. Because of the narrow localizer course width, overcontrolling of heading corrections is a common error.
   1) Unless the CDI shows a full deflection, heading corrections in 5° increments (or less) should keep you close to the centerline.
   2) The sooner you establish the correct wind correction angle, the easier it will be to track the localizer without chasing the CDI needle.
   3) Remember to fly a heading on your HI and then check the result on the CDI. DO NOT fly by the CDI needle.

c. When reintercepting the localizer after a procedure turn, you will need to cross-check the CDI frequently to prevent overshooting the course centerline.
   1) If your interception angle is 45°, start your turn to the inbound course as soon as the CDI moves from the full deflection position.

d. When you are being radar vectored to the localizer course, ATC will normally try to provide you with a 30° intercept angle.

e. During the interception of the localizer course, use other navigation aids, such as the ADF (if available) or GPS, to monitor the position of your airplane relative to the localizer course.

f. Once you have reached the localizer centerline, maintain the inbound heading until the CDI moves off center.

   1) Drift corrections should be small and reduced proportionally as the course narrows.

   2) By the time you reach the FAF, your wind correction angle should be established.

g. The final approach segment begins at the FAF.

   1) At this point, you should also have your airspeed stabilized at the approach airspeed.

   2) In the ILS RWY 28 example on page 298, the FAF is the glide slope intercept at the minimum altitude of 3,000 ft. MSL.

9. Apply the necessary adjustments to the published DH and visibility criteria for your airplane approach category when required.

a. FDC and Class II NOTAMs

   1) Flight Data Center (FDC) NOTAMs are regulatory in nature and inform you of amendments to IAPs or aeronautical charts prior to their normal publication.

      a) FDC NOTAMs are available from FSS and are published in the *Notices to Airmen Publication* (*NTAP*).

   2) The *NTAP* (formerly Class II NOTAMs) contains all FDC NOTAMs that are current at the time of publication and those NOTAMs (D) that are expected to remain in effect for 7 days after the issuance of the publication. The *NTAP* is issued every 28 days.

      a) During your preflight briefing, you must ask your FSS specialist for any published NOTAMs that are pertinent to your flight.

      b) Alternatively, you may view the NTAP web version at

         www.faa.gov/NTAP/default.htm

b. Inoperative airplane and ground navigation equipment

   1) A compass locator (i.e., LOM or LMM) or precision radar (PAR) may be substituted for the outer or middle marker.

      a) DME, VOR, or NDB fixes authorized in the IAP or surveillance radar (ASR) may be substituted for the outer marker.

   2) An inoperative glide slope would make the approach a LOC, which is a nonprecision approach with an MDA.

      a) The LOC approach is explained in detail beginning on page 265.

c. Inoperative visual aids associated with the landing environment

   1) An inoperative visual aid (e.g., approach light system) will not affect the DH but will increase the required visibility.

      a) NACO charts will have this information listed in the Inoperative Components Table located on the inside front cover.

      b) JEPP charts will have this information listed within the minimums section of the IAP chart.

d. National Weather Service (NWS) reporting factors and criteria

   1) On NACO IAP charts, this information is included below the minimums section.

10. Establish an initial rate of descent at the point where the electronic glide slope is intercepted, which approximates that required for your airplane to follow the glide slope.

a. The electronic glide slope of the ILS provides a 3° glide slope to the landing runway.

1) The glide slope width is only 1.4°.

a) The course width of a localizer is 3° to 6°; thus, the glide slope needle is approximately three times as sensitive as the localizer (CDI) needle and 12 times as sensitive as the VOR needle.

2) With a 3° glide slope and a glide slope course width of 0.7° (above and below the glide slope), the vertical width is

a) 350 ft. at 5 NM
b) 210 ft. at 3 NM
c) 70 ft. at 1 NM

b. At glide slope intercept, you should start your descent to the DH, tracking both the glide slope and the localizer.

1) Use the Rate of Descent table in the *Terminal Procedures Publication* (located on the inside back cover), applying the glide slope (GS) angle shown on the IAP chart (e.g., 3.00°) and your groundspeed.

a) JEPP charts have the rate-of-descent information with the timing information on the IAP chart.

2) Below is a general rule that can also be used:

Approximate rate of descent (fpm) = *Factor* × *Groundspeed (kt.)*

a) The factor is based on the glide slope angle:

i) 3° = 5
ii) 4.5° = 8
iii) 6° = 10

3) EXAMPLE: Your groundspeed is 90 kt., and the instrument approach indicates a 3° glide slope angle. Rate of descent = 5 x 90 = 450 fpm.

c. You will want to maintain a constant airspeed, whether in level flight or descending to the DH.

1) To accomplish this, all you will need to do initially is to have the power and trim set for your desired approach airspeed, e.g., 90 kt.

2) To descend, reduce power to a predetermined setting to establish the rate of descent desired.

a) As the power is reduced, the nose of the airplane will lower. All you may need to do is ensure that the nose does not pitch down too much.

b) Since the airplane is trimmed for your approach speed, no change of the trim or elevator control is required.

c) You will be able to adjust the rate of descent with minor power adjustments and still be able to maintain airspeed.

d. See Task IV.D., Rate Climbs and Descents, beginning on page 166.

e. Never begin your descent on the GS until you are established on the LOC.

11. *While on the final approach segment, allow no more than a three-quarter-scale deflection of either the localizer or the glide slope indications, and maintain the desired airspeed within 10 kt.*

a. The final approach segment is from the glide slope intercept to the DH.

b. Maintain the final approach course (i.e., localizer) and glide slope so that the indication of either is no greater than a three-quarter-scale deflection.

1) Use small heading changes (no more than 2° to 5° at a time) to track the LOC.

2) Adjust the descent rate (VSI) with power and pitch to stay on GS while maintaining airspeed.

c. Throughout the final approach segment, you must maintain your approach airspeed at all times, ±10 kt.

d. The heaviest demand on your piloting technique will occur during the descent from the glide slope intercept to the DH.

1) During this time, you will maintain the localizer with heading changes, adjust pitch to maintain the desired rate of descent, and adjust power to maintain airspeed.

2) Remember to make small adjustments and fly by the heading indicator, airspeed indicator, and vertical speed indicator. DO NOT fly the LOC/GS needles.

a) With the airplane properly trimmed for your approach speed, any changes in power will cause a pitch change. Thus, you can maintain both glide slope and airspeed with minor power adjustments.

i) Large corrections may require both pitch and power corrections.

12. Avoid descent below the DH before initiating a missed approach procedure or transitioning to a normal landing approach.

a. You will be very busy when you approach the DH because you must glance up from your instruments and determine if you can see the required visual references.

b. You will not level off at the DH. You will see it once as you start the climb on the missed approach procedure or as you descend below it on the way to your landing.

13. Initiate immediately the missed approach procedure when, at the DH, the required visual references for the intended runway are not distinctly visible and identifiable at the MAP.

a. The DH is the MAP for the ILS approach. If the required visual references are not visible and identified, immediately initiate the missed approach procedure.

1) Timing the approach from the outer marker (OM) is a good procedure.

a) It will allow you to continue the approach if the glide slope fails.

NOTE: While timing is a good operating procedure, during your practical test, it is not required. Also, timing is not required by the PTS.

b. To descend below the DH (FAR 91.175), you are required to

1) Have the runway environment in sight
2) Have visibility at or above the minimums for your approach category
3) Be in a position to make a normal descent to the intended runway

c. Runway environment is defined as any one of the following visual references required for descent below the DH:

1) Approach light system

a) However, you are not allowed to descend below 100 ft. above the touchdown zone elevation (TDZE) using the approach lights as a reference unless the red terminating bars or the red side row bars are also distinctly visible and identifiable.

2) The threshold
3) The threshold markings
4) The threshold lights
5) The runway end identifier lights
6) The visual approach slope indicator
7) The touchdown zone or touchdown zone markings
8) The touchdown zone lights
9) The runway or runway markings
10) The runway lights

d. You are required to execute the missed approach procedure when you cannot identify one of the required visual references at either of the following times:

1) Upon arrival at the DH
2) At any time that you are below the DH until touchdown

e. The missed approach procedure is written in the pilot briefing of NACO charts. Additionally, missed approach icons are used in the profile view.

1) In the example ILS RWY 28 IAP on page 298, the missed approach procedure is to execute a climb to 2,000 ft. MSL, then climbing left turn to 3,000 ft. MSL direct to AGC VOR/DME and hold.

f. Protected obstacle clearance areas for missed approach procedures are predicated on the assumption that the missed approach procedure is initiated at the DH, not before the DH. Reasonable buffers are provided for normal maneuvers.

1) When an early missed approach is executed (i.e., before the MAP), you should, unless otherwise cleared by ATC, fly the IAP as specified on the chart to the MAP at or above the glide slope before executing a turning maneuver.

14. Transition to a normal landing approach when your airplane is continuously in a position from which a descent to a landing on the intended runway can be made at a normal rate of descent using normal maneuvers.

a. See Task VI.E., Landing from a Straight-in or Circling Approach, beginning on page 332.

C. Common Errors during an ILS Instrument Approach

1. Failure to have essential knowledge of the information on the ILS instrument approach procedure chart.

a. Know your IAF and the way to arrive at the final approach course.

b. Know where you can expect to intercept the glide slope.

c. Know your minimum altitudes during each segment of the approach, including the DH.

d. Know the missed approach procedure.

2. Incorrect communication procedures or noncompliance with ATC clearances.

a. Always follow correct communication procedures.

b. You are required to follow all ATC clearances. If you cannot comply with a clearance, you must request an amended clearance from ATC.

3. Failure to accomplish checklist items.

a. You must complete the before-landing checklist for your airplane. The checklist is normally begun during the initial segment and completed by the start of the final approach segment.

b. An approach is a busy time in the cockpit. Attempt to complete your checklist as much as possible before glide slope intercept.

4. Faulty basic instrument flying technique.

a. It is a busy time, but you must continue your cross-check and instrument interpretation throughout the approach.

b. Remember to fly your airplane first, then track your course, and then talk.

c. Do not chase the needles, but fly a stabilized approach.

5. Inappropriate descent below the DH.

a. You can descend below the DH only when you have the required visual references and the visibility is equal to or better than published on the IAP chart.

b. You must judge your rate of descent so that you have time to look up, determine if you see the required visual references, and decide either to perform the missed approach procedure or to transition to a normal landing.

1) At the DH, you should be either climbing on the missed approach or descending for a landing.

D. ILS Approaches to Parallel Runways

1. ATC procedures permit ILS instrument approach operations to dual or triple parallel runway configurations.

a. ILS approaches to parallel runways are grouped into the following three classes:

1) Parallel (dependent) ILS approaches

2) Simultaneous parallel (independent) ILS approaches

3) Simultaneous close parallel (independent) ILS precision runway monitoring (PRM) approaches.

b. The classification of a parallel runway approach procedure is dependent on

1) Adjacent parallel runway centerline separation
2) ATC procedures
3) Airport ATC radar monitoring and communications capability

c. At some airports, one or more parallel localizer courses may be offset up to 3°.

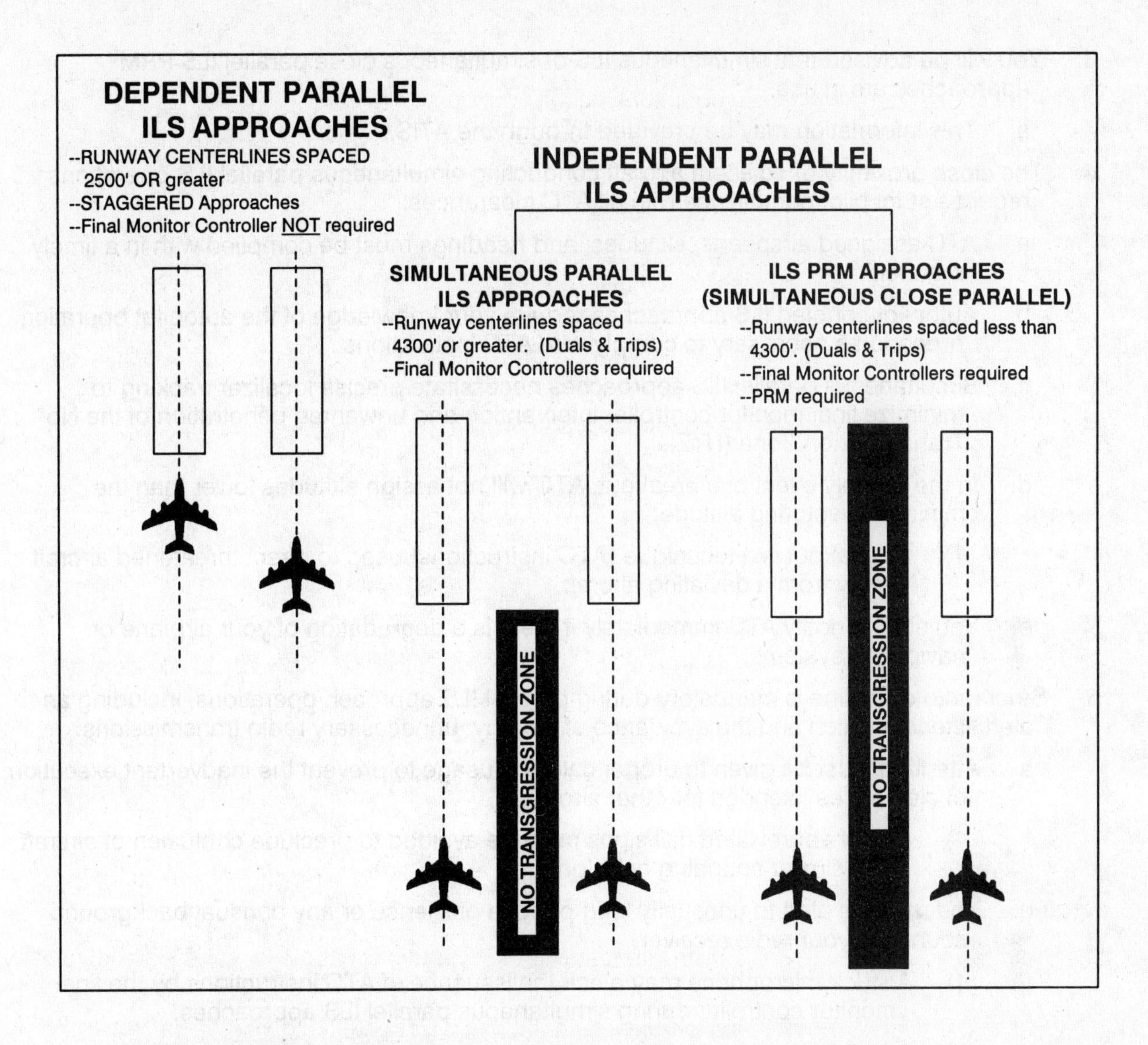

2. Parallel approach operations demand your heightened situational awareness. A thorough IAP chart review should be conducted with emphasis, as a minimum, on the following chart information:

   a. Name and number of the approach
   b. Localizer frequency
   c. Inbound localizer course
   d. Glide slope intercept altitude
   e. Decision height (DH)
   f. Missed approach instructions
   g. Special notes/procedures
   h. Assigned runway location/proximity to adjacent runways

3. You will be advised that simultaneous ILS or simultaneous close parallel ILS PRM approaches are in use.
    a. This information may be provided through the ATIS.
4. The close proximity of adjacent aircraft conducting simultaneous parallel ILS operations requires strict pilot compliance with all ATC clearances.
    a. ATC assigned airspeeds, altitudes, and headings must be complied with in a timely manner.
    b. Autopilot-coupled ILS approaches require your knowledge of the autopilot operation procedures necessary to comply with ATC instructions.
    c. Simultaneous parallel ILS approaches necessitate precise localizer tracking to minimize final monitor controller intervention and unwanted penetration of the No Transgression Zone (NTZ).
    d. In the unlikely event of a breakout, ATC will not assign altitudes lower than the minimum vectoring altitude.
        1) A breakout is a technique (ATC instructions) used to direct threatened aircraft away from a deviating aircraft.
    e. You should notify ATC immediately if there is a degradation of your airplane or navigation systems.
5. Strict radio discipline is mandatory during parallel ILS approach operations, including an alert listening watch and the avoidance of lengthy, unnecessary radio transmissions.
    a. Attention must be given to proper call sign usage to prevent the inadvertent execution of clearances intended for other aircraft.
        1) Use of abbreviated call signs must be avoided to preclude confusion of aircraft with similar sounding call signs.
    b. You must be alert to unusually long periods of silence or any unusual background sounds in your radio receiver.
        1) A stuck microphone may block the issuance of ATC instructions by the final monitor controller during simultaneous parallel ILS approaches.

E. Parallel ILS Approaches (Dependent)

1. Parallel ILS approaches may be conducted at airports having parallel runways separated by at least 2,500 ft. between centerlines.
2. A parallel (dependent) approach differs from a simultaneous (independent) approach in that
    a. The minimum distance between parallel runway centerlines is reduced.
    b. There is no requirement for radar monitoring or advisories.
    c. A staggered separation of aircraft on the adjacent localizer is required.
3. Whenever parallel ILS approaches are in progress, you will be informed that approaches to both runways are in use.

F. Simultaneous Parallel ILS Approaches (Independent)

1. Simultaneous parallel ILS approaches constitute an approach system permitting simultaneous ILS approaches to parallel runways that have centerlines separated by 4,300 to 9,000 ft. and are equipped with final monitor controllers.
   a. The IAP chart of each approach at an airport permitting simultaneous parallel ILS approaches will contain a note identifying the appropriate runways.
      1) EXAMPLE: "Simultaneous approaches authorized RWYs 14L and 14R."
   b. When advised that simultaneous parallel ILS approaches are in progress, you must inform approach control immediately if you have malfunctioning or inoperative receivers or if you do not desire a simultaneous parallel ILS approach.
2. Simultaneous parallel ILS approaches require radar monitoring to ensure separation between aircraft on the adjacent parallel approach course.
   a. Aircraft position is tracked by final monitor controllers who will issue instructions to aircraft that are observed deviating from the assigned localizer course.
      1) Staggered separation (as used in parallel ILS approaches) is not used in simultaneous parallel ILS approaches.
3. Radar monitoring service will be provided as follows:
   a. During a turn onto parallel final approach, aircraft will be provided 3-NM radar separation or a minimum of 1,000 ft. of vertical separation.
   b. Aircraft will not be vectored to intercept the final approach course at an angle greater than 30°.
   c. You will be instructed to monitor the tower frequency to receive advisories and instructions.
   d. Aircraft observed to overshoot the turn-on or to continue a track that will penetrate the No Transgression Zone (NTZ) will be instructed to return to the correct final approach course immediately. The final monitor controller may also issue missed approach or breakout instructions to the deviating aircraft.
      1) The NTZ is an area 2,000 ft. wide located equidistant between parallel final approach courses.
   e. If a deviating aircraft fails to respond to such instructions or is observed penetrating the NTZ, the aircraft on the adjacent final approach course may be instructed to alter course.

G. Simultaneous Close Parallel ILS PRM Approaches

1. The ILS precision runway monitor (PRM) system permits simultaneous ILS approaches to parallel runways with centerlines separated by less than 4,300 ft.
   a. The final controllers are equipped with high update radar and high resolution radar displays, collectively called a PRM system.
      1) The PRM system displays almost instantaneous radar information.
      2) Automated tracking software provides the controllers with aircraft identification, position, a 10-second projected position, as well as visual and aural alerts.
2. In addition to the required ILS components installed in your airplane, you must have the capability to listen to two radio frequencies simultaneously.

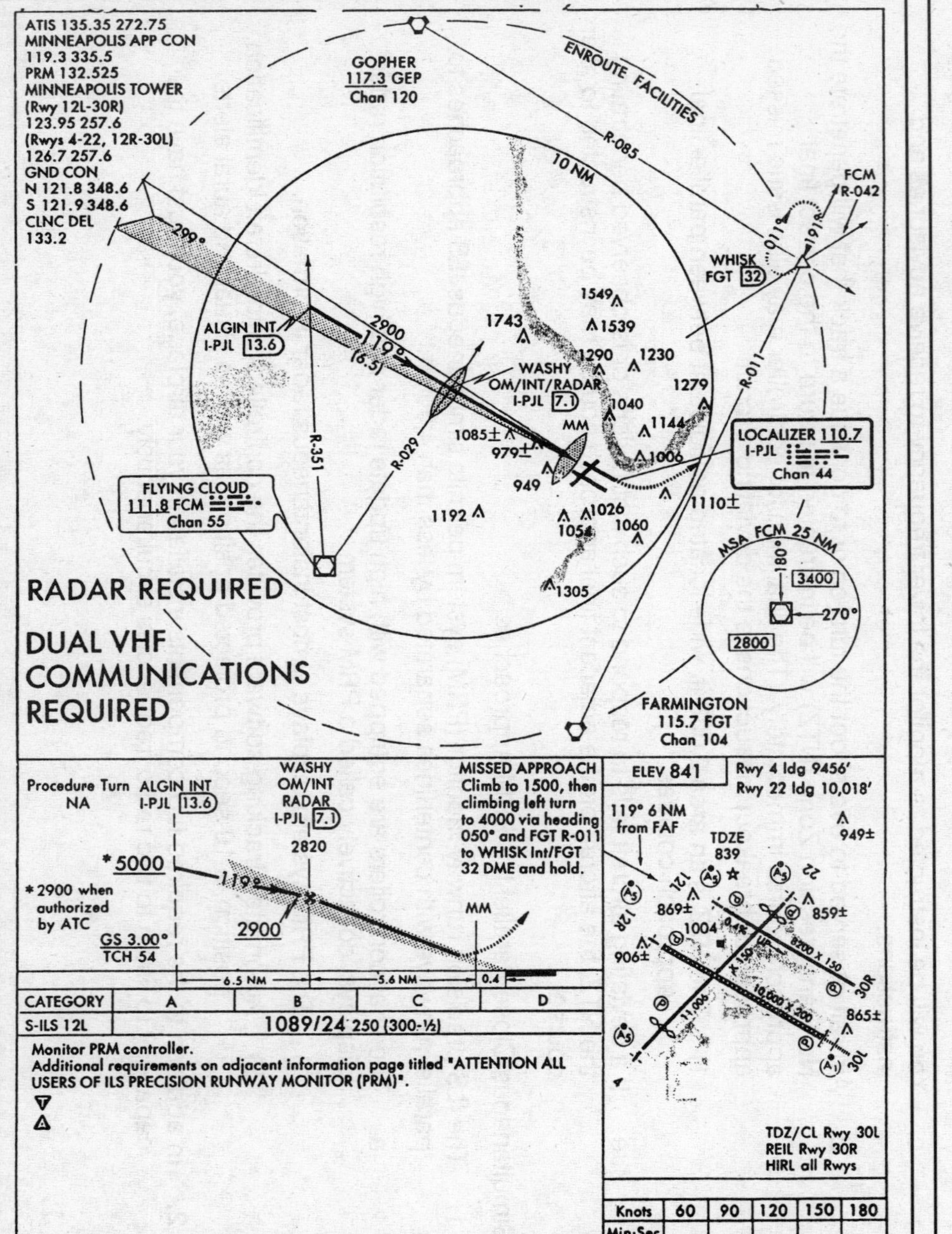

196 NC-1, 25 MAR 1999

ILS PRM RWY 12L Amdt 2A 99028
(SIMULTANEOUS CLOSE PARALLEL)
MINNEAPOLIS-ST PAUL INTL (WOLD-CHAMBERLAIN) (MSP)
AL-264 (FAA) MINNEAPOLIS, MINNESOTA

ATTENTION ALL USERS OF ILS PRECISION RUNWAY MONITOR (PRM)

Before initiating a simultaneous close parallel ILS PRM approach, PILOTS ARE EXPECTED TO HAVE VIEWED ONE OF THE TWO VIDEOS "RDU PRECISION RUNWAY MONITOR: A PILOTS APPROACH" or "ILS PRM APPROACH: INFORMATION FOR PILOTS" and MEET THE FOLLOWING REQUIREMENTS (if unable notify ATC);

1. **ATIS**: When the ATIS broadcast advises ILS PRM approaches are in progress, pilots will NOTIFY ATC on initial contact if they can not meet all the requirements on this information page.

2. **Dual VHF Communication required**: To avoid blocked transmissions, each runway will have two frequencies, a primary and a monitor frequency. The tower controller and monitor controller will transmit on both frequencies. Pilots will ONLY transmit on the primary frequency, but will listen to both frequencies. It is important that pilots do not switch to the monitor frequency until instructed by ATC to contact the tower. The volume levels should be set about the same on both radios so that the pilots will be able to hear transmissions on at least one frequency if the other is blocked.

3. **TCAS**: To avoid ATC giving instructions directly opposite to a TCAS RA, the TCAS will be placed in TA mode when entering the monitored PRM airspace at ALGIN INT. If a pilot is broken out from the ILS the TCAS should be reset to the TA/RA mode as soon as practicable.

4. **ALL ATC directed "breakouts" are to be HAND FLOWN**: Pilots, when directed to breakoff an approach, must assume that an aircraft is blundering toward their course and a breakout must be initiated immediately. The breakout must be hand flown to assure it is accomplished in the shortest amount of time. Controllers may give a descending breakout but in no case will the descent be below minimum vectoring altitude (MVA) which provides at least 1000' obstacle clearance. The MVA is 2,500' at Minneapolis-St Paul International.

5. **Phraseology - "TRAFFIC ALERT"**: If an aircraft enters the "NO TRANSGRESSION ZONE" (NTZ), the controller will breakout the threatened aircraft on the adjacent approach. The phraseology for the breakout will be:

"TRAFFIC ALERT, (aircraft call sign) TURN (left/right) IMMEDIATELY, HEADING (degrees), CLIMB/DESCEND AND MAINTAIN (altitude)".

NOTE: For additional information refer to ILS PRM in the Aeronautical Information Manual (AIM) or the Aeronautical Information Publication (AIP).

3. To ensure that separation is maintained and to avoid the development of a dangerous situation during simultaneous close ILS PRM approaches, you must immediately comply with final monitor controller instructions to prevent an imminent situation (called a breakout).
   a. A blunder, or deviation, must be recognized by one controller, the information must be passed on to another controller, and breakout instructions must be issued to the endangered aircraft.
   b. You will not have any warning that a breakout is imminent because the blundering, or deviating, aircraft will probably be on another frequency.
4. When you receive breakout instructions, it is important for you to assume that a deviating aircraft is headed into your approach course and to begin the breakout as soon as safety allows.
   a. Although you are encouraged to use the autopilot while flying an ILS PRM approach, the autopilot must be disengaged while flying a breakout.
   b. The controller may issue a descending breakout when there is a deviating aircraft from an adjacent approach course crossing your flight path.
      1) You must be aware that a descending breakout is a possibility.
      2) In no case will the controller instruct an aircraft to descend below the minimum vectoring altitude (MVA).
      3) You are not expected to exceed a 1,000-fpm rate of descent in a descending breakout.
5. Simultaneous close parallel ILS PRM approaches are identified by a separate IAP chart named ILS PRM (Simultaneous Close Parallel).
   a. See the previous page for the ILS PRM RWY 12L (Simultaneous Close Parallel) IAP at Minneapolis-St. Paul International and the information page.
6. When an airport is about to commission an ILS PRM IAP, general aviation pilots will be given information about these approaches at safety seminars and local FBOs.
   a. At the printing date of this book, the only ILS PRM IAPs are at the Minneapolis-St. Paul International Airport (KMSP).

END OF TASK

MISSED APPROACH

VI.C. TASK: MISSED APPROACH

REFERENCES: 14 CFR Parts 61, 91; AC 61-27; IAP; AIM.

Objective. To determine that the applicant:

1. Exhibits adequate knowledge of the elements related to missed approach procedures associated with standard instrument approaches.
2. Initiates the missed approach promptly by applying power, establishing a climb attitude, and reducing drag in accordance with the aircraft manufacturer's recommendations.
3. Reports to ATC beginning the missed approach procedure.
4. Complies with the published or alternate missed approach procedure.
5. Advises ATC or examiner anytime the aircraft is unable to comply with a clearance, restriction, or climb gradient.
6. Follows the recommended checklist items appropriate to the go-around procedure.
7. Requests, if appropriate, ATC clearance to the alternate airport, clearance limit, or as directed by the examiner.
8. Maintains the recommended airspeed within 10 kt.; heading, course, or bearing within 10°; and altitude(s) within 100 ft. (30 meters) during the missed approach procedure.

A. General Information

1. The objective of this task is to determine your knowledge and ability to perform missed approach procedures.
2. You will execute a missed approach when one of the following conditions exists:
   a. Arrival at the missed approach point (MAP) or the decision height (DH) and visual reference to the runway environment are insufficient to complete the landing.
   b. A safe landing is not possible.
   c. You are instructed to do so by ATC.

B. Task Objectives

1. **Exhibit adequate knowledge of the elements related to missed approach procedures associated with standard instrument approaches.**
   a. A missed approach procedure is designed for each instrument approach procedure. Thus, each procedure is unique.
      1) Protected obstacle clearance areas for missed approach are predicated on the assumption that the missed approach is initiated at the DH or at the MAP and not lower than the MDA.
         a) A climb of at least 200 feet per nautical mile (FPNM) is required, unless a higher climb gradient is published on the approach chart.
      2) If you decide, or are directed by ATC, to make a missed approach before the MAP, you need to remember that no obstacle clearance consideration is given to an early turn.
         a) Unless otherwise cleared by ATC, fly the approach as depicted on the IAP to the MAP at or above the MDA or DH before turning.
         b) There is no restriction on climbing early.

b. Missed approaches can be caused by many factors.
   1) The primary factor is weather below minimums, especially when the visibility decreases. Decreased visibility causes more missed approaches than the ceiling.
   2) Full-scale needle deflections experienced past the FAF require a missed approach.
      a) Once the needle indicates a full-scale deflection, you cannot measure how far off course you are.
   3) Although rare, equipment failures do happen. If your primary navigation equipment for the approach fails, you should execute a missed approach.
   4) Outside factors, such as traffic, aircraft spacing, turns to final approach course that are too sharp or too late, and controller errors, can also cause missed approaches.
   5) Pilot errors, such as setting the avionics wrong or failing to descend quickly enough, can lead to a missed approach.
   6) Any loss of the required visual references between the MDA (or DH) and touchdown can cause a missed approach.
   7) Above all else, if something seems wrong or you are uncertain, execute the missed approach. You may be averting a disaster.
c. When you are flying a circling approach, you must maintain visual references. If you lose them, execute the missed approach immediately.
   1) To become established on the missed approach course, you should make an initial climbing turn toward the landing runway and continue the turn until established on the missed approach course.
      a) This procedure will assure that you will remain within the circling and missed approach obstruction clearance areas.
d. The missed approach is potentially the most critical and dangerous maneuver after takeoff.
   1) It normally occurs at a very low altitude.
   2) Your airplane must transition from a descent to a climb promptly without a danger of stalling.
   3) You are transitioning or expecting to transition from instruments to visual, and suddenly you must to go back to instruments.
   4) The procedure requires precise airplane handling and execution.
e. You must read the missed approach procedures before you arrive at the FAF.
   1) Memorize at least the initial headings and altitudes.
   2) If possible, set your No. 2 NAV for the missed approach fix.
   3) If the procedure calls for a turn, use standard or one-half standard rate.
f. Every instrument approach should be flown with the full intention of executing the missed approach.
   1) Your having this mind-set will help you avoid any uncertainty at the MAP as to your course of action.
   2) Sight of the visual references should be treated as a pleasant surprise, not as an expected event.

2. Initiate the missed approach promptly by applying power, establishing a climb attitude, and reducing drag in accordance with your airplane manufacturer's recommendations.

a. At the MAP or anytime between the MAP and touchdown that the required visual references are not visible, you should execute a go-around procedure.

b. The first step is to add maximum allowable power promptly, but smoothly, and adjust the pitch attitude for V_X to establish a climb attitude.

1) In your airplane, V_X is ______.

2) This step assumes that you are executing the approach with approach flaps (not full flaps) and that V_X is the recommended go-around airspeed.

3) Use the airspeed recommended in your airplane's *POH*.

c. Next, reduce drag by following the procedures described in your *POH*.

1) Once you have established a positive rate of climb by checking the VSI and confirming it with the ALT, retract the landing gear (if retractable), and accelerate to V_Y.

a) In your airplane, V_Y is ______.

2) When reaching a safe altitude (e.g., 500 ft. AGL), retract the flaps.

3. Report to ATC that you are beginning the missed approach procedure.

a. Your first priority when beginning the missed approach procedure is to fly your airplane.

b. Once you have established a climb and have performed the initial tasks of applying power, climbing, and doing an initial cleanup of your airplane, inform ATC of your missed approach.

4. Comply with the published or alternate missed approach procedure.

a. You must perform the missed approach procedure that is written on the IAP chart.

1) At times, ATC may instruct you to perform an alternate missed approach procedure.

b. You should study the missed approach procedure before the FAF so that at the MAP you can concentrate on your flying without the need to look down to read instructions.

5. Advise ATC or your examiner anytime your airplane is unable to comply with a clearance, restriction, or climb gradient.

a. You must inform ATC (or your examiner) anytime your airplane's operating limitations forbid compliance with a clearance, restriction, or climb gradient.

1) Inform ATC (or your examiner) that you are unable to comply, and request an amended clearance.

6. Follow the recommended checklist items appropriate to the go-around procedure.

a. Initiating the missed approach procedure is identical to a go-around procedure. Thus, you should follow and complete the go-around (or balked landing) checklist in your *POH*.

7. **Request, if appropriate, ATC clearance to your alternate airport or a clearance limit, or follow your examiner's directions.**
   a. When you have missed the approach, you must decide what you will do next. If your approach was done in IMC and the weather is below minimums, your decision will be based on how much fuel is remaining and what is needed to reach your alternate airport.
   b. Normally you will take one of three steps:
      1) Request a clearance to your alternate airport. You may want to ask the controller if (s)he knows a closer airport than your alternate where aircraft are able to make successful approaches.
         a) If there is, request to go to that airport.
      2) Request a clearance to a fix (limit) and hold. This option is good, provided the weather is to improve shortly and you have enough fuel to attempt another approach later and, if necessary, fly to your alternate airport.
      3) Request to be sequenced into the traffic flow for another approach. This option is good, especially if other aircraft are able to complete the approach.
         a) Ask for the same approach, or maybe try a different approach to the same airport.
   c. During your practical test, your examiner may direct your actions.
8. ***Maintain the recommended airspeed within 10 kt.; heading, course, or bearing within 10°; and altitudes within 100 ft. during the missed approach procedure.***
   a. Maintain your basic instrument flying techniques throughout the missed approach procedure.
   b. This is important at the beginning since you must transition from looking outside for the runway environment to looking back inside the cockpit and reestablishing your instrument cross-check, instrument interpretation, and airplane control.

C. Common Errors during a Missed Approach

1. **Failure to have essential knowledge of the information on the instrument approach procedure chart.**
   a. While you study the IAP chart for your approach, you must also study the missed approach procedure.
   b. You should know the MAP and memorize at least the initial steps of the missed approach procedure.
   c. Know how to find the missed approach holding fix (if one exists).
2. **Failure to have the missed approach procedures committed to memory.**
   a. Be prepared mentally for a missed approach on each approach you make.
   b. Review your *POH* for the missed approach procedures (i.e., go-around).

3. **Failure to recognize conditions requiring a missed approach procedure.**
    a. You must execute a missed approach procedure anytime you arrive at the MAP (or DH) and do not have the required elements of the runway environment in sight and identified.
        1) Anytime after the MAP (or DH) to touchdown that you lose the runway visual contact with the required element, you must execute a missed approach.
    b. Execute a missed approach when instructed to do so by ATC.
4. **Failure to promptly initiate a missed approach procedure.**
    a. Protected obstacle clearance areas for missed approach are based on the assumption that the procedure is initiated at the MAP, not lower than the MDA or DH.
    b. Do not attempt to go even a little farther to attempt to see the runway.
    c. It is a good practice to announce aloud your descent to minimums; e.g., 500 ft. to MDA/DH, 200 ft. to MDA/DH, 100 ft. to MDA/DH.
5. **Failure to make the required report to ATC.**
    a. Once you complete the initial go-around checklist items and have your airplane stabilized on the missed approach procedure, you must inform ATC of your missed approach and the reason that it occurred.
6. **Failure to comply with the missed approach procedure.**
    a. You must comply with the published missed approach procedure or follow an alternative missed approach procedure provided by ATC.
    b. Inform ATC if you are unable to comply with any clearance, restriction, or climb gradient.
7. **Faulty basic instrument flying technique.**
    a. It is a busy time in the cockpit, but you must continue your cross-check and instrument interpretation throughout the missed approach procedure.
    b. Remember to fly your airplane first.

END OF TASK

CIRCLING APPROACH

VI.D. TASK: CIRCLING APPROACH

REFERENCES: 14 CFR Parts 61, 91; AC 61-27; IAP; AIM.

Objective. To determine that the applicant:

1. Exhibits adequate knowledge of the elements related to a circling approach procedure.
2. Selects and complies with the appropriate circling approach procedure considering turbulence and wind shear and considering the maneuvering capabilities of the aircraft.
3. Confirms the direction of traffic and adheres to all restrictions and instructions issued by ATC and the examiner.
4. Does not exceed the visibility criteria or descend below the appropriate circling altitude until in a position from which a descent to a normal landing can be made.
5. Maneuvers the aircraft, after reaching the authorized MDA, and maintains that altitude within +100 ft. (30 meters)/–0 ft. and a flight path that permits a normal landing on a runway at least 90° from the final approach course.

A. General Information

1. The objective of this task is to determine your knowledge and ability to perform a circling approach procedure.

B. Task Objectives

1. **Exhibit adequate knowledge of the elements related to a circling approach procedure.**
   a. A circling approach is required when the approach alignment or the runway and weather conditions do not warrant a straight-in landing.
      1) A circling approach is required if the final approach course deviates more than 30° (15° for GPS IAPs) from the centerline of the landing runway.
         a) In such cases, the approach will be designated with a letter (e.g., VOR-A, NDB-B), not with a runway number.
      2) You may perform a circling approach if you make an approach to one runway (e.g., ILS RWY 8) and desire to land on another runway (e.g., RWY 26) due to wind direction and speed.

b. The obstacle clearance protected areas for circling approach procedures are determined by aircraft approach category, as shown in the figure below.

1) The protected area is limited to 1.3 NM from the runway ends for Category A aircraft and to 4.5 NM for Category E.

a) Normally, the higher the category, the higher the MDA.

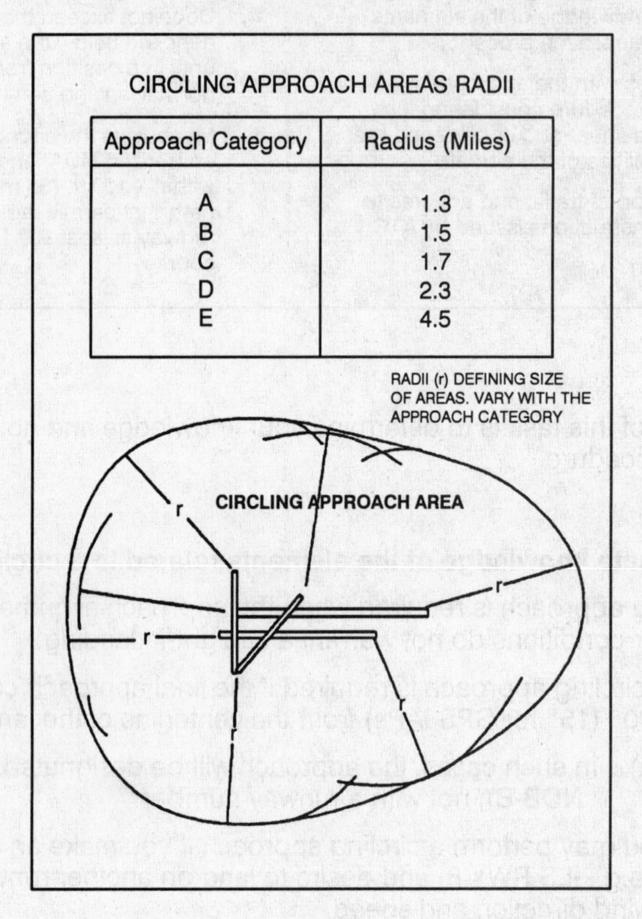

CIRCLING APPROACH AREAS RADII

Approach Category	Radius (Miles)
A	1.3
B	1.5
C	1.7
D	2.3
E	4.5

2) If you are operating in Category B, you must remain within 1.5 NM at the appropriate MDA.

a) This requirement applies even if the visibility minimum for the approach is greater than 1.5 NM.

b) Visibility higher than minimums does not allow you to leave the protected area.

2. Select and comply with the appropriate circling approach procedure considering turbulence and wind shear, and consider the maneuvering capabilities of your airplane.

a. Patterns that can be used for circling approaches are shown in the figure on page 240.

b. Pattern "A" can be flown when your final approach course intersects the runway centerline at less than a 90° angle and you sight the runway early enough to establish a base leg.

1) If you sight the runway too late to fly "A," you can circle as shown in "B."

c. Fly pattern "C" if it is desirable to land opposite the direction of the final approach and the runway is sighted in time for a turn downwind.

   1) If the runway is sighted too late, you can fly pattern "D."

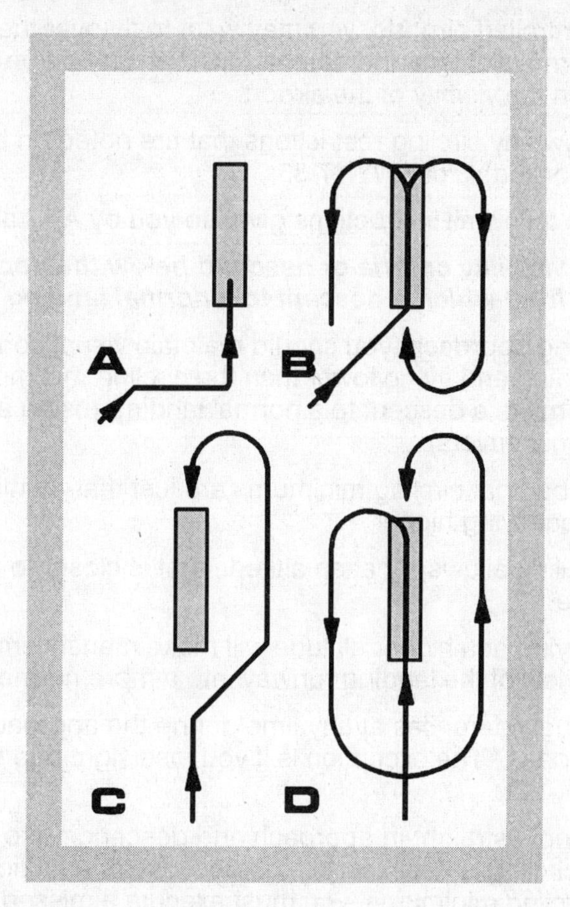

d. Sound judgment and knowledge of your capabilities and the performance of your airplane are the criteria for determining the pattern to be flown.

   1) Other factors that must be considered follow:

      a) Airport design
      b) Ceiling and visibility
      c) Wind direction and speed
      d) Turbulence or possible wind shear
      e) Final approach course alignment
      f) Distance from the FAF to the runway
      g) Altitude and airspeed
      h) ATC instructions

3. **Confirm the direction of traffic, and adhere to all restrictions and instructions issued by ATC and your examiner.**
   a. At controlled airports, ATC will inform you of the direction of traffic.
      1) At uncontrolled airports, you may want to fly over the airport to observe wind and turn indicators and other traffic that may be on the runway or may be flying in the vicinity of the airport.
   b. You must follow any circling restrictions that are noted on the IAP chart, e.g., circling NA (not authorized) E of RWY 17-35.
   c. You must also follow all instructions given to you by ATC and/or your examiner.
4. ***Do not exceed the visibility criteria or descend below the appropriate circling altitude until in a position from which a descent to a normal landing can be made.***
   a. During a circling approach, you should maintain visual contact with the runway of intended landing and fly no lower than the circling minimums (MDA) until you are in a position to make a descent to a normal landing (using a normal rate of descent and normal maneuvers).
      1) Remember that circling minimums are just that--minimums. Nothing prevents you from flying higher.
      2) If the ceiling allows, fly at an altitude that is closer to the VFR traffic pattern altitude.
         a) Flying at a higher altitude will make maneuvering safer and bring your view of the landing runway into a more normal perspective.
   b. If you lose visual references at any time during the approach, you must execute a missed approach. The exception is if you lose sight due to a normal bank of your airplane.
   c. If, after planning a straight-in approach and descending to straight-in minimums, you decide that circling is necessary (e.g., because of unanticipated winds), but you are below the circling minimums, you must execute a missed approach.
5. ***Maneuver your airplane, after reaching the authorized MDA, and maintain that altitude within +100 ft./–0 ft. and a flight path that permits a normal landing on a runway at least 90° from the final approach course.***
   a. This is a requirement for your practical test only.

C. Common Errors during a Circling Approach

1. **Failure to have essential knowledge of the circling approach information on the instrument approach procedure chart.**
   a. Besides knowing all the information required for the IAP, you must also know how you will approach the airport and decide which maneuver path is the shortest to the base or downwind leg of the desired landing runway.
   b. Know the limits of the protected obstacle clearance area.

2. **Failure to adhere to the published MDA and visibility criteria during the circling approach maneuver.**
   a. You must not descend below the MDA until you are in a position to land safely on the desired runway using a normal rate of descent and normal maneuvers.
      1) The protected obstacle clearance is at the MDA. The circling approach does not mean you can go below the MDA to remain clear of clouds.
   b. Visibility criteria must be maintained throughout the approach.
3. **Inappropriate pilot technique during the transition from the circling maneuver to the landing approach.**
   a. Avoid banks in excess of 30°, even during VFR flights during an approach.
   b. You should be circling at your approach speed. Maintain that airspeed throughout the transition. Do not dive down to the runway at high airspeeds.
   c. You should set up a normal descent angle and stabilize your descent and airspeed as soon as possible during the transition.

END OF TASK

LANDING FROM A STRAIGHT-IN OR CIRCLING APPROACH

VI.E. TASK: LANDING FROM A STRAIGHT-IN OR CIRCLING APPROACH

REFERENCES: 14 CFR Parts 61, 91; AC 61-27; AIM.

Objective. To determine that the applicant:

1. Exhibits adequate knowledge of the elements related to the pilot's responsibilities, and the environmental, operational, and meteorological factors which affect a landing from a straight-in or a circling approach.
2. Transitions at the DH, MDA, or VDP to a visual flight condition, allowing for safe visual maneuvering and a normal landing.
3. Adheres to all ATC (or examiner) advisories such as NOTAMs, wind shear, wake turbulence, runway surface, braking conditions, and other operational considerations.
4. Completes appropriate checklist items for the pre-landing and landing phase.
5. Maintains positive aircraft control throughout the complete landing maneuver.

A. General Information

1. The objective of this task is to determine your knowledge and ability to perform a landing from a straight-in or circling approach procedure.

B. Task Objectives

1. **Exhibit adequate knowledge of the elements related to your responsibilities, and the environmental, operational, and meteorological factors which affect a landing from a straight-in or circling approach.**
   a. You are responsible for the safe operation of your airplane at all times.
      1) You are the final authority to determine whether you can safely land your airplane from a straight-in or circling approach.
   b. Environmental factors can include items such as ATC and runway length, surface, or conditions.
   c. Operational factors are not only the components of an approach (e.g., VOR, approach light system, etc.) but also the interaction between you and the environment and the way the environment affects your operations.
   d. Meteorological factors are the current weather conditions. These conditions must allow you the opportunity to attempt a landing.
   e. All of these factors are tied together in any situation, and in this task the situation is a landing from a straight-in or circling approach.
      1) All factors are interrelated, and you must make decisions throughout the maneuver on how changes or current conditions will affect the outcome of the landing.

2. Transition at the DH, MDA, or VDP to a visual flight condition, allowing for safe visual maneuvering and a normal landing.

a. A **visual descent point (VDP)** is a defined point on the final approach course of some nonprecision straight-in approach procedures from which a normal descent from the MDA to the runway touchdown point may be commenced, provided the approach threshold of that runway, approach lights, or other markings identifiable with the approach end of that runway are clearly visible to you.

1) VDPs are intended to provide additional guidance, and no special technique is required to fly an IAP with a VDP.

a) You should not descend below the MDA prior to reaching the VDP and acquiring the necessary visual references.

2) The VDP will normally be identified by DME on VOR or LOC procedures.

a) If your airplane is not equipped to identify the VDP, you should fly the IAP as though no VDP has been provided.

3) The VDP is identified on the profile view of the IAP chart by a bold-faced **V**.

b. At the DH, MDA, or VDP, you will be required to transition from instruments to visually identifying the runway environment and to perform a normal landing using normal maneuvers.

1) You must mentally prepare yourself for this transition.

c. At the DH, MDA, or VDP, you may see only the initial lights of the approach lighting system. As you continue down, more light will appear, and eventually you will see the beginning of the runway. You must have your airplane stabilized and correctly aligned.

1) Landing after an approach to minimums requires you, in the remaining several hundred feet, to slow your airplane to the final landing approach speed, add the remainder of flaps (if appropriate), and touch down safely.

d. To make this procedure easier and safer, practice and know your airplane. Before you attempt any approaches, you must feel comfortable flying your airplane. You must also know how your airplane responds and what configuration is needed for all flight regimes.

3. Adhere to all ATC (or examiner) advisories, such as NOTAMs, wind shear, wake turbulence, runway surface, braking conditions, and other operational considerations.

a. NOTAMs may inform you of inoperative components, such as runway lighting systems. Visualize how the information will affect your transition to a visual approach.

b. Wind shear is the unexpected change in wind direction and/or windspeed. During an approach, it can cause severe turbulence and a possible decrease to your airspeed (when a headwind changes to a tailwind), causing your airplane to stall (and crash).

1) The best method of dealing with wind shear is avoidance. You should never conduct approaches through, or in close proximity to, an active thunderstorm. Thunderstorms provide visible signs of possible wind-shear activity.

2) Many airports that are served by the air carriers employ one of the following systems for wind shear detection:

   a) Terminal Doppler weather radar (TDWR) is designed to advise the controller of wind shear and microburst events impacting all runways and the areas ½ mile on either side of the extended centerline of the runways out to 3 miles on final approach and 2 miles on departure.

   b) The weather systems processor (WSP) provides the same products as the TDWR but uses the weather channel capabilities of the existing airport surveillance radar.

   c) The low-level wind shear alert system (LLWAS) employs wind sensors around the airport to warn ATC of the presence of hazardous wind shear and microbursts in the vicinity of the airport.

3) Elsewhere, pilot reports from airplanes preceding you on the approach can be very informational.

4) If you are conducting an approach with possible wind shear or a thunderstorm nearby, you should consider

   a) Using more power during the approach

   b) Flying the approach at a faster airspeed (general rule: adding one-half the gust factor to your airspeed)

   c) Staying as high as feasible on the approach until it is necessary to descend for a safe landing (unless a lower altitude is necessary due to low ceilings)

   d) Initiating a go-around at the first sign of a change in airspeed or an unexpected pitch change

      i) The most important factor is to use full power to get the airplane climbing.

      ii) Many accidents caused by wind shear are due to a severe downdraft (or a rapid change from headwind to tailwind) that pushes the aircraft into the ground. In extreme cases, even the power of an airliner is unable to counteract the descent.

c. Ensure that you understand how closely you are following heavy or large airplanes on the approach or what your distance is from departing airplanes on the runway.

   1) A small airplane landing behind a heavy airplane should be separated by 5 mi. (4 miles behind a large airplane).

   2) If you are concerned, immediately inform ATC of your concern, and request additional separation.

d. Many times, especially during inclement weather, ATIS will include the statement, "Braking action advisories are in effect." This statement means that you should expect deteriorating braking conditions and will be advised of the conditions by ATC.

   1) Braking action reports are received by ATC from the airport management, which uses special equipment, or from pilots who have already landed.

2) The quality of the braking action is described as

a) Good
b) Fair
c) Poor
d) Nil

3) If ATC does not volunteer a braking report, you should request it.

4) If you are inbound and receive a braking action report of nil, you should immediately advise ATC that you would like to proceed to an alternate airport. Except for aircraft with reversing capability, landing at an airport with nil braking is too dangerous.

5) If the braking action is reported as fair or poor, you should consider going elsewhere. If you elect to land, you should plan your approach accordingly.

a) Land at the very beginning of the usable portion of the runway.

b) Land at the minimum practicable airspeed.

c) Land directly on the centerline, with the aircraft's longitudinal axis parallel to it.

d) Avoid sudden or heavy braking, which could cause a skid.

e) Use aerodynamic braking as much as feasible.

4. Complete the appropriate checklist items for the pre-landing and landing phase.

a. Perform your pre-landing checklist (from your *POH*) no later than your arrival over the FAF.

1) Once you transition to visual reference and prior to touchdown, you should confirm that all checklist items are completed.

b. After you have landed and cleared the runway, you should perform your after-landing checklist from your *POH*.

5. Maintain positive airplane control throughout the complete landing maneuver.

a. Remember your flight is not complete until your airplane is properly parked, shut down, and secured on the ramp.

b. Your first priority is to maintain positive control of your airplane.

C. Common Errors during a Landing from a Straight-in or Circling Approach

1. Inappropriate division of attention during the transition from instrument to visual flight conditions.

a. During your transition from instrument to visual conditions, you should divide your attention between instrument and visual references.

1) You may find yourself in lower clouds again, and you need to be able to adjust quickly to a return to instrument conditions.

b. You need to pick up the visual references so you can adjust your descent in order to make an approach using a normal rate of descent and normal maneuvers.

1) When using an ILS, you should remain on the glide slope even when in visual conditions.

2. Failure to complete the required checklist items.

a. You must use your checklists as you do when flying VFR.

b. The checklist is the only way to ensure that your airplane is in the proper configuration for landing.

3. Failure to properly plan and perform the turn to final approach during a circling approach.

a. During a landing from a circling approach, you must maneuver within the protected area for obstacle clearance. To do this, you must properly plan your turn to final approach.

b. Your turn to final approach must allow you to make a normal descent to the runway.

1) If the runway is equipped with a visual approach slope indicator, you should remain on or above the visual approach slope.

4. Improper technique for wind shear, wake turbulence, and crosswind.

a. The best method of dealing with wind shear is avoidance.

1) If you conduct a landing, use more power and a higher airspeed.

b. ATC is responsible for wake turbulence separation. If you want more separation, make a request to ATC.

c. Apply normal crosswind landing procedures to correct for a crosswind during the landing phase of the approach.

5. Failure to maintain positive airplane control throughout the complete landing maneuver.

a. You are responsible for the safe operation of your airplane.

b. Remember that your responsibility is first to fly and maintain positive control of your airplane and then to accomplish other tasks, such as talk to ATC.

END OF TASK -- END OF CHAPTER

CHAPTER VII
EMERGENCY OPERATIONS

This chapter explains the two single-engine tasks (A-B) of Emergency Operations. These tasks include both knowledge and skill. Your examiner is required to test you on both tasks.

Since the scope of this book is for single-engine airplanes, we have omitted the following two multiengine tasks from our discussion:

1. One Engine Inoperative During Straight-and-Level Flight and Turns
2. One Engine Inoperative -- Instrument Approach

These tasks are reproduced in Appendix A, FAA Instrument Rating Practical Test Standards, on page 339.

LOSS OF COMMUNICATIONS

VII.A. TASK: LOSS OF COMMUNICATIONS

REFERENCES: 14 CFR Parts 61, 91; AIM.

Objective. To determine that the applicant exhibits adequate knowledge of the elements related to applicable loss of communications procedures to include:

1. Recognizing loss of communication.
2. Continuing to destination according to the flight plan.
3. When to deviate from the flight plan.
4. Timing for beginning an approach at destination.

A. General Information

1. The objective of this task is for you to demonstrate your knowledge and ability to operate under IFR with the loss of communications.
2. It is virtually impossible to provide regulations and procedures applicable to all possible situations associated with two-way radio communications failure.
   a. When confronted with a situation not covered in the FARs during two-way radio communications failure, you should exercise good judgment in whatever action you decide to take.
   b. Whether two-way radio communications failure constitutes an emergency depends on the circumstances and is a decision you have to make.
      1) You should not be reluctant to deviate from any rule to the extent necessary to meet the emergency (FAR 91.3).

B. Task Objectives

1. **Recognize the loss of communication.**
   a. The loss of communication can be either an ATC outage or a malfunction in your communication system due to an equipment problem (e.g., radio, speaker, microphone, headset) or an electrical failure.
   b. If you are not sure whether it is an ATC outage or your own radio problem, you should wait at least 1 minute before switching frequencies.
      1) This is normally the maximum amount of time needed for ATC to switch to a backup radio.
   c. You should attempt to reestablish contact by switching to either your previous frequency or one to which you expect to be switched.
      1) You should also attempt to contact an FSS or an Aeronautical Radio Incorporated (ARINC) station.
         a) ARINC is a commercial communications corporation that operates radios serving aviation interests and has the capability of relaying information to and from ATC facilities throughout the country.
      2) You should also monitor any NAVAID voice feature and even attempt to make contact on 121.5 Mhz.

d. If you can only receive, ATC will probably have you use the IDENT feature on your transponder to determine whether you can still receive their communications.

e. Once you have determined that you have a loss of communications, you should squawk 7600 on your transponder.

2. Continue to your destination according to your flight plan.

3. Determine when to deviate from your flight plan.

NOTE: Elements 2 and 3 are interrelated, and your author feels that it is easier to understand FAR 91.185, IFR Operations: Two-Way Radio Communications Failure, by incorporating those elements into the following discussion.

a. Unless otherwise authorized by ATC, each pilot who has two-way radio communications failure when operating under IFR shall comply with the rules of FAR 91.185, below.

b. **VFR conditions:** If the failure occurs in VFR weather conditions, or if VFR weather conditions are encountered after the failure, continue the flight under VFR and land as soon as practicable.

1) The primary objective of this provision is to prevent extended IFR operation in the ATC system in VFR weather conditions.

a) You should recognize that operations in VFR conditions may unnecessarily as well as adversely affect other users of the airspace, since ATC may be required to reroute or delay other users in order to protect the aircraft experiencing communications failure.

2) The requirement to "land as soon as practicable" does not mean "as soon as possible."

a) Use good judgment. You are not required to land at an unauthorized airport or an airport unsuitable for your airplane, and you do not have to land only minutes short of your destination.

c. **IFR conditions:** If the failure occurs in IFR conditions or if you cannot comply with b. above, continue the flight as follows:

1) **Route:**

a) By the route assigned in the last ATC clearance received;

b) If being radar vectored, by the direct route from the point of radio failure to the fix, route, or airway specified in the vector clearance;

c) In the absence of an assigned route, by the route that ATC has advised may be expected in a further clearance; or

d) In the absence of an assigned route or a route that ATC has advised may be expected in a further clearance, by the route filed in your IFR flight plan.

2) **Altitude:** At the highest of the following altitudes for the route segment being flown:

a) The altitude or flight level assigned in the last ATC clearance received;

b) The minimum altitude for IFR operations; or

c) The altitude or flight level ATC has advised may be expected in a further clearance.

d. The intent of the altitude provisions in FAR 91.185 is for you to select the appropriate altitude for the particular route segment being flown and make the necessary altitude adjustments for subsequent route segments.

1) If you received an "expect further clearance" containing a higher altitude to expect at a specified time or fix, maintain the highest of the following altitudes until that time/fix:

a) The last assigned altitude
b) The minimum altitude for IFR operations

2) Upon reaching the time/fix specified, you should commence climbing to the altitude advised to expect.

3) If the radio failure occurs after the time/fix specified, the altitude to be expected is not applicable, and you should maintain the highest of your last assigned altitude or the minimum IFR altitude for the route segment being flown.

4) If you receive an "expect further clearance" containing a lower altitude, you should maintain the highest of your last assigned altitude or the minimum IFR altitude for the route segment being flown until that time specified in item 4.b. on page 331.

5) EXAMPLES:

a) You experience a two-way radio failure after being cleared along a route at an assigned altitude of 7,000 ft. If the next route segment has an MEA of 9,000 ft., you should begin your climb to 9,000 ft. at the fix where the MEA rises, or prior to the fix if necessary to comply with an MCA at the fix. If the next route segment has an MEA of 5,000 ft., you should descend to 7,000 ft. (your last assigned altitude) because that altitude is higher than the MEA.

b) You experience two-way radio failure while being progressively descended to lower altitudes to begin an approach after receiving a clearance to maintain 2,700 ft. until crossing the VOR and then being cleared for the approach. The MOCA along the airway is 2,700 ft., the MEA is 4,000 ft., and you are within 22 NM of the VOR.

i) You should maintain 2,700 ft. until crossing the VOR because that altitude is the minimum IFR altitude for the route segment being flown.

c) The MEA between A and B is 5,000 ft.; between C and D is 11,000 ft.; and between D and E is 7,000 ft. You have been cleared via A, B, C, and D, to E. While flying between A and B, your assigned altitude was 6,000 ft. and ATC advised you to expect 8,000 ft. at B. Prior to receiving the higher altitude assignment (and before reaching B), you experience two-way radio failure.

i) You should maintain 6,000 ft. to B, then climb to 8,000 ft. (the altitude advised to expect).

ii) You should maintain 8,000 ft., then climb to 11,000 ft. at C, or prior to C if necessary to comply with an MCA at C.

iii) Upon reaching D, you should descend to 8,000 ft. (even though the MEA is 7,000 ft.) because 8,000 ft. was the highest of the altitude situations.

4. Know when to begin an approach at your destination.

a. If your destination airport has more than one published IAP, you must select the IAP you want to execute.

   1) If the approach you select has more than one IAF, you may select the most appropriate IAF.

b. You may not begin an approach unless certain conditions outlined in FAR 91.185 are met.

   1) When the clearance limit specified in your last ATC clearance is a fix from which an approach begins, commence descent, or descent and approach, as close as possible to the expect-further-clearance (EFC) time if one has been received or, if one has not been received, as close as possible to the estimated time of arrival (ETA), as calculated from the filed or amended (with ATC) estimated time en route (ETE).

   2) If the clearance limit is not a fix from which an approach begins, leave the clearance limit at the EFC time if one has been received or, if one has not been received, upon arrival over the clearance limit, and proceed to a fix from which an approach begins. Then commence descent, or descent and approach, as close as possible to the ETA, as calculated from the filed or amended (with ATC) ETE.

   3) If you have been cleared to your intended airport, proceed to a fix from which an approach begins, and commence descent and approach without delay.

END OF TASK

LOSS OF GYRO ATTITUDE AND/OR HEADING INDICATORS

VII.B. TASK: LOSS OF GYRO ATTITUDE AND/OR HEADING INDICATORS

REFERENCES: 14 CFR Part 61; AC 61-27; IAP.

Note: This approach shall count as one of the required nonprecision approaches.

Objective. To determine that the applicant:

1. Exhibits adequate knowledge of the elements relating to recognizing if attitude indicator and/or heading indicator is inaccurate or inoperative, and advises ATC or the examiner.
2. Advises ATC or examiner any time the aircraft is unable to comply with a clearance.
3. Demonstrates a nonprecision instrument approach without gyro attitude and heading indicators using the objectives of the nonprecision approach task (Area of Operation VI, Task A).

A. General Information

1. The objective of this task is for you to demonstrate your knowledge of recognizing a loss of your AI and/or HI and demonstrate your ability to safely conduct a nonprecision approach.

B. Task Objectives

1. **Exhibit adequate knowledge of the elements relating to recognizing if the attitude indicator (AI) and/or heading indicator (HI) is inaccurate or inoperative, and advise ATC or your examiner.**
   a. In most single-engine airplanes, the gyros for the HI and AI are powered by a vacuum system. The vacuum suction gauge indicates the operational status of this system.
      1) The correct operating range is found in your *POH*.
      2) In a few airplanes, the gyros for HI and/or AI may be electrically powered. Know the system for your airplane.
   b. Like any mechanical system, this system can malfunction suddenly or slowly.
      1) A slow decrease in vacuum pressure (as noted on the gauge) may indicate a dirty filter, dirty screens, a sticking regulator, a worn-out air pump, or a leak in the system.
      2) A sudden loss of pressure may indicate a sheared pump drive, pump failure, a collapsed line, or an inoperative gauge.
   c. A complete loss of vacuum pressure is noticeable immediately on the vacuum gauge or within minutes by incorrect AI and HI indications.
      1) A slow deterioration may lead to sluggish or incorrect readings, which may trick you if you are not constantly cross-checking and interpreting **all** instruments, including the vacuum pressure gauge.
         a) If conflicting information develops, use all of your instruments to recognize the inoperative instrument(s). Do not make control inputs based only on the information presented on the AI.

2) EXAMPLES:

a) If the AI shows a nose-high attitude, but the ASI is not decreasing and the ALT is not increasing, you should assume the AI is inoperative.

i) If the AI shows wings level, but the TC, HI, or magnetic compass shows a turn, you should assume a gyro failure.

ii) Compare all gyros to determine the one(s) in error.

b) You will most likely detect a HI failure when you have problems in continuing to track your intended course with reference to your navigation instrument.

i) Another method of detection is to notice whether the HI is moving while the TC is showing wings level with the ball centered.

d. Once you have identified the inoperative instrument(s), you should cover it (them) up.

e. IFR flight with an inoperative AI and HI is at least an urgent (i.e., semi-emergency) situation. Remember, you must decide whether it is an emergency situation and, if so, declare an emergency with ATC.

1) It is at least an urgent situation since you may not be able to comply immediately and accurately with all ATC clearances.

2) Inform ATC as soon as the situation occurs.

f. Partial-panel flight

1) The term "partial panel" usually means that one or both of your vacuum instruments (AI, HI) are lost. Generally, the failure lies in the vacuum pump, and both instruments will be inoperative.

2) With these instruments out, you will have to use the TC as primary for bank with the compass providing supplemental information. For pitch, you must use a combination of the ALT, ASI, and VSI.

3) Because of the importance of reliable bank information when the AI and HI have failed, it is preferable to have an aircraft equipped with a TC rather than a turn-and-slip indicator (T&SI).

a) Partial-panel flight is easier with the miniature aircraft than with the needle of the T&SI.

i) The TC indicates rate of roll in addition to rate of turn.

b) This is especially true in turbulence, when a T&SI is frequently unflyable due to its lack of stability.

4) Prior to training for partial-panel flight or for any recurrency training, you should reemphasize compass turns and timed turns. See Task IV.E., Timed Turns to Magnetic Compass Headings, beginning on page 175.

a) Both are essential for partial-panel flying.

5) Some high-performance aircraft are being equipped to eliminate the potential of being in a partial-panel situation.
   a) Multiengine aircraft have a vacuum pump on each engine; thus, it would take a double pump failure to cause a vacuum loss.
   b) Many single-engine aircraft have backup emergency pumps which, when selected, will provide vacuum pressure to the necessary flight instruments.
      i) Some also have a backup electrically powered AI.
6) Obtain a pad of 2-inch square Post-it™ notes (manufactured by 3M Company) at a local office supply store. Each sheet of paper in the pad contains adhesive along one edge. Alternatively, obtain the proper size non-transparent suction cups from a hardware or variety store.
   a) Keep Post-it™ notes or suction cups in your flight bag when flying IFR. THIS IS VERY IMPORTANT. If you are under IFR and lose your AI, you probably will be distracted.
      i) When the AI fails, the horizon flops over to the left or right, indicating a steep right or left turn.
      ii) As your horizon fails, you may need to cover the AI initially with your hand to avoid possible confusion.
      iii) As soon as practicable, cover the failed instrument with paper or a suction cup to free your hand.
7) During partial-panel flight, your first priority is to maintain wings level. With a TC and some practice, you can almost maintain wings level as well as you can with an operational AI.
   a) After a few minutes (especially if you use Post-it™ notes), your scan will start to return automatically to the TC to maintain wings level.
8) The next priority is to maintain the proper pitch attitude. Use a combination of the ALT, ASI, and VSI. By using information from all three instruments, you can maintain an accurate pitch attitude.
   a) If you notice a climb on the VSI and the altimeter, lower the nose slightly. The reverse is true if you notice a descent.
   b) You must remember to use all three instruments, but primarily the ALT. If you concentrate on only one instrument, you will tend to overcontrol and cause continuing altitude variations.
   c) One helpful hint is to make all moves slowly and deliberately. Large changes will cause instability and possible disorientation.
9) Once you have stabilized the aircraft in both pitch and roll, you can concentrate on navigation. Your navigation instruments should continue to provide course information. As for heading, you must revert to your magnetic compass.
   a) Remember to make use of ATC assistance. Since you have informed the controller of your situation, (s)he will do everything reasonable to clear your way, divert other traffic, and provide all the help possible. If you are having trouble maintaining a specific altitude, tell the controller, and (s)he will provide you with a block altitude.

10) Your primary objective is to reach VFR weather conditions. Ask ATC. It will be able to check the weather in the vicinity as well as ask other aircraft for PIREPs. VFR may be only a descent away.
    a) When you do start maneuvering using partial panel, you must remember to do everything slowly and deliberately. Keeping the aircraft stable and in control is of utmost importance.
    b) As long as you realize the navigation problems associated with a magnetic compass, it can be used with reasonable accuracy (periodic partial-panel practice is essential if you fly IFR!).
    c) If unable to get to VFR weather conditions within your fuel supply, you will be forced to execute an approach. It can be a regular approach (e.g., ILS, VOR, GPS) on partial panel or a no-gyro approach.
    d) The choice of approaches depends on availability and the weather conditions. On all approaches, use ATC radar (no-gyro) vectoring if possible. Also try to pick the best approach available (e.g., choose an ILS over a VOR). You want to execute one and only one approach.

2. **Advise ATC or your examiner any time your airplane is unable to comply with a clearance.**
    a. You must inform ATC (or your examiner) any time you cannot comply with a clearance.
        1) Request an amended clearance.
3. ***Demonstrate a nonprecision instrument approach without gyro attitude and heading indicators using the objectives of the nonprecision approach task.***
    a. See Task VI.A., Nonprecision Instrument Approach, beginning on page 232.

END OF TASK -- END OF CHAPTER

CHAPTER VIII
POSTFLIGHT PROCEDURES

This chapter explains the one task of Postflight Procedures. This task includes both knowledge and skill. Your examiner is required to test you on this task.

CHECKING INSTRUMENTS AND EQUIPMENT

VIII.A. TASK: CHECKING INSTRUMENTS AND EQUIPMENT

REFERENCES: 14 CFR Parts 61, 91.

Objective. To determine that the applicant:

1. Exhibits adequate knowledge of the elements relating to all navigation equipment for proper operation.
2. Notes all flight equipment for proper operation.
3. Notes all equipment and/or aircraft malfunctions and makes a written record of improper operation or failure of such equipment.

A. General Information
   1. The objective of this task is for you to demonstrate your ability to check instruments and equipment for proper operation.

B. Task Objectives
   1. **Exhibit adequate knowledge of the elements relating to all navigation equipment for proper operation.**
      a. To know whether the navigation equipment in your airplane is operating properly, you must first be thoroughly trained in the proper operation of the equipment.
      b. This task relates to the discussion in Task II.B., Aircraft Flight Instruments and Navigation Equipment, beginning on page 72.
   2. **Note all flight equipment for proper operation.**
      a. During your flight, you should know if the flight instruments and navigation equipment are operating properly.
      b. If you have problems with any flight equipment, you should make a note of it during flight.
   3. **Note all equipment and/or aircraft malfunctions and make a written record of improper operation or failure of such equipment.**
      a. You should make a written record of any equipment and/or airplane malfunctions so that those items can be repaired before your next flight.
         1) If you rent an airplane, there is usually an equipment and/or aircraft malfunction log (sometimes called a squawk sheet) to record any problems.
            a) A notation in the log will warn other pilots and maintenance personnel of problems.

END OF TASK -- END OF CHAPTER

APPENDIX A
FAA INSTRUMENT RATING PRACTICAL TEST STANDARDS (FAA-S-8081-4C with Changes 1 and 2)

The purpose of this appendix is to reproduce verbatim what you would get in PTS reprint books that are normally sold for $5.00 at FBOs. All of the tasks in this PTS are reproduced (and explained, discussed, and illustrated!!) elsewhere throughout this book.

FOREWORD

The Instrument Rating Practical Test Standards (PTS) book is published by the Federal Aviation Administration (FAA) to establish the standards for instrument rating certification practical tests for the airplane, helicopter, and powered lift, category and classes. These practical test standards shall also be used for the instrument portion of the commercial pilot-airship practical test. FAA inspectors and designated pilot examiners shall conduct practical tests in compliance with these standards. Flight instructors and applicants should find these standards helpful during training and when preparing for practical tests.

Richard O. Gordon
Acting Director, Flight Standards Service

INTRODUCTION

General Information

The Flight Standards Service of the Federal Aviation Administration (FAA) has developed this practical test standards book to be used by FAA inspectors and designated pilot examiners when conducting instrument rating: airplane, helicopter, and powered lift practical tests, and instrument proficiency checks for all aircraft. These practical test standards shall also be used for the instrument portion of the commercial pilot-airship practical test. Instructors are expected to use this book when preparing applicants for practical tests. Applicants should be familiar with this book and refer to these standards during their training.

This publication sets forth the practical test requirements for the addition of an instrument rating to a pilot certificate in airplanes, helicopters, and powered-lift aircraft.

Information considered directive in nature is described in this practical test standards book in terms, such as "shall" and "must," indicating the actions are mandatory. Guidance information is described in terms, such as "should" and "may," indicating the actions are desirable or permissive, but not mandatory.

The FAA gratefully acknowledges the valuable assistance provided by many individuals and companies who contributed their time and talent in assisting with the revision of these practical test standards.

These practical test standards may be accessed through the FedWorld Information System by computer modem at 703-321-3339. These standards may also be accessed on the Internet at http://www.fedworld.gov/pub/faa-att/faa-att.htm. This address accesses the index of training and testing files in the FAA-ATT Library on FedWorld. Subsequent changes to these standards, in accordance with AC 60-27, Announcement of Availability: Changes to Practical Test Standards, will be available through FedWorld and then later incorporated into a printed revision. For a listing of changes, AFS-600's Internet web site may be accessed at www.mmac.jccbi.gov/afs/afs600.

This publication may be purchased from the Superintendent of Documents, U.S. Government Printing Office, Washington, DC 20402.

Comments regarding this publication should be sent to:

U.S. Department of Transportation
Federal Aviation Administration
Flight Standards Service
Airman Testing Standards Branch, AFS-630
P.O. Box 25082
Oklahoma City, OK 73125

Practical Test Standard Concept

Title 14 of the Code of Federal Regulations (14 CFR) part 61 specifies the areas in which knowledge and skill must be demonstrated by the applicant before the issuance of an instrument rating. The CFR's provide the flexibility to permit the FAA to publish practical test standards containing specific TASKS in which pilot competency shall be demonstrated. The FAA will revise this book whenever it is determined that changes are needed in the interest of safety. Adherence to the provisions of the regulations and the practical test standards is mandatory for evaluation of instrument pilot applicants.

Practical Test Book Description

This test book contains the instrument rating practical test standards for airplane, helicopter, and powered lift. It also contains TASK requirements for the addition of airplane, helicopter, or powered lift, if an instrument rating is possessed by the applicant in at least one other aircraft category. Refer to the commercial pilot-airship practical test standard to determine the instrument TASKS required for that practical test. Required TASKS for instrument proficiency checks (PC) are also contained in these practical test standards.

Practical Test Standards Description

AREAS OF OPERATION are phases of the practical test arranged in a logical sequence within each standard. They begin with preflight preparation and end with postflight procedures. The examiner may conduct the practical test in any sequence that results in a complete and efficient test.

TASKS are titles of knowledge areas, flight procedures, or maneuvers appropriate to an AREA OF OPERATION.

The applicant who holds an airplane, helicopter, or powered lift instrument rating will not have to take the entire test when applying for an added rating. The TASKS required for each additional instrument rating are shown in the Rating Task Table on page 343.

Applicants for an instrument proficiency check required by 14 CFR section 61.57, must perform to the standards of the TASKS listed under PC in the Rating Task Table on page 343.

NOTE is used to emphasize special considerations required in the AREA OF OPERATION or TASK.

The REFERENCE identifies the publication(s) that describe(s) the TASK. Descriptions of TASKS are not included in the standards because this information can be found in the current issue of the listed references. Publications other than those listed may be used for references if their content conveys substantially the same meaning as the referenced publications. These practical test standards are based on the following references. The latest revision of these references shall be used.

14 CFR part 61	Certification: Pilots and Flight Instructors
14 CFR part 91	General Operating and Flight Rules
AC 00-6	Aviation Weather
AC 00-45	Aviation Weather Services
AC 60-28	English Language Skill Standards Required by 14 CFR parts 61, 63, and 65
AC 61-21	Flight Training Handbook
AC 61-23	Pilot's Handbook of Aeronautical Knowledge
AC 61-27	Instrument Flying Handbook
AC 61-84	Role of Preflight Preparation
AC 90-48	Pilot's Role in Collision Avoidance
AC 90-94	Guidelines for Using Global Positioning Systems
AIM	Aeronautical Information Manual
DP's	Instrument Departure Procedures
STAR's	Standard Terminal Arrivals
AFD	Airport Facility Directory
FDC NOTAM's	National Flight Data Center Notices to Airmen
IAP	Instrument Approach Procedures
Others	Pertinent Pilot's Operating Handbooks FAA-Approved Flight Manuals En Route Low Altitude Charts

The Objective lists the important elements that must be satisfactorily performed to demonstrate competency in a TASK. The Objective includes:

1. specifically what the applicant should be able to do;
2. the conditions under which the TASK is to be performed; and
3. the acceptable standards of performance.

Use of the Practical Test Standards Book

The instrument rating practical test standards are designed to evaluate competency in both knowledge and skill.

The FAA requires that all practical tests be conducted in accordance with the appropriate practical test standards and the policies set forth in the INTRODUCTION. Instrument rating applicants shall be evaluated in ALL TASKS included in the AREAS OF OPERATION of the appropriate practical test standard (unless instructed or noted otherwise).

In preparation for each practical test, the examiner shall develop a written "plan of action." The "plan of action" shall include all TASKS in each AREA OF OPERATION. If the elements in one TASK have already been evaluated in another TASK, they need not be repeated. For example: the "plan of action" need not include evaluating the applicant on complying with markings, signals, and clearances at the end of the flight if that element was sufficiently observed at the beginning of the flight. Any TASKS selected for evaluation during a practical test shall be evaluated in its entirety.

The TASKS apply to airplanes, helicopters, powered lift, and airships. In certain instances, NOTES describe differences in the performance of a TASK by an "airplane" applicant, "helicopter" applicant, or "powered lift" applicant. When using the practical test standards book, the examiner must evaluate the applicant's knowledge and skill in sufficient depth to determine that the standards of performance listed for all TASKS are met.

All TASKS in these practical test standards are required for the issuance of an instrument rating in airplanes, helicopters, and powered lift. However, when a particular element is not appropriate to the aircraft, its equipment, or operational capability, that element may be omitted. Examples of these element exceptions would be high altitude weather phenomena for helicopters, integrated flight systems for aircraft not so equipped, or other situations where the aircraft or operation is not compatible with the requirement of the element.

The examiner is not required to follow the precise order in which the AREAS OF OPERATION and TASKS appear in this book. The examiner may change the sequence or combine TASKS with similar Objectives to have an orderly and efficient flow of the practical test. For example, emergency descents may be combined with high altitude operations. The examiner's "plan of action" shall include the order and combination of TASKS to be demonstrated by the applicant in a manner that will result in an efficient and valid test.

Examiners shall place special emphasis upon areas of aircraft operation that are most critical to flight safety. Among these are precise aircraft control and sound judgment in Aeronautical Decision Making (ADM). Although these areas may or may not be shown under each TASK, they are essential to flight safety and shall receive careful evaluation throughout the practical test. If these areas are shown in the Objective, additional emphasis shall be placed on them. THE EXAMINER SHALL ALSO EMPHASIZE STALL/SPIN AWARENESS, WAKE TURBULENCE AVOIDANCE, LOW LEVEL WIND SHEAR, COLLISION AVOIDANCE, RUNWAY INCURSION AVOIDANCE, AND CHECKLIST USAGE.

Practical Test Prerequisites: Instrument Rating

An applicant for an instrument rating practical test is required by 14 CFR part 61 to:

1. hold at least a current private pilot certificate with an aircraft rating appropriate to the instrument rating sought;
2. pass the appropriate instrument rating knowledge test since the beginning of the 24th month before the month in which the practical test is taken;
3. obtain the applicable instruction and aeronautical experience prescribed for the instrument rating sought;
4. hold at least a current third-class medical certificate;
5. be able to read, speak, write, and understand the English language; and
6. obtain a written statement from an authorized flight instructor certifying that the applicant has been given flight instruction in preparation for the practical test within 60 days preceding the date of application. The statement shall also state that the instructor finds the applicant competent to pass the practical test and that the applicant has satisfactory knowledge of the subject area(s) in which a deficiency was indicated by the airman knowledge test report.

If there are questions concerning English language requirements, refer to AC 60-28, English Language Skill Standards Required by 14 CFR parts 61, 63, and 65, or your local Flight Standards District Office (FSDO). English language requirements should be determined to be met prior to beginning the practical test.

Aircraft and Equipment Required for the Practical Test

The instrument rating applicant is required by 14 CFR part 61 to provide an airworthy, certificated aircraft for use during the practical test. Its operating limitations must not prohibit the TASKS required on the practical test. Flight instruments are those required for controlling the aircraft without outside references. The required radio equipment is that which is necessary for communications with air traffic control (ATC), and for the performance of two of the following nonprecision approaches: (VOR, NDB, GPS, LOC, LDA, SDF) and one precision approach: (glide slope, localizer, marker beacon, and approach lights).

To obtain an **instrument rating with multiengine privileges**, an applicant must demonstrate competency in a multiengine airplane not limited to center thrust. The multiengine airplane that is used to obtain unlimited multiengine privileges must have a V_{MC} speed established by the manufacturer, and produce an asymmetrical thrust configuration with the loss of one or more engines. If an instrument flight test is conducted in a multiengine airplane limited to center thrust, a limitation shall be placed on the applicant's certificate: (INSTRUMENT RATING, AIRPLANE MULTIENGINE, LIMITED TO CENTER THRUST).

Use of FAA-Approved Flight Simulator or Flight Training Device

An airman applicant for instrument rating certification is authorized to use an FAA-qualified and approved flight simulator or flight training device, to complete certain flight TASK requirements listed in this practical test standard.

When flight TASKS are accomplished in an aircraft, certain TASK elements may be accomplished through "simulated" actions in the interest of safety and practicality, but when accomplished in a flight simulator or flight training device, these same actions would not be "simulated." For example, when in an aircraft, a simulated engine fire may be addressed by retarding the throttle to idle, simulating the shutdown of the engine, simulating the discharge of the fire suppression agent, if applicable, simulating the disconnect of associated electrical, hydraulic, and pneumatics systems, etc. However, when the same emergency condition is addressed in a flight simulator or flight training device, all TASK elements must be accomplished as would be expected under actual circumstances.

Similarly, safety of flight precautions taken in the aircraft for the accomplishment of a specific maneuver or procedure (such as limiting altitude in an approach to stall or setting maximum airspeed for an engine failure expected to result in a rejected takeoff) need not be taken when a flight simulator or flight training device is used.

It is important to understand that whether accomplished in an aircraft, flight simulator or flight training device, all TASKS and elements for each maneuver or procedure shall have the same performance standards applied equally for determination of overall satisfactory performance.

The applicant must demonstrate all of the instrument approach procedures required by 14 CFR part 61. At least one instrument approach procedure must be demonstrated in an airplane, helicopter, or powered lift as appropriate. At least one precision and one nonprecision approach not selected for actual flight demonstration may be performed in flight simulators or flight training devices that meet the requirements of appendix 1 of this practical test standard.

Examiner[1] Responsibility

The examiner conducting the practical test is responsible for determining that the applicant meets the acceptable standards of knowledge and skill of each TASK within the appropriate practical test standard. Since there is no formal division between the "oral" and "skill" portions of the practical test, this becomes an ongoing process throughout the test. To avoid unnecessary distractions, oral questioning should be used judiciously at all times, especially during the flight portion of the practical test.

Examiners shall test to the greatest extent practicable the applicant's correlative abilities rather than mere rote enumeration of facts throughout the practical test.

During the flight portion of the practical test, the examiner shall evaluate the applicant's use of visual scanning, and collision avoidance procedures, when appropriate. Except for takeoff and landing, all TASKS shall be conducted solely by reference to instruments under actual or simulated instrument flight conditions.

The examiner may not assist the applicant in the management of the aircraft, radio communications, navigational equipment, and navigational charts. In the event the test is conducted in an aircraft operation requiring a crew of two, the examiner may assume the duties of the second in command. Most helicopters certified for IFR operations must be flown using two pilots or a single pilot with an approved autopilot or a stability augmentation system (SAS). Therefore, when conducting practical tests in a helicopter (without autopilot, SAS, or copilot), examiners may act as an autopilot (e.g., hold heading and altitude), when requested, to allow applicants to tune radios, select charts, etc. Examiners may perform the same functions as an autopilot but should not act as a copilot performing more extensive duties. The examiner shall remain alert for other traffic at all times. The examiner shall use proper ATC terminology when simulating ATC clearances.

Satisfactory Performance

Satisfactory performance to meet the requirements for certification is based on the applicant's ability to safely:

1. perform the approved AREAS OF OPERATION for the certificate or rating sought within the approved standards;
2. demonstrate mastery of the aircraft with the successful outcome of each TASK performed never seriously in doubt;
3. demonstrate satisfactory proficiency and competency within the approved standards;
4. demonstrate sound judgment and ADM; and
5. demonstrate single-pilot competence if the aircraft is type certificated for single-pilot operations.

---

[1] The word "examiner" is used throughout the standard to denote either the FAA inspector or FAA designated pilot examiner who conducts an official practical test.

Unsatisfactory Performance

If, in the judgment of the examiner, the applicant does not meet the standards of performance of any TASK performed, the associated AREA OF OPERATION is failed and therefore, the practical test is failed. The examiner or applicant may discontinue the test at any time after the failure of an AREA OF OPERATION which makes the applicant ineligible for the certificate or rating sought. The test will be continued only with the consent of the applicant. If the test is either continued or discontinued, the applicant is entitled to credit for only those AREAS OF OPERATION satisfactorily performed. However, during the retest and at the discretion of the examiner, any TASK may be reevaluated including those previously passed.

Typical areas of unsatisfactory performance and grounds for disqualification are:

1. Any action or lack of action by the applicant that requires corrective intervention by the examiner to maintain safe flight.
2. Failure to use proper and effective visual scanning techniques, when applicable, to clear the area before and while performing maneuvers.
3. Consistently exceeding tolerances stated in the Objectives.
4. Failure to take prompt corrective action when tolerances are exceeded.

When a disapproval notice is issued, the examiner shall record the applicant's unsatisfactory performance in terms of AREA OF OPERATIONS appropriate to the practical test conducted.

Letter of Discontinuance

When a practical test is discontinued for reasons other than unsatisfactory performance (i.e., equipment failure, weather, illness), FAA Form 8710.1, Airman Certificate and/or Rating Application, and, if applicable, AC Form 8080-2, Airman Knowledge Test Report, shall be returned to the applicant. The examiner at that time should prepare, sign, and issue a Letter of Discontinuance to the applicant. The Letter of Discontinuance should identify the AREAS OF OPERATION of the practical test that were successfully completed. The applicant shall be advised that the Letter of Discontinuance shall be presented to the examiner when the practical test is resumed, and made part of the certification file.

Crew Resource Management (CRM)

CRM " ...refers to the effective use of ALL available resources; human resources, hardware, and information." Human resources "...includes all other groups routinely working with the cockpit crew (or pilot) who are involved in decisions that are required to operate a flight safely. These groups include, but are not limited to: dispatchers, cabin crewmembers, maintenance personnel, and air traffic controllers." CRM is not a single TASK, it is a set of skill competencies that must be evident in all TASKS in this practical test standard as applied to either single pilot or a crew operation. Examiners are required to exercise proper CRM competencies in conducting tests as well as expecting the same from applicants.

Applicant's Use of Checklists

Throughout the practical test, the applicant is evaluated on the use of an appropriate checklist. Proper use is dependent on the specific TASK being evaluated. The situation may be such that the use of the checklist, while accomplishing elements of an Objective, would be either unsafe or unfeasible, especially in a single-pilot operation. In this case, the method might demand the need to review the checklist after the elements have been met. In any case, use of a checklist must consider proper scanning vigilance and division of attention at all times.

Use of Distractions During Practical Tests

Numerous studies indicate that many accidents have occurred when the pilot has been distracted during critical phases of flight. To evaluate the pilot's ability to utilize proper control technique while dividing attention both inside and outside the cockpit, the examiner shall cause a realistic distraction during the flight portion of the practical test to evaluate the applicant's ability to divide attention while maintaining safe flight.

Metric Conversion Initiative

To assist the pilots in understanding and using the metric measurement system, the practical test standards refer to the metric equivalent of various altitudes throughout. The inclusion of meters is intended to familiarize pilots with its use. The metric altimeter is arranged in 10 meter increments; therefore, when converting from feet to meters, the exact conversion, being too exact for practical purposes, is rounded to the nearest 10 meter increment or even altitude as necessary.

Positive Exchange of Flight Controls

During flight, there must always be a clear understanding between pilots of who has control of the aircraft. Prior to flight, a briefing should be conducted that includes the procedure for the exchange of flight controls. A positive three-step process in the exchange of flight controls between pilots is a proven procedure and one that is strongly recommended.

When one pilot wishes to give the other pilot control of the aircraft, he or she will say "You have the flight controls." The other pilot acknowledges immediately by saying, "I have the flight controls." The first pilot again says "You have the flight controls." When control is returned to the first pilot, follow the same procedure. A visual check is recommended to verify that the exchange has occurred. There should never be any doubt as to who is flying the aircraft.

Flight Instructor Responsibility

An appropriately rated flight instructor is responsible for training the instrument rating pilot applicant to acceptable standards in all subject matter areas, procedures, and maneuvers included in the TASKS within the appropriate instrument rating pilot practical test standard. Because of the impact of their teaching activities in developing safe, proficient pilots, flight instructors should exhibit a high level of knowledge, skill, and the ability to impart that knowledge and skill to students. Additionally, the flight instructor must certify that the applicant is able to perform safely as an instrument pilot and is competent to pass the required practical test.

Throughout the applicant's training, the flight instructor is responsible for emphasizing the performance of effective visual scanning, collision avoidance, and runway incursion avoidance procedures. These areas are covered, in part, in AC 90-48, Pilot's Role in Collision Avoidance; AC 61-21, Flight Training Handbook; AC 61-23, Pilot's Handbook of Aeronautical Knowledge; and the Aeronautical Information Manual.

Emphasis on Attitude Instrument Flying and Partial-Panel Skills

The FAA is concerned about numerous fatal aircraft accidents involving spatial disorientation of instrument-rated pilots who have attempted to control and maneuver their aircraft in clouds with inoperative gyroscopic heading and attitude indicators.

Many of the light aircraft operated in instrument meteorological conditions (IMC) are not equipped with dual, independent, gyroscopic heading or attitude indicators and in many cases are equipped with only a single-vacuum source. Therefore, the FAA has stressed that it is imperative for instrument pilots to acquire and maintain adequate partial-panel instrument skills and that they be cautioned not to be overly reliant upon the gyro-instrument systems.

The instrument rating practical test standards place increased emphasis on basic attitude instrument flying and require the demonstration of partial-panel, nonprecision instrument approach procedures.

Applicants may have an unfair advantage during partial-panel TASKS during an instrument approach due to the location of the magnetic compass in some aircraft. When cross-checking the magnetic compass heading, a view of the runway or other visual clue may be sighted. It is the examiner's responsibility to determine if the applicant is receiving visual clues from outside the cockpit. If an examiner feels that the applicant is receiving outside visual clues, the examiner may devise other options to limit the applicant's view. By no means shall the examiner limit his or her view as the safety pilot.

AREA OF OPERATION IV requires the performance of basic instrument flight TASKS under both full-panel and partial-panel conditions. These TASKS are described in detail in AC 61-27, Instrument Flying Handbook. The TASKS require a knowledge of attitude instrument flying procedures and a demonstration of the skills to perform the basic instrument maneuvers with full-instrument-panel and with certain instruments inoperative. The attitude instrument flying system of teaching is described in AC 61-27 and is recommended by the FAA because it requires specific knowledge and interpretation of each individual instrument during training. The Instrument Flight Instructor Lesson Guide in AC 61-27 also provides a course of training which is designed to develop the student's partial-panel skills.

A nonprecision partial-panel approach is considered one of the most demanding situations that could be encountered. If applicants can master this situation, they can successfully complete a less difficult precision approach. **If an actual partial-panel approach in IMC becomes necessary, a less difficult precision approach should be requested, if available. Sound judgment would normally dictate such requests. However, this TASK during the instrument practical test requires that a nonprecision approach be performed.**

Examiners should determine that the applicant demonstrates competency in either the PRIMARY and SUPPORTING or the CONTROL and PERFORMANCE CONCEPT method of instrument flying.

RATING TASK TABLE

ADDITIONAL INSTRUMENT RATING DESIRED				
AREA OF OPERATION	Required TASKS are indicated by either the TASK letter(s) that apply(s) or an indication that all or none of the TASKS must be tested.			
	IA	IH	IPL	PC
I	NONE	NONE	NONE	NONE
II	A,C	A,C	A,C	A,B,C
III	NONE	NONE	NONE	A,B,C
IV	A,B,C, D,F,G	A,B,C, D,F,G	A,B,C, D,F,G	A,B,C, D,G
V	NONE	NONE	NONE	ALL
VI	ALL	A,B,C,E	A,B,C,E	ALL*
VII	ALL**	ALL**	ALL**	ALL**
VIII	ALL	ALL	ALL	ALL

LEGEND
IA - Instrument airplane
IH - Instrument helicopter
IPL - Instrument powered lift
PC - Proficiency check

NOTE: Except as noted, all TASKS are required for *initial issuance* of an instrument rating.

*TASK D, Circling Approach, is applicable *only* to the *airplane* category.

**TASKS B and C are applicable *only* to *multiengine airplanes*.

CONTENTS

APPLICANT'S PRACTICAL TEST CHECKLIST

APPOINTMENT WITH EXAMINER:

EXAMINER'S NAME ____________________

LOCATION ____________________

DATE/TIME ____________________

ACCEPTABLE AIRCRAFT

- ☐ View-limiting device
- ☐ Aircraft Documents: Airworthiness Certificate
- ☐ Registration Certificate
- ☐ Rating Limitations
- ☐ Aircraft Maintenance Records: Airworthiness Inspections

Personal Equipment

- ☐ Current Aeronautical Charts
- ☐ Computer and Plotter
- ☐ Flight Plan Form
- ☐ Flight Logs
- ☐ Current AIM

Personal Records

- ☐ Identification - Photo/Signature ID
- ☐ Pilot Certificate
- ☐ Medical Certificate
- ☐ Completed FAA Form 8710-1, Application for an Airman Certificate and/or Rating
- ☐ Airman Knowledge Test Report
- ☐ Logbook with Instructor's Endorsement
- ☐ Notice of Disapproval (if applicable)
- ☐ Approved School Graduation Certificate (if applicable)
- ☐ Examiner's Fee (if applicable)

EXAMINER'S PRACTICAL TEST CHECKLIST

APPLICANT'S NAME ____________________

LOCATION ____________________

DATE/TIME ____________________

I. PREFLIGHT PREPARATION

- ☐ **A.** Weather Information
- ☐ **B.** Cross-Country Flight Planning

II. PREFLIGHT PROCEDURES

- ☐ **A.** Aircraft Systems Related to IFR Operations
- ☐ **B.** Aircraft Flight Instruments and Navigation Equipment
- ☐ **C.** Instrument Cockpit Check

III. AIR TRAFFIC CONTROL CLEARANCES AND PROCEDURES

- ☐ **A.** Air Traffic Control Clearances
- ☐ **B.** Compliance with Departure, En Route, and Arrival Procedures and Clearances
- ☐ **C.** Holding Procedures

IV. FLIGHT BY REFERENCE TO INSTRUMENTS

- ☐ **A.** Straight-and-Level Flight
- ☐ **B.** Change of Airspeed
- ☐ **C.** Constant Airspeed Climbs and Descents
- ☐ **D.** Rate Climbs and Descents
- ☐ **E.** Timed Turns to Magnetic Compass Headings
- ☐ **F.** Steep Turns
- ☐ **G.** Recovery from Unusual Flight Attitudes

V. NAVIGATION SYSTEMS

- ☐ Intercepting and Tracking Navigational Systems and DME Arcs

VI. INSTRUMENT APPROACH PROCEDURES

- ☐ **A.** Nonprecision Instrument Approach
- ☐ **B.** Precision ILS Instrument Approach
- ☐ **C.** Missed Approach
- ☐ **D.** Circling Approach
- ☐ **E.** Landing from a Straight-in or Circling Approach

VII. EMERGENCY OPERATIONS

- ☐ **A.** Loss of Communications
- ☐ **B.** One Engine Inoperative During Straight-and-Level Flight and Turns **(Multiengine Airplane)**
- ☐ **C.** One Engine Inoperative -- Instrument Approach **(Multiengine Airplane)**
- ☐ **D.** Loss of Gyro Attitude and/or Heading Indicators

VIII. POSTFLIGHT PROCEDURES

- ☐ Checking Instruments and Equipment

I. AREA OF OPERATION: PREFLIGHT PREPARATION

A. TASK: WEATHER INFORMATION

REFERENCES: 14 CFR part 61; AC 00-6, AC 00-45; AIM.

NOTE: Where current weather reports, forecasts, or other pertinent information is not available, this information will be simulated by the examiner in a manner which will adequately measure the applicant's competence.

Objective. To determine that the applicant:

1. Exhibits adequate knowledge of the elements related to aviation weather information by obtaining, reading, and analyzing the applicable items, such as--
   a. weather reports and forecasts.
   b. pilot and radar reports.
   c. surface analysis charts.
   d. radar summary charts.
   e. significant weather prognostics.
   f. winds and temperatures aloft.
   g. freezing level charts.
   h. stability charts.
   i. severe weather outlook charts.
   j. tables and conversion graphs.
   k. SIGMET's and AIRMET's.
   l. ATIS reports.
2. Correctly analyzes the assembled weather information pertaining to the proposed route of flight and destination airport, and determines whether an alternate airport is required, and, if required, whether the selected alternate airport meets the regulatory requirement.

B. TASK: CROSS-COUNTRY FLIGHT PLANNING

REFERENCES: 14 CFR parts 61, 91; AC 61-27, AC 61-23, AC 90-94; AFD; AIM.

Objective. To determine that the applicant:

1. Exhibits adequate knowledge of the elements by presenting and explaining a preplanned cross-country flight, as previously assigned by the examiner (preplanning is at examiner's discretion). It should be planned using real time weather and conform to the regulatory requirements for instrument flight rules within the airspace in which the flight will be conducted.
2. Exhibits adequate knowledge of the aircraft's performance capabilities by calculating the estimated time en route and total fuel requirement based upon factors, such as:
   a. power settings.
   b. operating altitude or flight level.
   c. wind.
   d. fuel reserve requirements.

3. Selects and correctly interprets the current and applicable en route charts, instrument departure procedures (DP's), Standard Terminal Arrival (STAR), and Standard Instrument Approach Procedure Charts (IAP).
4. Obtains and correctly interprets applicable NOTAM information.
5. Determines the calculated performance is within the aircraft's capability and operating limitations.
6. Completes and files a flight plan in a manner that accurately reflects the conditions of the proposed flight. (Does not have to be filed with ATC.)
7. Demonstrates adequate knowledge of Global Positioning Systems (GPS) and Receiver Autonomous Integrity Monitoring (RAIM) capability, when aircraft is so equipped.

II. AREA OF OPERATION: PREFLIGHT PROCEDURES

A. TASK: AIRCRAFT SYSTEMS RELATED TO IFR OPERATIONS

REFERENCES: 14 CFR parts 61, 91; AC 61-27, AC 61-84.

Objective. To determine that the applicant exhibits adequate knowledge of the elements related to applicable aircraft anti-icing/deicing system(s) and their operating methods to include:

1. Airframe.
2. Propeller/intake.
3. Fuel.
4. Pitot-static.

B. TASK: AIRCRAFT FLIGHT INSTRUMENTS AND NAVIGATION EQUIPMENT

REFERENCES: 14 CFR parts 61, 91; AC 61-27, AC 61-84, AC 90-48.

Objective. To determine that the applicant:

1. Exhibits adequate knowledge of the elements related to applicable aircraft flight instrument system(s) and their operating characteristics to include-
   a. pitot-static.
   b. altimeter.
   c. airspeed indicator.
   d. vertical speed indicator.
   e. attitude indicator.
   f. horizontal situation indicator.
   g. magnetic compass.
   h. turn-and-slip indicator/turn coordinator.
   i. heading indicator.
   j. electrical systems.
   k. vacuum systems.
2. Exhibits adequate knowledge of the applicable aircraft navigation system(s) and their operating characteristics to include-
   a. VHF omnirange (VOR).
   b. distance measuring equipment (DME).
   c. instrument landing system (ILS).
   d. marker beacon receiver/indicators.
   e. transponder/altitude encoding.
   f. automatic direction finder (ADF).
   g. global positioning system (GPS).

C. TASK: INSTRUMENT COCKPIT CHECK

REFERENCES: 14 CFR parts 61, 91; AC 61-27.

Objective. To determine that the applicant:

1. Exhibits adequate knowledge of the elements related to preflighting instruments, avionics, and navigation equipment cockpit check by explaining the reasons for the check and how to detect possible defects.
2. Performs the preflight on instruments, avionics, and navigation equipment cockpit check by following the checklist appropriate to the aircraft flown.
3. Determines that the aircraft is in condition for safe instrument flight including-
   a. radio communications equipment.
   b. radio navigation equipment including the following, as appropriate to the aircraft flown:
      (1) VOR/VORTAC.
      (2) ADF.
      (3) ILS.
      (4) GPS.
      (5) LORAN.
   c. magnetic compass.
   d. heading indicator.
   e. attitude indicator.
   f. altimeter.
   g. turn-and-slip indicator/turn coordinator.
   h. vertical speed indicator.
   i. airspeed indicator.
   j. clock.
   k. power source for gyro-instruments.
   l. pitot heat.
4. Notes any discrepancies and determines whether the aircraft is safe for instrument flight or requires maintenance.

III. AREA OF OPERATION: AIR TRAFFIC CONTROL CLEARANCES AND PROCEDURES

NOTE: The ATC clearance may be an actual or simulated ATC clearance based upon the flight plan.

A. TASK: AIR TRAFFIC CONTROL CLEARANCES

REFERENCES: 14 CFR parts 61, 91; AC 61-27; AIM.

Objective. To determine that the applicant:

1. Exhibits adequate knowledge of the elements related to ATC clearances and pilot/controller responsibilities to include tower en route control and clearance void times.
2. Copies correctly, in a timely manner, the ATC clearance as issued.
3. Determines that it is possible to comply with ATC clearance.
4. Interprets correctly the ATC clearance received and, when necessary, requests clarification, verification, or change.
5. Reads back correctly, in a timely manner, the ATC clearance in the sequence received.
6. Uses standard phraseology when reading back clearance.
7. Sets the appropriate communication and navigation frequencies and transponder codes in compliance with the ATC clearance.

B. TASK: COMPLIANCE WITH DEPARTURE, EN ROUTE, AND ARRIVAL PROCEDURES AND CLEARANCES

REFERENCES: 14 CFR parts 61, 91; AC 61-27; DP's; En Route Low Altitude Charts; STAR's.

Objective. To determine that the applicant:

1. Exhibits adequate knowledge of the elements related to DP's, En Route Low Altitude Charts, STAR's, and related pilot/controller responsibilities.
2. Uses the current and appropriate navigation publications for the proposed flight.
3. Selects and uses the appropriate communication frequencies; selects and identifies the navigation aids associated with the proposed flight.
4. Performs the appropriate aircraft checklist items relative to the phase of flight.
5. Establishes two-way communications with the proper controlling agency, using proper phraseology.
6. Complies, in a timely manner, with all ATC instructions and airspace restrictions.
7. Exhibits adequate knowledge of two-way radio communication failure procedures.
8. Intercepts, in a timely manner, all courses, radials, and bearings appropriate to the procedure, route, or clearance.
9. Maintains the applicable airspeed within 10 knots; headings within 10°; altitude within 100 feet (30 meters); and tracks a course, radial, or bearing.

C. TASK: HOLDING PROCEDURES

REFERENCES: 14 CFR parts 61, 91; AC 61-27; AIM.

NOTE: Any reference to DME will be disregarded if the aircraft is not so equipped.

Objective. To determine that the applicant:

1. Exhibits adequate knowledge of the elements related to holding procedures.
2. Changes to the holding airspeed appropriate for the altitude or aircraft when 3 minutes or less from, but prior to arriving at, the holding fix.
3. Explains and uses an entry procedure that ensures the aircraft remains within the holding pattern airspace for a standard, nonstandard, published, or nonpublished holding pattern.
4. Recognizes arrival at the holding fix and initiates prompt entry into the holding pattern.
5. Complies with ATC reporting requirements.
6. Uses the proper timing criteria, where applicable, as required by altitude or ATC instructions.
7. Complies with pattern leg lengths when a DME distance is specified.
8. Uses proper wind correction procedures to maintain the desired pattern and to arrive over the fix as close as possible to a specified time.
9. Maintains the airspeed within 10 knots; altitude within 100 feet (30 meters); headings within 10°; and tracks a selected course, radial, or bearing.

IV. AREA OF OPERATION: FLIGHT BY REFERENCE TO INSTRUMENTS

NOTE: The examiner shall require the performance of all TASKS. At least two of the TASKS, A through E as selected by the examiner, shall be performed without the use of the attitude and heading indicators. TASK F shall be performed using all available instruments; TASK G shall be performed without the use of the attitude indicator.

A. TASK: STRAIGHT-AND-LEVEL FLIGHT

REFERENCES: 14 CFR part 61; AC 61-27.

Objective. To determine that the applicant:

1. Exhibits adequate knowledge of the elements related to attitude instrument flying during straight-and-level flight.
2. Maintains straight-and-level flight in the aircraft configuration specified by the examiner.
3. Maintains the heading within 10°, altitude within 100 feet (30 meters), and airspeed within 10 knots.
4. Uses proper instrument cross-check and interpretation, and applies the appropriate pitch, bank, power, and trim corrections.

B. TASK: CHANGE OF AIRSPEED

REFERENCES: 14 CFR part 61; AC 61-27.

Objective. To determine that the applicant:

1. Exhibits adequate knowledge of the elements relating to attitude instrument flying during change of airspeeds in straight-and-level flight and in turns.
2. Establishes a proper power setting when changing airspeed.
3. Maintains the heading within 10°, angle of bank within 5° when turning, altitude within 100 feet (30 meters), and airspeed within 10 knots.
4. Uses proper instrument cross-check and interpretation, and applies the appropriate pitch, bank, power, and trim corrections.

C. TASK: CONSTANT AIRSPEED CLIMBS AND DESCENTS

REFERENCES: 14 CFR part 61; AC 61-27.

Objective. To determine that the applicant:

1. Exhibits adequate knowledge of the elements relating to attitude instrument flying during constant airspeed climbs and descents.
2. Demonstrates climbs and descents at a constant airspeed between specific altitudes in straight or turning flight as specified by the examiner.
3. Enters constant airspeed climbs and descents from a specified altitude, airspeed, and heading.
4. Establishes the appropriate change of pitch and power to establish the desired climb and descent performance.
5. Maintains the airspeed within 10 knots, heading within 10° or, if in a turning maneuver, within 5° of the specified bank angle.
6. Performs the level-off within 100 feet (30 meters) of the specified altitude.
7. Uses proper instrument cross-check and interpretation, and applies the appropriate pitch, bank, power, and trim corrections.

D. TASK: RATE CLIMBS AND DESCENTS

REFERENCES: 14 CFR part 61; AC 61-27.

Objective. To determine that the applicant:

1. Exhibits adequate knowledge of the elements relating to attitude instrument flying during rate climbs and descents.
2. Demonstrates climbs and descents at a constant rate between specific altitudes in straight or turning flight as specified by the examiner.
3. Enters rate climbs and descents from a specified altitude, airspeed, and heading.
4. Establishes the appropriate change of pitch, bank, and power to establish the specified rate of climb or descent.
5. Maintains the specified rate of climb and descent within 100 feet per minute, airspeed within 10 knots, heading within 10°, or if in a turning maneuver, within 5° of the specified bank angle.
6. Performs the level-off within 100 feet (30 meters) of the specified altitude.
7. Uses proper instrument cross-check and interpretation, and applies the appropriate pitch, bank, power, and trim corrections.

E. TASK: TIMED TURNS TO MAGNETIC COMPASS HEADINGS

REFERENCES: 14 CFR part 61; AC 61-27.

NOTE: If the aircraft has a turn and slip indicator, the phrase "miniature aircraft of the turn coordinator" applies to the turn needle.

Objective. To determine that the applicant:

1. Exhibits adequate knowledge of elements and procedures relating to calibrating the miniature aircraft of the turn coordinator, the operating characteristics and errors of the magnetic compass, and the performance of timed turns to specified compass headings.
2. Establishes indicated standard rate turns, both right and left.
3. Applies the clock correctly to the calibration procedure.
4. Changes the miniature aircraft position, as necessary, to produce a standard rate turn.
5. Makes timed turns to specified compass headings.
6. Maintains the altitude within 100 feet (30 meters), airspeed within 10 knots, bank angle 5° of a standard or half-standard rate turn, and rolls out on specified headings within 10°.

F. TASK: STEEP TURNS

REFERENCES: 14 CFR part 61; AC 61-27.

Objective. To determine that the applicant:

1. Exhibits adequate knowledge of the factors relating to attitude instrument flying during steep turns.
2. Enters a turn using a bank of approximately 45° for an airplane and 30° for a helicopter.
3. Maintains the specified angle of bank for either 180° or 360° of turn, both left and right.
4. Maintains altitude within 100 feet (30 meters), airspeed within 10 knots, 5° of specified bank angle, and rolls out within 10° of the specified heading.
5. Uses proper instrument cross-check and interpretation, and applies the appropriate pitch, bank, power, and trim corrections.

G. TASK: RECOVERY FROM UNUSUAL FLIGHT ATTITUDES

REFERENCES: 14 CFR part 61; AC 61-27.

NOTE: Any intervention by the examiner to prevent the aircraft from exceeding any operating limitations, or entering an unsafe flight condition, shall be disqualifying.

Objective. To determine that the applicant:

1. Exhibits adequate knowledge of the elements relating to attitude instrument flying during recovery from unusual flight attitudes (both nose-high and nose-low).
2. Uses proper instrument cross-check and interpretation, and applies the appropriate pitch, bank, and power corrections in the correct sequence to return the aircraft to a stabilized level flight attitude.

V. AREA OF OPERATION: NAVIGATION SYSTEMS

TASK: INTERCEPTING AND TRACKING NAVIGATIONAL SYSTEMS AND DME ARCS

REFERENCES: 14 CFR parts 61, 91; AC 61-27; AIM.

NOTE: Any reference to DME arcs, ADF, or GPS shall be disregarded if the aircraft is not equipped with these specified navigational systems.

Objective. To determine that the applicant:

1. Exhibits adequate knowledge of the elements related to intercepting and tracking navigational systems and DME arcs.
2. Tunes and correctly identifies the navigation facility.
3. Sets and correctly orients the radial to be intercepted into the course selector or correctly identifies the radial on the RMI.
4. Intercepts the specified radial at a predetermined angle, inbound or outbound from a navigational facility.
5. Maintains the airspeed within 10 knots, altitude within 100 feet (30 meters), and selected headings within 5°.
6. Applies proper correction to maintain a radial, allowing no more than three-quarter-scale deflection of the CDI or within 10° in case of an RMI.
7. Determines the aircraft position relative to the navigational facility or from a waypoint in the case of GPS.
8. Intercepts a DME arc and maintains that arc within 1 nautical mile.
9. Recognizes navigational receiver or facility failure, and when required, reports the failure to ATC.

VI. AREA OF OPERATION: INSTRUMENT APPROACH PROCEDURES

A. TASK: NONPRECISION INSTRUMENT APPROACH

REFERENCES: 14 CFR parts 61, 91; AC 61-27; IAP; AIM.

NOTE: Any reference to DME arcs, ADF, or GPS shall be disregarded if the aircraft is not equipped with the above specified navigational systems. If the aircraft is equipped with any of the above navigational systems, the examiner may ask the applicant to demonstrate those types of approaches. The examiner shall select two nonprecision approaches utilizing different approach systems.

Objective. To determine that the applicant:

1. Exhibits adequate knowledge of the elements related to an instrument approach procedure.
2. Selects and complies with the appropriate instrument approach procedure to be performed.

3. Establishes two-way communications with ATC, as appropriate, to the phase of flight or approach segment, and uses proper radio communication phraseology and technique.
4. Selects, tunes, identifies, and confirms the operational status of navigation equipment to be used for the approach procedure.
5. Complies with all clearances issued by ATC or the examiner.
6. Recognizes if heading indicator and/or attitude indicator is inaccurate or inoperative, advises controller, and proceeds with approach.
7. Advises ATC or examiner anytime the aircraft is unable to comply with a clearance.
8. Establishes the appropriate aircraft configuration and airspeed considering turbulence and wind shear, and completes the aircraft checklist items appropriate to the phase of the flight.
9. Maintains, prior to beginning the final approach segment, altitude within 100 feet (30 meters), heading within 10° and allows less than a full-scale deflection of the CDI or within 10° in the case of an RMI, and maintains airspeed within 10 knots.
10. Applies the necessary adjustments to the published MDA and visibility criteria for the aircraft approach category when required, such as-
    a. FDC and Class II NOTAM's.
    b. inoperative aircraft and ground navigation equipment.
    c. inoperative visual aids associated with the landing environment.
    d. National Weather Service (NWS) reporting factors and criteria.
11. Establishes a rate of descent and track that will ensure arrival at the MDA prior to reaching the MAP with the aircraft continuously in a position from which descent to a landing on the intended runway can be made at a normal rate using normal maneuvers.
12. Allows, while on the final approach segment, no more than a three-quarter-scale deflection of the CDI or within 10° in case of an RMI, and maintains airspeed within 10 knots.
13. Maintains the MDA, when reached, within +100 feet (30 meters), –0 feet to the MAP.
14. Executes the missed approach procedure when the required visual references for the intended runway are not distinctly visible and identifiable at the MAP.
15. Executes a normal landing from a straight-in or circling approach when instructed by the examiner.

B. TASK: PRECISION ILS INSTRUMENT APPROACH

REFERENCES: 14 CFR parts 61, 91; AC 61-27; IAP; AIM.

Objective. To determine that the applicant:

1. Exhibits adequate knowledge of the elements of an ILS instrument approach procedure.
2. Selects and complies with the appropriate ILS instrument approach procedure to be performed.
3. Establishes two-way communications with ATC, as appropriate to the phase of flight or approach segment, and uses proper radio communications phraseology and technique.
4. Selects, tunes, identifies, and confirms the operational status of ground and aircraft navigation equipment to be used for the approach procedure.
5. Complies with all clearances issued by ATC or the examiner.
6. Advises ATC or examiner anytime the aircraft is unable to comply with a clearance.
7. Establishes the appropriate aircraft configuration and airspeed, considering turbulence and wind shear, and completes the aircraft checklist items appropriate to the phase of flight.
8. Maintains, prior to beginning the final approach segment, specified altitude within 100 feet (30 meters), heading or course within 10°, and airspeed within 10 knots.
9. Applies the necessary adjustments to the published DH and visibility criteria for the aircraft approach category when required, such as--
    a. FDC and Class II NOTAM's.
    b. inoperative aircraft and ground navigation equipment.
    c. inoperative visual aids associated with the landing environment.
    d. National Weather Service (NWS) reporting factors and criteria.
10. Establishes an initial rate of descent at the point where the electronic glide slope is intercepted, which approximates that required for the aircraft to follow the glide slope to DH.
11. Allows, while on the final approach segment, no more than three-quarter-scale deflection of either the localizer or glide slope indications, and maintains the specified airspeed within 10 knots.
12. Avoids descent below the DH before initiating a missed approach procedure or transitioning to a normal landing approach.
13. Initiates immediately the missed approach procedure when, at the DH, the required visual references for the intended runway are not distinctly visible and identifiable.
14. Transitions to a normal landing approach when the aircraft is continuously in a position from which a descent to a landing on the intended runway can be made at a normal rate of descent using normal maneuvers.

C. TASK: MISSED APPROACH

REFERENCES: 14 CFR parts 61, 91; AC 61-27; IAP; AIM.

Objective. To determine that the applicant:

1. Exhibits adequate knowledge of the elements related to missed approach procedures associated with standard instrument approaches.
2. Initiates the missed approach promptly by applying power, establishing a climb attitude, and reducing drag in accordance with the aircraft manufacturer's recommendations.
3. Reports to ATC beginning the missed approach procedure.
4. Complies with the published or alternate missed approach procedure.
5. Advises ATC or examiner anytime the aircraft is unable to comply with a clearance, restriction, or climb gradient.
6. Follows the recommended checklist items appropriate to the go-around procedure.
7. Requests, if appropriate, ATC clearance to the alternate airport, clearance limit, or as directed by the examiner.
8. Maintains the recommended airspeed within 10 knots; heading, course, or bearing within 10°; and altitude(s) within 100 feet (30 meters) during the missed approach procedure.

D. TASK: CIRCLING APPROACH

REFERENCES: 14 CFR parts 61, 91; AC 61-27; IAP; AIM.

Objective. To determine that the applicant:

1. Exhibits adequate knowledge of the elements related to a circling approach procedure.
2. Selects and complies with the appropriate circling approach procedure considering turbulence and wind shear and considering the maneuvering capabilities of the aircraft.
3. Confirms the direction of traffic and adheres to all restrictions and instructions issued by ATC and the examiner.
4. Does not exceed the visibility criteria or descend below the appropriate circling altitude until in a position from which a descent to a normal landing can be made.
5. Maneuvers the aircraft, after reaching the authorized MDA and maintains that altitude within +100 feet (30 meters), –0 feet and a flightpath that permits a normal landing on a runway at least 90° from the final approach course.

E. TASK: LANDING FROM A STRAIGHT-IN OR CIRCLING APPROACH

REFERENCES: 14 CFR parts 61, 91; AC 61-27; AIM.

Objective. To determine that the applicant:

1. Exhibits adequate knowledge of the elements related to the pilot's responsibilities, and the environmental, operational, and meteorological factors which affect a landing from a straight-in or a circling approach.
2. Transitions at the DH, MDA, or VDP to a visual flight condition, allowing for safe visual maneuvering and a normal landing.
3. Adheres to all ATC (or examiner) advisories, such as NOTAM's, wind shear, wake turbulence, runway surface, braking conditions, and other operational considerations.
4. Completes appropriate checklist items for the pre-landing and landing phase.
5. Maintains positive aircraft control throughout the complete landing maneuver.

VII. AREA OF OPERATION: EMERGENCY OPERATIONS

A. TASK: LOSS OF COMMUNICATIONS

REFERENCES: 14 CFR parts 61, 91; AIM.

Objective. To determine that the applicant exhibits adequate knowledge of the elements related to applicable loss of communication procedures to include:

1. Recognizing loss of communication.
2. Continuing to destination according to the flight plan.
3. When to deviate from the flight plan.
4. Timing for beginning an approach at destination.

***B. TASK: ONE ENGINE INOPERATIVE DURING STRAIGHT-AND-LEVEL FLIGHT AND TURNS (MULTIENGINE AIRPLANE)**

REFERENCES: 14 CFR part 61; AC 61-21, AC 61-27.

Objective. To determine that the applicant:

1. Exhibits adequate knowledge of the procedures used if engine failure occurs during straight-and-level flight and turns while on instruments.
2. Recognizes engine failure simulated by the examiner during straight-and-level flight and turns.
3. Sets all engine controls, reduces drag, and identifies and verifies the inoperative engine.
4. Establishes the best engine-inoperative airspeed and trims the aircraft.
5. Verifies the accomplishment of prescribed checklist procedures for securing the inoperative engine.
6. Establishes and maintains the recommended flight attitude, as necessary, for best performance during straight-and-level and turning flight.
7. Attempts to determine the reason for the engine failure.
8. Monitors all engine control functions and makes necessary adjustments.
9. Maintains the specified altitude within 100 feet (30 meters), (if within the aircraft's capability), airspeed within 10 knots, and the specified heading within 10°.
10. Assesses the aircraft's performance capability and decides an appropriate action to ensure a safe landing.
11. Avoids loss of aircraft control, or attempted flight contrary to the engine-inoperative operating limitations of the aircraft.

***C. TASK: ONE ENGINE INOPERATIVE -- INSTRUMENT APPROACH (MULTIENGINE AIRPLANE)**

REFERENCES: 14 CFR part 61; AC 61-21, AC 61-27; IAP.

Objective. To determine that the applicant:

1. Exhibits adequate knowledge of the elements by explaining the procedures used during an instrument approach in a multiengine aircraft with one engine inoperative.
2. Recognizes promptly, engine failure simulated by the examiner.
3. Sets all engine controls, reduces drag, and identifies and verifies the inoperative engine.
4. Establishes the best engine-inoperative airspeed and trims the aircraft.
5. Verifies the accomplishment of prescribed checklist procedures for securing the inoperative engine.
6. Establishes and maintains the recommended flight attitude and configuration for the best performance for all maneuvering necessary for the instrument approach procedures.
7. Attempts to determine the reason for the engine failure.
8. Monitors all engine control functions and makes necessary adjustments.
9. Requests and receives an actual or a simulated ATC clearance for an instrument approach.
10. Follows the actual or a simulated ATC clearance for an instrument approach.
11. Establishes a rate of descent that will ensure arrival at the MDA prior to reaching the MAP with the aircraft continuously in a position from which descent to a landing on the intended runway can be made straight-in or circling.
12. Maintains, where applicable, the specified altitude within 100 feet (30 meters), the airspeed within 10 knots if within the aircraft's capability, and the heading within 10°.
13. Sets the navigation and communication equipment used during the approach and uses the proper communications technique.

*Multiengine tasks which are not explained in this book.

14. Avoids loss of aircraft control, or attempted flight contrary to the engine-inoperative operating limitations of the aircraft.

15. Complies with the published criteria for the aircraft approach category when circling.

16. Allows, while on final approach segment, no more than three-quarter-scale deflection of either the localizer or glide slope or GPS indications, or within 10° of the nonprecision final approach course.

17. Completes a safe landing.

D. TASK: LOSS OF GYRO ATTITUDE AND/OR HEADING INDICATORS

REFERENCES: 14 CFR part 61; AC 61-27; IAP.

Note: This approach shall count as one of the required nonprecision approaches.

Objective. To determine that the applicant:

1. Exhibits adequate knowledge of the elements relating to recognizing if attitude indicator and/or heading indicator is inaccurate or inoperative, and advises ATC or the examiner.
2. Advises ATC or examiner anytime the aircraft is unable to comply with a clearance.
3. Demonstrates a nonprecision instrument approach without gyro attitude and heading indicators using the objectives of the nonprecision approach TASK (AREA OF OPERATION VI, TASK A).

VIII. AREA OF OPERATION: POSTFLIGHT PROCEDURES

TASK: CHECKING INSTRUMENTS AND EQUIPMENT

REFERENCES: 14 CFR parts 61, 91.

Objective. To determine that the applicant:

1. Exhibits adequate knowledge of the elements relating to all instrument and navigation equipment for proper operation.
2. Notes all flight equipment for proper operation.
3. Notes all equipment and/or aircraft malfunctions and makes a written record of improper operation or failure of such equipment.

Appendix 1 - Levels of Simulation Devices
TASK VS. SIMULATION DEVICE CREDIT

Examiners conducting the instrument rating practical tests with flight simulation devices should consult appropriate documentation to ensure that the device has been approved for training, testing, or checking. The documentation for each device should reflect that the following activities have occurred:

1. The device must be evaluated, determined to meet the appropriate standards, and assigned the appropriate qualification level by the National Simulator Program Manager. The device must continue to meet qualification standards through continuing evaluations as outlined in the appropriate advisory circular (AC). For airplane flight training devices (FTDs), AC 120-45 (as amended), Airplane Flight Training Device Qualifications, will be used. For simulators, AC 120-40 (as amended), Airplane Simulator Qualification, will be used.
2. The FAA must approve the device for training, testing, and checking the specific flight TASKS listed in this appendix.
3. The device must continue to support the level of student or applicant performance required by this practical test standard.

NOTE: Users of the following chart are cautioned that use of the chart alone is incomplete. The description and Objective of each TASK as listed in the body of the practical test standard, including all NOTES, must also be incorporated for accurate simulation device use.

USE OF CHART

X Creditable.

A Creditable if appropriate systems are installed and operating.

NOTE: 1. Level 1 FTDs that have been issued a letter authorizing their use by the FAA Administrator and placed in service on or prior to August 2, 1996, may continue to be used for only those TASKS originally found acceptable. Use of Level 1, 2, or 3 FTDs may not be used for aircraft requiring a type rating.

2. If a FTDs or a simulator is used for the practical test, the instrument approach procedures conducted in that FTD or simulator are limited to one precision and one nonprecision approach procedure.

3. Postflight procedures means, closing flight plans, checking for discrepancies and malfunctions, and noting them on a log or maintenance form.

	FLIGHT SIMULATION DEVICE LEVEL											
FLIGHT TASK **Areas of Operation**	1	2	3	4	5	6	7		A	B	C	D
II. Preflight Procedures												
C. Instrument Cockpit Check*	—	A	X	A	A	X	X		X	X	X	X
III. Air Traffic Control Clearances and Procedures												
A. Air Traffic Control Clearances*	—	A	X	A	A	X	X		X	X	X	X
B. Departure, En Route and Arrival Clearances*	—	—	X	—	—	X	X		X	X	X	X
C. Holding Procedures	—	—	X	—	—	X	X		X	X	X	X
IV. Flight by Reference to Instruments												
A. Straight-and-Level Flight	—	—	X	—	—	X	X		X	X	X	X
B. Changes in Airspeed	—	—	X	—	—	X	X		X	X	X	X
C. Constant Airspeed Climbs and Descents	—	—	X	—	—	X	X		X	X	X	X
D. Rate Climbs and Descents	—	—	X	—	—	X	X		X	X	X	X
E. Times Turns to Magnetic Headings	—	—	X	—	—	X	X		X	X	X	X
F. Steep Turns	—	—	X	—	—	X	X		X	X	X	X
G. Unusual Flight Attitudes	—	—	—	—	—	—	X		X	X	X	X
V. Navigation Systems												
Intercepting and Tracking Course of Navigation Systems and DME ARCs	—	A	X	—	A	X	X		X	X	X	X
VI. Instrument Approach Procedures												
A. Nonprecision Approach Procedure (VOR, NDB, GPS, LOC, LDA, SDF), if equipped	—	—	X	—	—	X	X		X	X	X	X
B. Precision Approach Procedures (ILS)	—	—	X	—	—	X	X		X	X	X	X
C. Missed Approach Procedures	—	—	X	—	—	X	X		X	X	X	X
D. Circling Approach Procedures (NA Helicopters)	—	—	—	—	—	—	—		—	—	X	X
E. Landing from a Straight-in or Circling Approach	—	—	—	—	—	—	—		—	—	X	X
VII. Emergency Operations (ME) & (SE)**												
A. Loss of Communications	—	X	X	—	—	X	X		X	X	X	X
B. One Engine Inoperative, Straight-and-Level Flight and Turns	—	—	X	—	—	X	X		X	X	X	X
C. One Engine Inoperative, Instrument Approach	—	—	—	—	—	—	—		X	X	X	X
D. Loss of Gyro Attitutde and/or Heading Indicators	—	—	—	—	—	—	X		X	X	X	X
VIII. Postflight Procedures												
Checking Instruments and Equipment	—	A	X	—	A	X	X		X	X	X	X

*Aircraft required for those items that cannot be checked using a flight training device or flight simulator.
**Multiengine - Single Engine

APPENDIX B
FLIGHT TRAINING SYLLABUS

This appendix contains a reprint of the 28 flight training lessons from Gleim's ***Instrument Pilot Syllabus***. The Gleim system focuses on the 24 PTS tasks (single-engine airplane) required by the FAA on your practical test. It is designed to help you develop "PTS proficiency" as quickly and easily as possible.

Stage One (lessons 1 through 13) of the flight syllabus is designed to provide you with a foundation of good attitude instrument flying skills.

Stage Two (lessons 14 through 23) of the flight syllabus is designed to provide you with the knowledge and skills required to use the navigation systems in your airplane, including holding procedures and instrument approach procedures.

Stage Three (lessons 24 through 28) of the flight syllabus is designed to provide you with practical experience in operating under IFR during cross-country flights. Additionally, this stage includes the final preparation for your practical test.

This syllabus is intended specifically to meet Part 141 requirements. As such, the 50 hr. of cross-country flight as pilot in command is not included in this syllabus. Furthermore, our suggested time for completion is 36 hr., while Part 61 requires 40 hr. of actual or simulated instrument time.

The following table lists each lesson with the minimum time we have allotted for its completion:

Lesson	Topic	Minimum Time
	Stage One	
1	Attitude Instrument Flying	1.0
2	Instrument Takeoff, Steep Turns, and Airspeed Changes	1.0
3	Rate Climbs/Descents, Timed Turns, and Magnetic Compass Turns	1.0
4	Partial-Panel Flying	1.0
5	Attitude Instrument Flying (review)	1.0
6	Partial-Panel Flying (review)	1.0
7	Basic Instrument Flight Patterns	1.5
8	VOR/VORTAC Procedures	1.0
9	VOR Time/Distance to Station and DME Arcs	1.0
10	GPS and ADF Procedures	1.0
11	GPS Procedures and ADF Time/Distance to Station	1.0
12	Tracking the Localizer	1.0
13	Stage One Check	1.5
	Stage Two	
14	VOR Holding	1.0
15	GPS and ADF Holding	1.0
16	Localizer Holding	1.0
17	DME and Intersection Holding	1.0
18	VOR Instrument Approach	1.0
19	GPS and NDB Instrument Approaches	1.5
20	Localizer Instrument Approach	1.5
21	ILS Instrument Approach	1.5
22	Instrument Approaches (review)	1.5
23	Stage Two Check	1.0
	Stage Three	
24	Cross-Country and Emergency Procedures	2.0
25	Cross-Country Procedures	2.0
26	Cross-Country Procedures	3.5
27	Maneuvers Review	1.0
28	Stage Three Check	1.5
	TOTAL	36.0

The following is a brief description of the parts of each flight lesson in this syllabus:

Objective: We open each lesson with an objective, usually a sentence or two, to help you gain perspective and understand the goal for that particular lesson.

Text References: This section tells you which reference books you will need to study or refer to while mastering the tasks within the lesson. Abbreviations are given to facilitate the cross-referencing process.

Content: Each lesson contains a list of the tasks required to be completed before moving to the next lesson. A task may be listed as a "review item" (a task that was covered in a previous lesson) or as a "new item" (a task that is introduced to you for the first time). Each task is preceded by three blank "checkoff" boxes which may be used by your CFII to keep track of your progress and to indicate that each task was completed.

There are three boxes because it may take more than one flight to complete the lesson. Your CFII may mark the box(es) next to each task in one of the following methods (or any other method desired):

	*	
✓ - task completed to lesson completion standards	D - demonstrated by CFII A - accomplished by you S - safe/satisfactory P - meets PTS standards	1 - above lesson standard 2 - meets lesson standard 3 - below lesson standard

Most tasks are followed by book and page references that tell you where to find the information you need to study to accomplish the task successfully.

Completion Standards: Based on these standards, your CFII determines how well you have met the objective of the lesson in terms of knowledge and skill.

Instructor's Comments and Lesson Assignment: Space is provided for your CFII's critique of the lesson, which you can refer to later. Your CFII may also write any specific assignment for the next lesson.

Lesson Certification: This section will be signed by you and your CFII as a record that the lesson was completed.

*System suggested by the U.S. Air Force flying clubs

STAGE ONE

Stage One Objective

The pilot will be able to precisely control the airplane using basic attitude instrument flying skills and the airplane's navigation systems.

Stage One Completion Standards

The stage will be completed when the pilot satisfactorily passes the Stage One check and is able to precisely control the airplane using basic attitude instrument flying skills under full and partial panel instrument panel conditions. Additionally, the pilot will be able to fly a predetermined course using the airplane's navigation systems.

FLIGHT LESSON 1: Attitude Instrument Flying

Objective

To review basic attitude instrument flying and to develop the pilot's proficiency and confidence in flying the airplane solely by reference to instruments.

Text References

Instrument Pilot Flight Maneuvers and Practical Test Prep (FM)

Instrument Pilot Syllabus (SYL)*

Pilot's Operating Handbook (POH)

Content

1. Preflight briefing
2. Review items
   - ☐☐☐ Attitude instrument flying - FM 137-140
   - ☐☐☐ Airplane flight instruments - FM 72-83; POH-7
   - ☐☐☐ Straight-and-level flight - FM 141-145
   - ☐☐☐ Change of airspeed - FM 146-153
   - ☐☐☐ Constant airspeed climbs and descents - FM 154-165
   - ☐☐☐ Standard rate turns - CFII
   - ☐☐☐ Maneuvering during slow flight - SYL*
   - ☐☐☐ Power-off stall - SYL*
   - ☐☐☐ Power-on stall - SYL*
   - ☐☐☐ Recovery from unusual flight attitudes - FM 185-190
   - ☐☐☐ Use of checklists - POH-4
   - ☐☐☐ Radio communications - CFII
   - ☐☐☐ Normal takeoff and landing - CFII
3. Postflight critique and preview of next lesson

*You can download these Additional Instrument Flight Maneuvers at (www.gleim.com/Aviation/Updates/books/ipfm/) or purchase Gleim's *Instrument Pilot Syllabus*.

Completion Standards

The lesson will have been successfully completed when the pilot demonstrates an understanding of attitude instrument flying as related to airplane control. The pilot will maintain airplane control at all times and maintain the desired altitude, ±200 ft.; airspeed, ±10 kt.; and heading, ±15°.

Instructor's comments: ______________________________

____

____

____

____

____

____

____

____

____

____

Lesson assignment: ______________________________

____

Lesson certification:

______________________________ ______________________________

(Pilot) | Instructor | Certificate No. | Exp. Date

FLIGHT LESSON 2: Instrument Takeoff, Steep Turns, and Airspeed Changes

Objective

To introduce the pilot to IFR preflight, instrument takeoff, change of airspeed, steep turns, and postflight procedures and to increase the pilot's proficiency in attitude instrument flying.

Text References

Instrument Pilot Flight Maneuvers and Practical Test Prep (FM)

Instrument Pilot Syllabus (SYL)*

Pilot's Operating Handbook (POH)

Content

1. Flight Lesson 1 complete? Yes ____ Copy placed in pilot's folder? Yes ____
2. Preflight briefing
3. Review items
   - ☐☐☐ Airplane flight instruments - FM 72-83; POH-7
   - ☐☐☐ Use of checklists - POH-4
   - ☐☐☐ Radio communications - CFII
   - ☐☐☐ Attitude instrument flying - FM 137-140
   - ☐☐☐ Straight-and-level flight - FM 141-145
   - ☐☐☐ Change of airspeed - FM 146-153
   - ☐☐☐ Standard rate turns - CFII
   - ☐☐☐ Constant airspeed climbs and descents - FM 154-165
   - ☐☐☐ Recovery from unusual flight attitudes - FM 185-190
   - ☐☐☐ Power-on stall - SYL*

   ____
4. New items
   - ☐☐☐ IFR preflight inspection - CFII
   - ☐☐☐ Airplane systems related to IFR operations - FM 68-71
   - ☐☐☐ Airplane navigation equipment - FM 84-93; POH-9
   - ☐☐☐ Instrument cockpit check - FM 94-99
   - ☐☐☐ Instrument takeoff - SYL*
   - ☐☐☐ Steep turns - FM 181-184
   - ☐☐☐ IFR postflight procedures - FM 338
5. Postflight critique and preview of next lesson

*You can download these Additional Instrument Flight Maneuvers at (www.gleim.com/Aviation/Updates/books/ipfm/) or purchase Gleim's *Instrument Pilot Syllabus*.

Completion Standards

The lesson will have been successfully completed when the pilot displays an understanding of IFR preflight and postflight procedures, the instrument cockpit check, and the instrument takeoff procedures. The pilot will demonstrate increased proficiency in attitude instrument flying. The pilot will be able to maintain the desired altitude, ±200 ft.; airspeed, ±10 kt.; and heading, ±15°.

Instructor's comments: ______________________________

____

____

____

____

____

Lesson assignment: ______________________________

____

Lesson certification:

______________________________ ______________________________
(Pilot) Instructor Certificate No. Exp. Date

FLIGHT LESSON 3: Rate Climbs/Descents, Timed Turns, and Magnetic Compass Turns

Objective

To introduce the pilot to constant rate climbs and descents, timed turns to magnetic compass headings, and magnetic compass turns. Additionally, this lesson will increase the pilot's proficiency in attitude instrument flying.

Text References

Instrument Pilot Flight Maneuvers and Practical Test Prep (FM)

Instrument Pilot Syllabus (SYL)*

Pilot's Operating Handbook (POH)

Content

1. Flight Lesson 2 complete? Yes ____ Copy placed in pilot's folder? Yes ____
2. Preflight briefing
3. Review items

   ☐☐☐ IFR preflight inspection - CFII
   ☐☐☐ Airplane systems related to IFR operations - FM 68-71
   ☐☐☐ Airplane flight instruments and navigation equipment - FM 72-93; POH-9
   ☐☐☐ Instrument cockpit check - FM 94-99
   ☐☐☐ Straight-and-level flight - FM 141-145
   ☐☐☐ Standard rate turns - CFII
   ☐☐☐ Recovery from unusual flight attitudes - FM 185-190
   ☐☐☐ Slow flight - SYL*
   ☐☐☐ Power-off stall - SYL*
   ☐☐☐ Steep turns - FM 181-184
   ☐☐☐ IFR postflight procedures - FM 338

   ____

4. New items

   ☐☐☐ Rate climbs and descents - FM 166-174
   ☐☐☐ Timed turns to magnetic compass headings - FM 175-180
   ☐☐☐ Magnetic compass turns - FM 176

5. Postflight critique and preview of next lesson

*You can download these Additional Instrument Flight Maneuvers at (www.gleim.com/Aviation/Updates/books/ipfm/) or purchase Gleim's *Instrument Pilot Syllabus*.

Completion Standards

The lesson will have been successfully completed when the pilot displays an understanding of how to perform rate climbs and descents, calibrate the turn coordinator for timed turns, and make turns to headings by use of a timed turn or the magnetic compass. Additionally, the pilot will demonstrate increased proficiency in attitude instrument flying. The pilot will be able to maintain the desired altitude, ±150 ft.; airspeed, ±10 kt.; and heading, ±15°.

Instructor's comments: ____

____

____

____

____

____

____

____

Lesson assignment: ____

____

Lesson certification:

____________________ ____

(Pilot) Instructor Certificate No. Exp. Date

FLIGHT LESSON 4: Partial-Panel Flying

Objective

To introduce the pilot to partial-panel attitude instrument flying and to increase the pilot's proficiency in constant rate climbs and descents, timed turns, and magnetic compass turns.

Text References

Instrument Pilot Flight Maneuvers and Practical Test Prep (FM)

Content

1. Flight Lesson 3 complete? Yes ____ Copy placed in pilot's folder? Yes ____
2. Preflight briefing
3. Review items (full panel)
   - ☐☐☐ Rate climbs and descents - FM 166-174
   - ☐☐☐ Timed turns to magnetic compass headings - FM 175-180
   - ☐☐☐ Magnetic compass turns - FM 176
4. New items ________________________
   - ☐☐☐ Loss of gyro attitude and/or heading indicators - FM 332-335
   - ☐☐☐ Straight-and-level (partial panel) - FM 141-145
   - ☐☐☐ Magnetic compass turns (partial panel) - FM 176
   - ☐☐☐ Constant airspeed climbs and descents (partial panel) - FM 154-165
   - ☐☐☐ Rate climbs and descents (partial panel) - FM 166-174
5. Postflight critique and preview of next lesson

Completion Standards

The lesson will have been successfully completed when the pilot demonstrates an understanding of the skills required to maintain airplane control while conducting partial-panel attitude instrument flight. Additionally, the pilot will demonstrate increased proficiency in constant rate climbs and descents, timed turns, and magnetic compass turns. The pilot will be able to maintain the desired altitude, ±150 ft.; airspeed, ±10 kt.; heading, ±15°; and climb/descent rate, ±200 fpm; and to roll out on the desired heading, ±15°.

Instructor's comments: ____

____

____

____

____

____

____

____

____

____

____

____

____

____

____

Lesson assignment: ____

____

Lesson certification:

(Pilot) ____________________ Instructor ____________________ Certificate No. ____________ Exp. Date ____________

FLIGHT LESSON 5: Attitude Instrument Flying (review)

Objective

To review previous lessons to gain proficiency in full- and partial-panel attitude instrument flying.

Text References

Instrument Pilot Flight Maneuvers and Practical Test Prep (FM)

Instrument Pilot Syllabus (SYL)*

Content

1. Flight Lesson 4 complete? Yes ____ Copy placed in pilot's folder? Yes ____
2. Preflight briefing
3. Review items
   a. Full-panel attitude instrument flight
      - ☐☐☐ Instrument cockpit check - FM 94-99
      - ☐☐☐ Instrument takeoff - SYL*
      - ☐☐☐ Straight-and-level - FM 141-145
      - ☐☐☐ Constant airspeed climbs and descents - FM 154-165
      - ☐☐☐ Rate climbs and descents - FM 166-174
      - ☐☐☐ Standard rate turns - CFII
      - ☐☐☐ Steep turns - FM 181-184
      - ☐☐☐ Change of airspeed - FM 146-153
      - ☐☐☐ Maneuvering during slow flight - SYL*
      - ☐☐☐ Stalls - SYL*
      - ☐☐☐ Recovery from unusual flight attitudes - FM 185-190

   b. Partial-panel attitude instrument flight
      - ☐☐☐ Straight-and-level - FM 141-145
      - ☐☐☐ Magnetic compass turns - FM 176
      - ☐☐☐ Constant airspeed climbs and descents - FM 154-165
      - ☐☐☐ Rate climbs and descents - FM 166-174
4. Postflight critique and preview of next lesson

*You can download these Additional Instrument Flight Maneuvers at (www.gleim.com/Aviation/Updates/books/ipfm/) or purchase Gleim's *Instrument Pilot Syllabus*.

Completion Standards

The lesson will have been successfully completed when the pilot demonstrates smooth, coordinated control of the airplane during full-panel attitude instrument flight. Additionally, the pilot will demonstrate increased understanding of partial-panel operations. The pilot will be able to maintain the desired altitude, ±150 ft.; airspeed, ±10 kt.; heading, ±15°; angle of bank, ±5°; and climb/descent rate, ±200 fpm; to roll out on the desired heading, ±15°; and to level off at the desired altitude, ±150 ft. Recovery procedures from stalls and unusual attitudes should be done correctly with the successful outcome never seriously in doubt.

Instructor's comments: ____

____

____

____

____

____

____

____

Lesson assignment: ____

____

Lesson certification:

________________ ____

(Pilot) Instructor Certificate No. Exp. Date

FLIGHT LESSON 6: Partial-panel Flying (review)

Objective

To increase the pilot's proficiency in partial-panel attitude instrument flying and to introduce the pilot to more complex partial-panel procedures.

Text References

Instrument Pilot Flight Maneuvers and Practical Test Prep (FM)

Instrument Pilot Syllabus (SYL)*

Content

1. Flight Lesson 5 complete? Yes ____ Copy placed in pilot's folder? Yes ____
2. Preflight briefing
3. Review items (partial panel)
   - ☐☐☐ Straight-and-level - FM 141-145
   - ☐☐☐ Magnetic compass turns - FM 176
   - ☐☐☐ Constant airspeed climbs and descents - FM 154-165

   ____
4. New items (partial panel)
   - ☐☐☐ Timed turns to magnetic compass headings - FM 175-181
   - ☐☐☐ Maneuvering during slow flight - SYL*
   - ☐☐☐ Stalls - SYL*
   - ☐☐☐ Recovery from unusual attitudes - FM 185-190
5. Postflight critique and preview of next lesson

*You can download these Additional Instrument Flight Maneuvers at (www.gleim.com/Aviation/Updates/books/ipfm/) or purchase Gleim's *Instrument Pilot Syllabus*.

Completion Standards

The lesson will have been successfully completed when the pilot displays an understanding of the correct recovery procedures during stalls and unusual attitudes without overcontrolling during partial-panel operations. Additionally, the pilot will demonstrate increased proficiency in partial-panel attitude instrument flight.

Instructor's comments: ______________________________

____

Lesson assignment: ______________________________

____

Lesson certification:

______________________________ ______________________________

(Pilot) Instructor Certificate No. Exp. Date

FLIGHT LESSON 7: Basic Instrument Flight Patterns

Objective

To further develop the pilot's ability to precisely control the airplane during attitude instrument flying by combining previously learned maneuvers. This will be accomplished by introducing the pilot to the FAA's various basic instrument flight patterns.

Text References

Instrument Pilot Syllabus (SYL)*

Content

1. Flight Lesson 6 complete? Yes ____ Copy placed in pilot's folder? Yes ____
2. Preflight briefing
3. New items
   - ☐☐☐ Pattern "A" - SYL*
   - ☐☐☐ Pattern "B" - SYL*
   - ☐☐☐ Vertical S - SYL*
   - ☐☐☐ Vertical S-1 - SYL*
   - ☐☐☐ Vertical S-2 - SYL*
   - ☐☐☐ 80/260 procedure turn - SYL*
   - ☐☐☐ Standard procedure turn - SYL*
   - ☐☐☐ Teardrop holding pattern entry - SYL*
   - ☐☐☐ Holding pattern - SYL*
   - ☐☐☐ Patterns applicable to circling approaches - SYL*
4. Postflight critique and preview of next lesson

*You can download these Additional Instrument Flight Maneuvers at (www.gleim.com/Aviation/Updates/books/ipfm/) or purchase Gleim's *Instrument Pilot Syllabus*.

Completion Standards

The lesson will have been successfully completed when the pilot displays an understanding of the various basic instrument flight patterns. The pilot will be able to maintain the desired altitude, ±150 ft.; airspeed, ±10 kt.; and heading, ±15°; to roll out on the specified heading, ±15°; and to maintain the desired rate of climb/descent, ±200 fpm.

Instructor's comments: ____

____

____

____

____

____

____

____

____

____

____

____

____

____

Lesson assignment: ____

____

Lesson certification:

(Pilot)	Instructor	Certificate No.	Exp. Date

FLIGHT LESSON 8: VOR/VORTAC Procedures

Note: For the remainder of this syllabus, the term VOR will be used to include VOR, VORTAC, and VOR/DME stations.

Objective

To develop the pilot's ability to determine the airplane's position in relation to a VOR station and to intercept and track a predetermined radial.

Text References

Instrument Pilot Flight Maneuvers and Practical Test Prep (FM)

Instrument Pilot Syllabus (SYL)*

Content

1. Flight Lesson 7 complete? Yes ____ Copy placed in pilot's folder? Yes ____
2. Preflight briefing
3. Review items
   - ☐☐☐ Constant airspeed climbs and descents - FM 154-165
   - ☐☐☐ Pattern "A" (partial panel) - SYL*
   - ☐☐☐ Vertical S - SYL*
   - ☐☐☐ Teardrop holding pattern entry - SYL*
   - ☐☐☐ Recovery from unusual flight attitudes (partial panel) - FM 185-190
4. New items
   - ☐☐☐ VOR accuracy test - FM 96-97
   - ☐☐☐ VOR orientation - FM 193, 196, 198
   - ☐☐☐ Intercepting and tracking a VOR radial - FM 194-197, 201-202
5. Postflight critique and preview of next lesson

*You can download these Additional Instrument Flight Maneuvers at (www.gleim.com/Aviation/Updates/books/ipfm/) or purchase Gleim's *Instrument Pilot Syllabus*.

Completion Standards

The lesson will have been successfully completed when the pilot displays an understanding of VOR orientation and of intercepting and tracking predetermined radials. Additionally, the pilot will demonstrate increased proficiency in attitude instrument flight. The pilot will be able to maintain the desired altitude, ±150 ft.; airspeed, ±10 kt.; and heading, ±10°.

Instructor's comments: ________________________________

Lesson assignment: ________________________________

Lesson certification:

(Pilot) Instructor Certificate No. Exp. Date

FLIGHT LESSON 9: VOR Time/Distance to Station and DME Arcs

Objective

To introduce the pilot to VOR time and distance calculations and the interception and tracking of DME arcs, if the airplane is DME equipped. Additionally, the pilot will gain more proficiency in VOR orientation, radial interception, and tracking.

Text References

Instrument Pilot Flight Maneuvers and Practical Test Prep (FM)

Instrument Pilot Syllabus (SYL)*

Content

1. Flight Lesson 8 complete? Yes ____ Copy placed in pilot's folder? Yes ____
2. Preflight briefing
3. Review items
   - ☐☐☐ Pattern "B" - SYL*
   - ☐☐☐ Vertical S-2 - SYL*
   - ☐☐☐ Holding pattern - SYL*
   - ☐☐☐ VOR orientation - FM 193, 196, 198
   - ☐☐☐ Intercepting and tracking a VOR radial - FM 194-197, 201-202

   ____
4. New items
   - ☐☐☐ VOR time and distance calculations - SYL*
   - ☐☐☐ Intercepting and tracking DME arcs (if the airplane is DME equipped) - FM 198-202
5. Postflight critique and preview of next lesson

*You can download these Additional Instrument Flight Maneuvers at (www.gleim.com/Aviation/Updates/books/ipfm/) or purchase Gleim's *Instrument Pilot Syllabus*.

Completion Standards

The lesson will have been successfully completed when the pilot can demonstrate an understanding of VOR time and distance calculations and how to intercept and track a DME arc (if the airplane is DME equipped). Additionally, the pilot will demonstrate increased proficiency in VOR orientation, intercepting and tracking VOR radials, and, attitude instrument flight. The pilot will be able to maintain the desired altitude, ±100 ft.; airspeed, ±10 kt.; and heading, ±10°, and to track a radial allowing no more than a 3/4-scale deflection of the course deviation indicator (CDI) or remaining within 10° in the case of an RMI.

Instructor's comments: ______________________________

____

____

____

____

____

____

____

____

____

____

Lesson assignment: ______________________________

____

Lesson certification:

____

(Pilot) Instructor Certificate No. Exp. Date

FLIGHT LESSON 10: GPS and ADF Procedures

Note: In this lesson and throughout the remainder of this syllabus, any reference to DME arcs, GPS, or ADF shall be disregarded if the airplane is not equipped with these specified navigational systems.

Objective

To introduce the pilot to GPS and ADF orientation, the method used to track a GPS course, and the method used to intercept and track an NDB bearing. Additionally, the pilot will increase proficiency in VOR procedures.

Text References

Instrument Pilot Flight Maneuvers and Practical Test Prep (FM)

Instrument Pilot Syllabus (SYL)*

Content

1. Flight Lesson 9 complete? Yes ____ Copy placed in pilot's folder? Yes ____
2. Preflight briefing
3. Review items

☐	☐	☐	Basic instrument flight patterns (as directed by your instructor) - SYL*
☐	☐	☐	VOR orientation - FM 193, 196, 198
☐	☐	☐	VOR time and distance calculations - SYL*
☐	☐	☐	Intercepting and tracking VOR radials and DME arcs - FM 194-197, 198-202

4. New items

☐	☐	☐	GPS and ADF orientation - FM 202-203 (ADF), 210 (GPS)
☐	☐	☐	Tracking a GPS course - FM 208-210
☐	☐	☐	Intercepting and tracking NDB bearings - FM 203-208

5. Postflight critique and preview of next lesson

*You can download these Additional Instrument Flight Maneuvers at (www.gleim.com/Aviation/Updates/books/ipfm/) or purchase Gleim's *Instrument Pilot Syllabus*.

Completion Standards

The lesson will have been successfully completed when the pilot can demonstrate an understanding of GPS and ADF orientation, the method used to track a GPS course, and the method used to intercept and track a predetermined NDB bearing. Additionally, the pilot will demonstrate increased proficiency in VOR procedures. The pilot will be able to maintain the desired altitude, ±100 ft.; airspeed, ±10 kt.; heading, ±10°; and DME arc, ±2 NM, and to track a VOR radial with no more than a 3/4-scale deflection of the CDI or within 10° in the case of an RMI.

Instructor's comments: ______________________________

____

Lesson assignment: ______________________________

Lesson certification:

(Pilot) ____________ Instructor ____________ Certificate No. ____________ Exp. Date

FLIGHT LESSON 11: GPS Procedures and ADF Time/Distance to Station

Objective

To introduce the pilot to ADF time and distance calculations. Additionally, the pilot will increase proficiency in tracking GPS courses and in ADF orientation and intercepting and tracking NDB bearings.

Text References

Instrument Pilot Flight Maneuvers and Practical Test Prep (FM)

Instrument Pilot Syllabus (SYL)*

Content

1. Flight Lesson 10 complete? Yes ____ Copy placed in pilot's folder? Yes ____
2. Preflight briefing
3. Review items
   - ☐☐☐ Basic instrument flight patterns (as directed by your instructor) - SYL*
   - ☐☐☐ GPS and ADF orientation - FM 202-203 (ADF), 210 (GPS)
   - ☐☐☐ Tracking a GPS course - FM 208-210
   - ☐☐☐ Intercepting and tracking NDB bearings - FM 203-208
   - ______________________________
4. New item
   - ☐☐☐ ADF time and distance calculations - SYL*
5. Postflight critique and preview of next lesson

*You can download these Additional Instrument Flight Maneuvers at (www.gleim.com/Aviation/Updates/books/ipfm/) or purchase Gleim's *Instrument Pilot Syllabus*.

Completion Standards

The lesson will have been successfully completed when the pilot can demonstrate an understanding of ADF time and distance calculations. The pilot will also demonstrate increased proficiency in GPS and ADF orientation, tracking a GPS course, and intercepting and tracking a predetermined NDB bearing. The pilot will be able to maintain the desired altitude, ±100 ft.; airspeed, ±10 kt.; heading, ±10°; GPS CDI with no more than 3/4-scale deflection; and an NDB bearing, ±10°.

Instructor's comments: ______________________________

____

____

____

____

____

____

____

____

____

____

____

____

____

____

Lesson assignment: ______________________________

____

Lesson certification:

______________________________ ______________________________

(Pilot) Instructor Certificate No. Exp. Date

FLIGHT LESSON 12: Tracking the Localizer

Note: For the remainder of this syllabus, the term "localizer" will be used to include localizer, simplified directional facility (SDF), and localizer-type directional aid (LDA).

Objective

To introduce the pilot to the procedures to track the front and back course localizer. The pilot will also gain more proficiency in VOR, DME, GPS, and ADF procedures.

Text References

Instrument Pilot Flight Maneuvers and Practical Test Prep (FM)

Instrument Pilot Syllabus (SYL)*

Content

1. Flight Lesson 11 complete? Yes ____ Copy placed in pilot's folder? Yes ____
2. Preflight briefing
3. Review items
   - ☐☐☐ Intercepting and tracking VOR radials - FM 193-196, 201-202
   - ☐☐☐ Intercepting and tracking DME arcs - FM 198-202
   - ☐☐☐ Intercepting and tracking NDB bearings - FM 202-208
   - ☐☐☐ Tracking GPS courses - FM 208-210
   - ☐☐☐ VOR and ADF time and distance calculations - SYL*

   ____
4. New items
   - ☐☐☐ Tracking the localizer front course - FM 86, 270
   - ☐☐☐ Tracking the localizer back course - FM 86, 270
5. Postflight critique and preview of next lesson

*You can download these Additional Instrument Flight Maneuvers at (www.gleim.com/Aviation/Updates/books/ipfm/) or purchase Gleim's *Instrument Pilot Syllabus*.

Completion Standards

The lesson will have been successfully completed when the pilot can demonstrate an understanding of front and back course localizer tracking. The pilot will be able to maintain the desired altitude, ±100 ft.; airspeed, ±10 kt.; heading, ±10°; DME arc, ±2 NM; and NDB bearing, ±10°, and to track a VOR radial or a GPS course with no more than a 3/4-scale deflection of the CDI or within 10° in the case of an RMI.

Instructor's comments: ______________________________

____

Lesson assignment: ______________________________

____

Lesson certification:

____

(Pilot) | Instructor | Certificate No. | Exp. Date

FLIGHT LESSON 13: Stage One Check

Objective

During this stage check, an authorized instructor will determine if the pilot is proficient in attitude instrument flying and in the use of navigation equipment.

Text References

Instrument Pilot Flight Maneuvers and Practical Test Prep (FM)

Pilot's Operating Handbook (POH)

Content

1. Flight Lesson 12 complete? Yes ____ Copy placed in pilot's folder? Yes ____
2. Preflight briefing
3. Review items
   a. Preflight procedures
   - ☐☐☐ Airplane systems related to IFR operations - FM 68-71
   - ☐☐☐ Airplane flight instruments and navigation equipment - FM 72-93, POH-9
   - ☐☐☐ Instrument cockpit check - FM 94-99

   b. Attitude instrument flying -- full and partial panel unless otherwise indicated
   - ☐☐☐ Straight-and-level flight - FM 141-145
   - ☐☐☐ Change of airspeed - FM 146-153
   - ☐☐☐ Constant airspeed climbs and descents - FM 154-165
   - ☐☐☐ Rate climbs and descents - FM 166-174
   - ☐☐☐ Timed turns to magnetic compass headings - FM 175-181
   - ☐☐☐ Magnetic compass turns (partial panel only) - FM 176
   - ☐☐☐ Steep turns (full panel only) - FM 181-184
   - ☐☐☐ Recovery from unusual flight attitudes - FM 185-190

   c. Navigation systems
   - ☐☐☐ Intercept and track VOR radials - FM 193-196, 198, 201-202
   - ☐☐☐ Intercept and track DME arcs - FM 198-202
   - ☐☐☐ Intercept and track NDB bearings - FM 202-208
   - ☐☐☐ Track a GPS course - FM 208-210
   - ☐☐☐ Track a localizer - FM 270
4. Postflight critique and preview of next lesson

Completion Standards

The lesson will have been successfully completed when the pilot can demonstrate proficiency in attitude instrument flight (full and partial panel) and navigation procedures. The pilot will be able to maintain the desired altitude, ±100 ft.; airspeed, ±10 kt.; heading, ±5°; rate of climb, ±100 fpm; DME arc, ±2 NM; and NDB bearing, ±10°; to roll out on the desired heading, ±10°; and to track a VOR radial or a GPS course with no more than a 3/4-scale deflection of the CDI or within 10° in the case of an RMI.

Instructor's comments: ________________________________

____

____

____

____

____

____

____

Lesson assignment: ________________________________

____

Lesson certification:

________________ ________________________________

(Pilot) Instructor Certificate No. Exp. Date

STAGE TWO

Stage Two Objective

The pilot will be introduced to holding procedures and instrument approach procedures, including missed approaches.

Stage Two Completion Standards

The stage will be completed when the pilot demonstrates proficiency in holding procedures using various navigation systems and all types of instrument approach procedures.

Lesson	Topic
14.	VOR Holding
15.	GPS and ADF Holding
16.	Localizer Holding
17.	DME and Intersection Holding
18.	VOR Instrument Approach
19.	GPS and NDB Instrument Approaches
20.	Localizer Instrument Approach
21.	ILS Instrument Approach
22.	Instrument Approaches (Review)
23.	Stage Two Check

FLIGHT LESSON 14: VOR Holding

Objective

To introduce the pilot to holding procedures at a VOR station, including ATC clearances and holding instructions and holding pattern entry procedures.

Text References

Instrument Pilot Flight Maneuvers and Practical Test Prep (FM)

Instrument Pilot Syllabus (SYL)*

Content

1. Flight Lesson 13 complete? Yes ____ Copy placed in pilot's folder? Yes ____
2. Preflight briefing
3. Review items
   - ☐☐☐ Airplane systems related to IFR operations - FM 68-71
   - ☐☐☐ Instrument cockpit check - FM 94-99
   - ☐☐☐ Instrument takeoff - SYL*
   - ☐☐☐ Intercept and track a VOR radial - FM 193-196, 198, 201-202
   - ☐☐☐ Loss of gyro attitude and/or heading indicators - FM 332-335
   - ______________________________
4. New items
   - ☐☐☐ Track a VOR radial (partial panel) - FM 193-196, 198, 201-202
   - ☐☐☐ ATC clearances and holding instructions - FM 122-123
   - ☐☐☐ Holding pattern entry procedures - FM 124-129
     - ☐☐☐ Parallel procedure
     - ☐☐☐ Teardrop procedure
     - ☐☐☐ Direct entry procedure
   - ☐☐☐ VOR holding procedures - FM 129-132
     - ☐☐☐ Standard pattern
     - ☐☐☐ Nonstandard pattern
5. Postflight critique and preview of next lesson

*You can download these Additional Instrument Flight Maneuvers at (www.gleim.com/Aviation/Updates/books/ipfm/) or purchase Gleim's *Instrument Pilot Syllabus*.

Completion Standards

The lesson will have been successfully completed when the pilot can demonstrate an understanding of ATC holding instructions, holding pattern entry procedures, and the procedures to remain in the holding pattern. Additionally, the pilot will demonstrate increased knowledge in systems related to IFR operations and proficiency in the instrument cockpit check and partial-panel instrument flight. The pilot will be able to maintain the desired altitude, ±100 ft.; airspeed, ±10 kt.; and heading, ±10°.

Instructor's comments: ______________________________

____

____

____

____

____

____

____

____

Lesson assignment: ______________________________

____

Lesson certification:

______________________________ ______________________________

(Pilot) Instructor Certificate No. Exp. Date

FLIGHT LESSON 15: GPS and ADF Holding

Objective

To introduce the pilot to holding procedures at an NDB station with an ADF and/or GPS holding procedures. Additionally, the pilot will increase proficiency in VOR holding and complying with ATC clearances and holding instructions.

Text References

Instrument Pilot Flight Maneuvers and Practical Test Prep (FM)

Content

1. Flight Lesson 14 complete? Yes ____ Copy placed in pilot's folder? Yes ____
2. Preflight briefing
3. Review items
   - ☐☐☐ ATC clearances and holding instructions - FM 123
   - ☐☐☐ Holding pattern entry procedures - FM 124-129
     - ☐☐☐ Parallel procedure
     - ☐☐☐ Teardrop procedure
     - ☐☐☐ Direct entry procedure
   - ☐☐☐ VOR holding procedures - FM 129-132
   - ☐☐☐ Loss of gyro attitude and/or heading indicators - FM 332-335

   ____
4. New items
   - ☐☐☐ Track a GPS course or an NDB bearing (partial panel) - FM 202-210
   - ☐☐☐ ADF holding procedures (if the airplane is so equipped) - FM 129-132, 134-135
     - ☐☐☐ Standard pattern
     - ☐☐☐ Nonstandard pattern
   - ☐☐☐ GPS holding procedures (if the airplane is so equipped) - FM 129-132, 136
     - ☐☐☐ Standard pattern
     - ☐☐☐ Nonstandard pattern
5. Postflight critique and preview of next lesson

Completion Standards

The lesson will have been successfully completed when the pilot can demonstrate an understanding of GPS and/or ADF holding procedures while maintaining orientation at all times. Additionally, the pilot will demonstrate increased proficiency in copying and complying with ATC holding instructions, entries to holding, and VOR holding procedures. The pilot will be able to maintain the desired altitude, ±100 ft.; airspeed, ±10 kt.; and heading, ±10°, and to track a selected course, radial, or bearing.

Instructor's comments: ______________________________

____

____

____

____

____

____

____

____

Lesson assignment: ______________________________

____

Lesson certification:

______________________________ ______________________________

(Pilot) Instructor Certificate No. Exp. Date

FLIGHT LESSON 16: Localizer Holding

Objective

To introduce the pilot to localizer holding procedures. Additionally, the pilot will increase proficiency in holding pattern procedures is increased through the practice of VOR, GPS, and ADF holding.

Text References

Instrument Pilot Flight Maneuvers and Practical Test Prep (FM)

Content

1. Flight Lesson 15 complete? Yes ____ Copy placed in pilot's folder? Yes ____
2. Preflight briefing
3. Review items
   - ☐☐☐ ATC holding instructions - FM 123
   - ☐☐☐ Holding pattern entry procedures - FM 124-129
   - ☐☐☐ VOR holding procedures - FM 129-132
   - ☐☐☐ GPS holding procedures - FM 129-132
   - ☐☐☐ ADF holding procedures - FM 129-132, 134-135
   - ______________________________
4. New items
   - ☐☐☐ Localizer holding procedures - FM 129-132
     - ☐☐☐ Standard pattern
     - ☐☐☐ Nonstandard pattern
5. Postflight critique and preview of next lesson

Completion Standards

The lesson will have been successfully completed when the pilot can demonstrate an understanding of localizer holding procedures. Additionally, the pilot will demonstrate greater proficiency in holding pattern entry procedures and VOR, GPS, and ADF holding procedures. The pilot will be able to maintain the desired altitude, ±100 ft.; airspeed, ±10 kt.; and heading, ±10°, and to track a selected course, bearing, or radial.

Instructor's comments: ______________________________

Lesson assignment: ______________________________

Lesson certification:

(Pilot) ______________ Instructor ______________ Certificate No. ______________ Exp. Date

FLIGHT LESSON 17: DME and Intersection Holding

Objective

To introduce the pilot to DME holding and intersection holding procedures. Additionally, the pilot's proficiency in holding procedures is increased through practice.

Text References

Instrument Pilot Flight Maneuvers and Practical Test Prep (FM)

Content

1. Flight Lesson 16 complete? Yes ____ Copy placed in pilot's folder? Yes ____
2. Preflight briefing
3. Review
   - ☐☐☐ ATC holding instructions - FM 123
   - ☐☐☐ Holding pattern entry procedures - FM 124-129
   - ☐☐☐ VOR holding procedures - FM 129-132
   - ☐☐☐ GPS holding procedures - FM 129-132
   - ☐☐☐ ADF holding procedures - FM 129-132, 134-135
   - ☐☐☐ Localizer holding procedures - FM 129-132
   - ____________________
4. New items
   - ☐☐☐ DME holding procedures - FM 129-132
     - ☐☐☐ Holding course away from the VORTAC or VOR/DME
     - ☐☐☐ Holding course toward the VORTAC or VOR/DME
   - ☐☐☐ Intersection holding - FM 133-134
     - ☐☐☐ Intersection defined by two VOR radials
     - ☐☐☐ Intersection defined by a VOR radial and an NDB bearing
5. Postflight critique and preview of next lesson

Completion Standards

The lesson will have been successfully completed when the pilot can demonstrate an understanding of DME holding and intersection holding. Additionally, the pilot should demonstrate increased proficiency in VOR, GPS, ADF, and localizer holding procedures. The pilot will be able to maintain the desired altitude, ±100 ft.; airspeed, ±10 kt.; and heading, ±10°, and track the desired radial, course, or bearing.

Instructor's comments: ____________________

____

____

____

____

____

____

____

____

____

____

Lesson assignment: ____________________

____

Lesson certification:

(Pilot)	Instructor	Certificate No.	Exp. Date

FLIGHT LESSON 18: VOR Instrument Approach

Objective

To introduce the pilot to VOR instrument approach procedures, including ATC clearances that pertain to the approach. Additionally, the pilot will be able to properly enter the holding pattern depicted for the approach(es).

Text References

Instrument Pilot Flight Maneuvers and Practical Test Prep (FM)

Content

1. Flight Lesson 17 complete? Yes ____ Copy placed in pilot's folder? Yes ____
2. Preflight briefing
3. Review item
   - ☐☐☐ Holding procedures - FM 122-136
4. New items
   - ☐☐☐ Segments of an instrument approach - FM 211-212
   - ☐☐☐ Instrument approach procedure charts - FM 214-219
   - ☐☐☐ Airplane approach category - FM 219-220
   - ☐☐☐ Procedure turn - FM 220-222
   - ☐☐☐ Advance information on instrument approach - FM 213-214
   - ☐☐☐ Full approach - CFII
   - ☐☐☐ Radar vectoring - FM 212-213
   - ☐☐☐ ATC approach clearances - FM 218-219
   - ☐☐☐ Inoperative airplane and ground navigation equipment - FM 239
   - ☐☐☐ Inoperative visual aids associated with the landing environment - FM 239
   - ☐☐☐ National Weather Service (NWS) reporting factors and criteria - FM 239
   - ☐☐☐ VOR, VOR/DME, VORTAC instrument approach procedures (full approach) - FM 233-242
     - ☐☐☐ DME arc - FM 198-201
   - ☐☐☐ Missed approach - FM 312-316
   - ☐☐☐ Circling approach - FM 317-321
   - ☐☐☐ Landing from a straight-in or circling approach - FM 322-326
5. Postflight critique and preview of next lesson

Completion Standards

The lesson will have been successfully completed when the pilot can explain and use the information on the approach charts, understand and comply with ATC approach clearances, and execute the approach and missed approach procedures. The pilot will be able to maintain altitude, ±100 ft.; airspeed, ±10 kt.; and heading, ±10°, with no more than a 3/4-scale deflection of the CDI, or within 10° in the case of an RMI, throughout the approach. Additionally, the pilot will demonstrate the proper holding pattern entry procedure and proper timing criteria or leg lengths, as appropriate.

Instructor's comments: ____________________

____

____

____

____

____

____

____

Lesson assignment: ____________________

____

Lesson certification:

____________________ ____________________

(Pilot) Instructor Certificate No. Exp. Date

FLIGHT LESSON 19: GPS and NDB Instrument Approaches

Objective

To introduce the pilot to GPS and NDB approach procedures. Additionally, the pilot will increase proficiency in VOR approach procedures.

Text References

Instrument Pilot Flight Maneuvers and Practical Test Prep (FM)

Content

1. Flight Lesson 18 complete? Yes ____ Copy placed in pilot's folder? Yes ____
2. Preflight briefing
3. Review items
   - ☐☐☐ Segments of an instrument approach - FM 211-212
   - ☐☐☐ GPS receiver autonomous integrity monitoring (RAIM) - FM 91-93
   - ☐☐☐ ATC approach clearances - FM 221
   - ☐☐☐ Holding procedures - FM 122-136
   - ☐☐☐ VOR instrument approach - FM 233-242
   - ☐☐☐ Circling approach - FM 317-321
   - ☐☐☐ Missed approach - FM 312-316
   - ☐☐☐ Landing from a straight-in or circling approach - FM 322-326
   - ☐☐☐ ____________________
4. New items
   - ☐☐☐ GPS instrument approach design concepts - FM 251-254
   - ☐☐☐ GPS instrument approach procedures - FM 254-264
     - ☐☐☐ Full approach
     - ☐☐☐ Radar vectors
     - ☐☐☐ Missed approach
   - ☐☐☐ NDB instrument approach procedures (full approach) - FM 242-251
     - ☐☐☐ Missed approach
5. Postflight critique and preview of next lesson

Completion Standards

The lesson will have been successfully completed when the pilot can demonstrate the proper GPS and/or NDB approach and the missed approach procedures. The pilot will demonstrate an increased proficiency in holding, VOR approaches, circling approach procedures, and landing from a straight-in or circling approach. The pilot will be able to maintain the desired altitude, ±100 ft.; airspeed, ±10 kt.; heading, ±10°; and NDB bearing, ±10°, and to maintain the GPS course with no more than a 3/4-scale deflection of the CDI.

Instructor's comments: ____________________

____

____

____

____

____

____

____

____

Lesson assignment: ____________________

____

Lesson certification:

____________________ ____________________

(Pilot) Instructor Certificate No. Exp. Date

FLIGHT LESSON 20: Localizer Instrument Approach

Objective

To introduce the pilot to localizer instrument approach procedures. Additionally, the pilot will increase proficiency in VOR, GPS, and NDB approaches.

Text References

Instrument Pilot Flight Maneuvers and Practical Test Prep (FM)

Content

1. Flight Lesson 19 complete? Yes ____ Copy placed in pilot's folder? Yes ____
2. Preflight briefing
3. Review items
   - ☐☐☐ VOR instrument approach (partial panel) - FM 233-242, 332-335
   - ☐☐☐ GPS instrument approach - FM 251-264
   - ☐☐☐ NDB instrument approach - FM 242-251
   - ☐☐☐ Missed approach - FM 312-316
   - ☐☐☐ Circling approach - FM 317-321
   - ☐☐☐ Landing from a straight-in or circling approach - FM 322-326
   - ________________________________
4. New items
   - ☐☐☐ Localizer instrument approach - FM 265-275
     - ☐☐☐ Front course
     - ☐☐☐ Back course (if available)
     - ☐☐☐ Missed approach
   - ☐☐☐ LDA instrument approach (if available) - FM 275-284
   - ☐☐☐ SDF instrument approach (if available) - FM 285-294
5. Postflight critique and preview of next lesson

Completion Standards

The lesson will have been successfully completed when the pilot can demonstrate knowledge in localizer instrument approach procedures. The pilot will be able to maintain the desired airspeed, ±10 kt.; heading, ±10°; GPS CDI with no more than a 3/4-scale deflection; and the desired NDB bearing, ±10° (NDB approach). Prior to beginning the final approach segment, the pilot will maintain altitude, ±100 ft., and, while on the final approach segment, descend and maintain the MDA, +100/–50 ft.

Instructor's comments: ________________________________

____

____

____

____

____

____

____

____

____

____

____

Lesson assignment: ________________________________

____

Lesson certification:

____

(Pilot) Instructor Certificate No. Exp. Date

FLIGHT LESSON 21: ILS Instrument Approach

Objective

To introduce the pilot to the ILS instrument approach procedures. Additionally, the pilot will increase proficiency in GPS, NDB, and localizer instrument approaches using full and partial panel.

Text References

Instrument Pilot Flight Maneuvers and Practical Test Prep (FM)

Content

1. Flight Lesson 20 complete? Yes ____ Copy placed in pilot's folder? Yes ____
2. Preflight briefing
3. Review items

☐☐☐ GPS instrument approach (partial panel) - FM 251-264, 332-335
☐☐☐ NDB instrument approach (partial panel) - FM 242-251, 332-335
☐☐☐ Localizer instrument approach - FM 265-275
  ☐☐☐ Front course
  ☐☐☐ Back course
☐☐☐ Recovery from unusual flight attitudes (without attitude indicator) - FM 185-190

____

4. New items

☐☐☐ ILS instrument approach - FM 295-311
  ☐☐☐ Missed approach - FM 312-316
  ☐☐☐ Circling approach - FM 317-321

5. Postflight critique and preview of next lesson

Completion Standards

The lesson will have been successfully completed when the pilot can demonstrate knowledge in ILS instrument approach procedures. The pilot will display an increase in proficiency in conducting GPS, NDB, and localizer instrument approaches, including partial-panel approaches. During the nonprecision approaches, the pilot will be able to

1. Maintain, prior to the beginning of the final approach segment, altitude, ±100 ft.; heading, ±10°; and airspeed, ±10 kt., with less than a full-scale deflection of the CDI or within 10° in the case of an RMI or ADF.
2. Arrive at the MDA prior to reaching the MAP.
3. Maintain, while on the final approach segment, no more than a 3/4-scale deflection of the CDI or within 10° in the case of RMI or ADF; airspeed, ±10 kt.; altitude, +100 ft./–50 ft.

Additionally, the pilot will display proficiency in recovering from an unusual flight attitudes.

Instructor's comments: ________________________________

____

____

____

____

____

____

____

____

____

____

Lesson assignment: ________________________________

____

Lesson certification:

________________ ________________________________

(Pilot) Instructor Certificate No. Exp. Date

FLIGHT LESSON 22: Instrument Approaches (review)

Objective

To review previous lessons to gain proficiency in instrument approaches (full and partial panel). Additionally, the pilot will be introduced to radar approaches and a timed approach from a holding fix.

Text References

Instrument Pilot Flight Maneuvers and Practical Test Prep (FM)

Content

1. Flight Lesson 21 complete? Yes ____ Copy placed in pilot's folder? Yes ____
2. Preflight briefing
3. Review items
   - ☐☐☐ VOR instrument approach - FM 233-242
     - ☐☐☐ Full panel
     - ☐☐☐ Partial panel
   - ☐☐☐ GPS instrument approach - FM 251-264
     - ☐☐☐ Full panel
     - ☐☐☐ Partial panel
   - ☐☐☐ NDB instrument approach - FM 242-251
     - ☐☐☐ Full panel
     - ☐☐☐ Partial panel
   - ☐☐☐ Localizer instrument approach - FM 265-275
     - ☐☐☐ Full panel
     - ☐☐☐ Partial panel
   - ☐☐☐ ILS instrument approach - FM 295-311
     - ☐☐☐ Full panel
     - ☐☐☐ Partial panel

---

4. New items
   - ☐☐☐ Timed approaches from a holding fix - FM 223
   - ☐☐☐ Radar approaches - FM 224-226
     - ☐☐☐ Precision approach (PAR)
     - ☐☐☐ Surveillance approach (ASR)
     - ☐☐☐ No-gyro approach (partial panel)
   - ☐☐☐ Side-step maneuver - FM 226
5. Postflight critique and preview of next lesson

Completion Standards

The lesson will have been successfully completed when the pilot demonstrates an understanding of timed approaches from a holding fix, radar approaches, and the side-step maneuver. Additionally, the pilot will demonstrate an increased proficiency in performing instrument approaches, under both full- and partial-panel operations.

Instructor's comments: ______________________________

____

____

____

Lesson assignment: ______________________________

____

Lesson certification:

______________________________ ______________________________

(Pilot) Instructor Certificate No. Exp. Date

FLIGHT LESSON 23: Stage Two Check

Objective

During this stage check an authorized instructor will determine if the pilot is proficient in holding procedures and instrument approach procedures.

Text References

Instrument Pilot Flight Maneuvers and Practical Test Prep (FM)

Content

1. Flight Lesson 22 complete? Yes ____ Copy placed in pilot's folder? Yes ____
2. Preflight briefing
3. Review items
   - ☐☐☐ Holding procedures using navigation equipment in airplane - FM 122-136
   - ☐☐☐ ATC instructions and clearances relating to holding and approaches - FM 123, 221
   - ☐☐☐ Nonprecision approaches (instructor to select at least one to be conducted under partial panel) - FM 232-294
     - ☐☐☐ VOR approach - FM 233-242
     - ☐☐☐ GPS approach - FM 251-264
     - ☐☐☐ NDB approach - FM 242-251
     - ☐☐☐ Localizer (back course) approach - FM 265-275
   - ☐☐☐ ILS approach - FM 295-311
   - ☐☐☐ Circling approach - FM 317-321
   - ☐☐☐ Missed approach - FM 312-316
   - ☐☐☐ Landing from a straight-in or circling approach - FM 322-326
4. Postflight critique and preview of next lesson

Completion Standards

The lesson will have been successfully completed when the pilot can demonstrate proficiency in holding procedures and instrument approach procedures to the standards listed in the current FAA Instrument Rating Practical Test Standards.

Instructor's comments: ______________________________

Lesson assignment: ______________________________

Lesson certification:

(Pilot) ____________ Instructor ____________ Certificate No. ____________ Exp. Date ____________

STAGE THREE

Stage Three Objective

The pilot will be instructed in the conduct of cross-country flights in an airplane while operating under IFR within the U.S. National Airspace System. Additionally, the pilot will be instructed in the procedures to be used in the event of loss of communications. Finally, the pilot will receive instruction in preparation for the instrument rating (airplane) practical test.

Stage Three Completion Standards

The stage will be completed when the pilot demonstrates the ability to conduct cross-country flights in an airplane while operating under IFR, including the loss of communications procedures. Finally, the pilot will demonstrate proficiency in all tasks of the instrument rating (airplane) practical test and meet or exceed the minimum acceptable standards for the instrument rating.

Lesson	Topic
24.	Cross-Country and Emergency Procedures
25.	Cross-Country Procedures
26.	Cross-Country Procedures
27.	Maneuvers Review
28.	Stage Three Check

FLIGHT LESSON 24: Cross-Country and Emergency Procedures

Objective

To introduce the pilot to IFR cross-country procedures that include flight planning. Additionally, the pilot will be introduced to departure, en route, arrival, and loss of communications procedures.

Text References

Instrument Pilot Flight Maneuvers and Practical Test Prep (FM)

Pilot's Operating Handbook (POH)

Content

1. Flight Lesson 23 complete? Yes ____ Copy placed in pilot's folder? Yes ____
2. Preflight briefing
3. Review items
   - ☐☐☐ Airplane systems related to IFR operations - FM 68-71
   - ☐☐☐ Airplane flight instruments and navigation systems - FM 72-93, POH-9
   - ☐☐☐ Instrument cockpit check - FM 94-99
   - ☐☐☐ Instrument takeoff - SYL
   - ☐☐☐ Intercepting and tracking navigational systems - FM 192-210
   - ☐☐☐ Holding procedures - FM 122-136
   - ☐☐☐ Arrival procedures - FM 211-229
   - ☐☐☐ Instrument approach procedures - FM 232-326
   - ☐☐☐ Loss of gyro attitude and/or heading indicators -FM 332-335
   - ☐☐☐ Postflight procedures - FM 338

   ____

4. New items
   - ☐☐☐ Weather information - FM 48-57
   - ☐☐☐ Cross-country flight planning - FM 58-65
   - ☐☐☐ Air traffic control clearances - FM 102-110
   - ☐☐☐ Compliance with departure, en route, and arrival procedures and clearances - FM 111-121
   - ☐☐☐ Emergency operations
     - ☐☐☐ Loss of communications - FM 328-331
     - ☐☐☐ Instrument, system, equipment failures - POH-3, 9
     - ☐☐☐ Icing and turbulence - POH-3
     - ☐☐☐ Engine failure - POH-3
     - ☐☐☐ Low fuel status - CFII, POH-3
   - ☐☐☐ Calculating ETEs and ETAs - CFII
   - ☐☐☐ En route course changes - CFII
5. Postflight critique and preview of next lesson

Completion Standards

The lesson will have been successfully completed when the pilot demonstrates an understanding of IFR flight planning; copying, readback, and compliance with ATC clearances; IFR departures and arrivals; and emergency operations. Additionally, the pilot will be able to calculate ETEs and ETAs and understand the reasons for course changes that are issued by ATC or requested due to weather conditions. The pilot will be able to maintain the desired altitude, ±100 ft.; airspeed, ±10 kt.; and heading, ±10°, and to track a course, radial, or bearing.

Instructor's comments: ____

____

____

____

Lesson assignment: ____

____

Lesson certification:

______________________ ____

(Pilot) Instructor Certificate No. Exp. Date

FLIGHT LESSON 25: Cross-Country Procedures

Objective

To further increase the pilot's ability to conduct IFR cross-country operations. This flight should include an airport that is at least a straight-line distance of more than 50 NM from the departure point and, if possible, an airport that has a radar approach.

Text References

Instrument Pilot Flight Maneuvers and Practical Test Prep (FM)

Content

1. Flight Lesson 24 complete? Yes ____ Copy placed in pilot's folder? Yes ____
2. Preflight briefing
3. Review items
   - ☐☐☐ Weather information - FM 48-57
   - ☐☐☐ Cross-country flight planning - FM 58-65
   - ☐☐☐ Compliance with departure, en route, and arrival procedures and clearances - FM 111-121
   - ☐☐☐ Holding procedures - FM 122-136
   - ☐☐☐ Instrument approach procedures - FM 211
     - ☐☐☐ Partial-panel nonprecision instrument approach - FM 232-294
     - ☐☐☐ Missed approach - FM 312-316
     - ☐☐☐ Circling approach - FM 317-321
     - ☐☐☐ Radar or no-gyro approach, if available - FM 227-228
   - ☐☐☐ Emergency operations
     - ☐☐☐ Loss of communications - FM 328-331
     - ☐☐☐ Loss of gyro attitude and/or heading indicators - FM 332-335
4. Postflight critique and preview of next lesson

Completion Standards

The lesson will have been successfully completed when the pilot demonstrates an increased proficiency in IFR flight planning, understanding and complying with ATC clearances, and conducting a cross-country flight under IFR. The pilot will be able to maintain the desired altitude, ±100 ft.; airspeed, ±10 kt.; and heading, ±10°, and to track a course, radial, or bearing.

Instructor's comments: ______________________________

____

Lesson assignment: ______________________________

Lesson certification:

(Pilot) ____________ Instructor ____________ Certificate No. ____________ Exp. Date ____________

FLIGHT LESSON 26: Cross-Country Procedures

Objective

To increase the pilot's proficiency in IFR cross-country operations. This cross-country flight must be performed under IFR and must be at least 250 NM along airways or ATC-directed routing with at least one segment of the flight consisting of at least a straight-line distance of 100 NM between airports. Additionally, it must involve an instrument approach at each airport and three different kinds of approaches with the use of navigation systems.

Text References

Instrument Pilot Flight Maneuvers and Practical Test Prep (FM)

Pilot's Operating Handbook (POH)

Content

1. Flight Lesson 25 complete? Yes ____ Copy placed in pilot's folder? Yes ____
2. Preflight briefing
3. Review items
   - ☐☐☐ Weather information - FM 48-57
   - ☐☐☐ Cross-country flight planning - FM 58-65
   - ☐☐☐ Instrument takeoff - SYL
   - ☐☐☐ Compliance with ATC clearances - FM 111-121
   - ☐☐☐ Holding procedures - FM 122-136
   - ☐☐☐ ILS instrument approach - FM 295-311
   - ☐☐☐ Nonprecision instrument approach (at least two different kinds of approaches) - FM 232-294
   - ☐☐☐ Missed approach - FM 312-316
   - ☐☐☐ Landing from a straight-in or circling approach - FM 322-326
   - ☐☐☐ Emergency operations (simulate) - POH-3
     - ☐☐☐ Loss of communications - FM 328-331
     - ☐☐☐ Loss of gyro attitude and/or heading indicators - FM 332-335
     - ☐☐☐ Low fuel supply - CFII, POH-3
     - ☐☐☐ Engine failure - POH-3
4. Postflight critique and preview of next lesson

Completion Standards

The lesson will have been successfully completed when the pilot demonstrates a thorough understanding of IFR procedures. The pilot will complete each task to the standards specified in the current FAA Instrument Rating Practical Test Standards.

Instructor's comments: ____

____

____

____

____

____

____

____

____

____

____

Lesson assignment: ____

____

Lesson certification:

________________________ ____

(Pilot) Instructor Certificate No. Exp. Date

FLIGHT LESSON 27: Maneuvers Review

Objective

To review procedures and maneuvers covered previously.

Text References

Instrument Pilot Flight Maneuvers and Practical Test Prep (FM)

Content

1. Flight Lesson 26 complete? Yes ____ Copy placed in pilot's folder? Yes ____
2. Preflight briefing
3. Review items
   - ☐☐☐ Airplane systems related to IFR operations - FM 68-71
   - ☐☐☐ Airplane flight instruments and navigation equipment - FM 72-93, POH-9
   - ☐☐☐ Timed turns to magnetic compass headings - FM 175-181
   - ☐☐☐ Steep turns - FM 181-184
   - ☐☐☐ Recovery from unusual flight attitudes (without attitude indicator) - FM 185-190
   - ☐☐☐ Maneuvers as assigned by the instructor - CFII
4. Postflight critique and preview of next lesson

Completion Standards

The lesson will have been successfully completed when the pilot demonstrates proficiency in the maneuvers performed. The pilot will complete each task to the standards specified in the current FAA Instrument Rating Practical Test Standards.

Instructor's comments: ____________________

Lesson assignment: ____________________

Lesson certification:

(Pilot) Instructor Certificate No. Exp. Date

FLIGHT LESSON 28: Stage Three Check

Objective

The pilot will be able to demonstrate the required proficiency of an instrument-rated pilot by using the current FAA Instrument Rating Practical Test Standards.

Text References

Instrument Pilot Flight Maneuvers and Practical Test Prep (FM)

Pilot's Operating Handbook (POH)

Content

1. Flight Lesson 27 complete? Yes ____ Copy placed in pilot's folder? Yes ____
2. Stage Check Tasks
   - ☐☐☐ Weather information - FM 48-57
   - ☐☐☐ Cross-country flight planning - FM 58-65
   - ☐☐☐ Airplane systems related to IFR operations - FM 68-71
   - ☐☐☐ Airplane flight instruments and navigation equipment - FM 72-93, POH-1, 7, 9
   - ☐☐☐ Instrument cockpit check - FM 94-99
   - ☐☐☐ Air traffic control clearances - FM 102-110
   - ☐☐☐ Compliance with departure, en route, and arrival procedures and clearances - FM 111-121
   - ☐☐☐ Holding procedures - FM 122-136
   - ☐☐☐ Straight-and-level flight - FM 141-145
   - ☐☐☐ Change of airspeed - FM 146-153
   - ☐☐☐ Constant airspeed climbs and descents - FM 154-165
   - ☐☐☐ Rate climbs and descents - FM 166-174
   - ☐☐☐ Timed turns to magnetic compass headings - FM 175-181
   - ☐☐☐ Steep turns - FM 181-184
   - ☐☐☐ Recovery from unusual flight attitudes - FM 185-190
   - ☐☐☐ Intercepting and tracking navigational systems and DME arcs - FM 192-210
   - ☐☐☐ Nonprecision instrument approach (perform two, one using partial panel) - FM 211-294
   - ☐☐☐ Precision ILS instrument approach - FM 295-311
   - ☐☐☐ Missed approach - FM 312-316
   - ☐☐☐ Circling approach - FM 317-321
   - ☐☐☐ Landing from a straight-in or circling approach - FM 322-326
   - ☐☐☐ Loss of communications - FM 328-331
   - ☐☐☐ Loss of gyro attitude and/or heading indicators - FM 332-335
   - ☐☐☐ Checking instruments and equipment - FM 338
3. Postflight critique
4. Flight lesson 28 complete? Yes ____
   Copy of lesson and graduation certificate placed in pilot's folder? Yes ____

Completion Standards

This lesson will have been successfully completed when the pilot demonstrates the required level of proficiency in all tasks of the current Instrument Rating Practical Test Standards. If additional instruction is necessary, the chief flight instructor will assign the additional training. If the flight is satisfactory, the chief instructor will complete the pilot's training records and issue a graduation certificate.

Instructor's comments: ______________________________

____

____

____

____

____

____

____

Lesson assignment: ______________________________

____

Lesson certification:

(Pilot)	Instructor	Certificate No.	Exp. Date

APPENDIX C
FAA FLIGHT INSTRUCTOR-INSTRUMENT PRACTICAL TEST STANDARDS (FAA-S-8081-9A Reprinted) AIRPLANE ONLY

The purpose of this appendix is to reproduce verbatim what you would get in PTS reprint books that are normally sold for $5.00 at FBOs. All of these PTSs are reproduced (and explained, discussed, and illustrated!!) elsewhere throughout this book.

FOREWORD

The Flight Instructor—Instrument Practical Test Standards (PTS) book has been published by the Federal Aviation Administration (FAA) to establish the standards for flight instructor certification and instrument rating practical tests for airplanes and helicopters. FAA inspectors and designated pilot examiners shall conduct practical tests in compliance with these standards. Flight instructors and applicants should find these standards helpful during training and when preparing for the practical test.

Joseph K. Tintera, Manager
Regulatory Support Division
Flight Standards Service.

CONTENTS

INTRODUCTION

General Information

The Flight Standards Service of the Federal Aviation Administration (FAA) has developed this practical test book as the standard that shall be used by FAA inspectors and designated pilot examiners when conducting flight instructor—instrument (airplane and helicopter) practical tests. Flight instructors are expected to use this book when preparing applicants for practical tests. Applicants should be familiar with this book and refer to these standards during their training. Information considered directive in nature is described in this practical test book in terms such as "shall" and "must" indicating the actions are mandatory. Guidance information is described in terms such as "should" and "may" indicating the actions are desirable or permissive, but not mandatory.

The FAA gratefully acknowledges the valuable assistance provided by many individuals and organizations throughout the aviation community who contributed their time and talent in assisting with the revision of these practical test standards.

This practical test standard may be downloaded from the Regulatory Support Division's, AFS-600, web site at http://afs600.faa.gov. Subsequent changes to this standard, in accordance with AC 60-27, Announcement of Availability: Changes to Practical Test Standards, will also be available on AFS-600's web site and then later incorporated into a printed revision.

This publication may be purchased from the Superintendent of Documents, U.S. Government Printing Office, Washington, DC 20402.

Comments regarding this publication should be sent to:

U.S. Department of Transportation
Federal Aviation Administration
Flight Standards Service
Airman Testing Standards Branch, AFS-630
P.O. Box 25082
Oklahoma City, OK 73125

Practical Test Standard Concept

Title 14 of the Code of Federal Regulations (14 CFR) part 61 specifies the areas in which knowledge and skill must be demonstrated by the applicant before the issuance of a flight instructor certificate with the associated category and class ratings. The CFR's provide the flexibility to permit the FAA to publish practical test standards containing the AREAS OF OPERATION and specific TASKS in which competency shall be demonstrated. The FAA shall revise this book whenever it is determined that changes are needed in the interest of safety. Adherence to the provisions of the regulations and the practical test standards is mandatory for the evaluation of flight instructor applicants.

Test Book Description

This test book contains the practical test standards for flight instructor—instrument (airplane and helicopter).

Practical Test Standards Description

AREAS OF OPERATION are phases of the practical test arranged in a logical sequence within each standard. They begin with Fundamentals of Instructing and end with Postflight Procedures. The examiner, however, may conduct the practical test in any sequence that will result in a complete and efficient test. Not withstanding the above, the oral portion of the practical test shall be completed prior to the flight portion.

TASKS are titles of knowledge areas, flight procedures, or maneuvers appropriate to an AREA OF OPERATION.

NOTE is used to emphasize special considerations required in the AREA OF OPERATION or TASK.

REFERENCE identifies the publication(s) that describe(s) the TASK. Descriptions of TASKS and maneuver tolerances are not included in these standards because this information can be found in the current issue of the listed reference. Publications other than those listed may be used for references if their content conveys substantially the same meaning as the referenced publications.

These practical test standards are based on the following references:

14 CFR part 1	Definitions and Abbreviations
14 CFR part 61	Certification: Pilots and Flight Instructors
14 CFR part 91	General Operating and Flight Rules
14 CFR part 95	IFR Altitudes
14 CFR part 97	Standard Instrument Approach Procedures
NTSB Part 830	Notification and Reporting of Aircraft Accidents and Incidents
AC 00-2	Advisory Circular Checklist
AC 00-6	Aviation Weather
AC 00-45	Aviation Weather Services
AC 61-23	Pilot's Handbook of Aeronautical Knowledge
AC 61-65	Certification: Pilots and Flight Instructors
AC 61-67	Stall and Spin Awareness Training
AC 61-84	Role of Preflight Preparation
AC 61-98	Currency and Additional Qualification Requirements for Certificated Pilots
AC 90-48	Pilots' Role in Collision Avoidance
AC 90-94	Guidelines for Using Global Positioning Systems
FAA-H-8083-9	Aviation Instructor's Handbook
FAA-H-8083-15	Instrument Flying Handbook
FAA-S-8081-4	Instrument Rating Practical Test Standards
AIM	Aeronautical Information Manual
IAP's	Instrument Approach Procedures
DP's	Departure Procedures
STAR's	Standard Terminal Arrivals
AFD	Airport/Facility Directory
NOTAM's	Notices to Airmen
	Enroute Low Altitude Charts
	Appropriate Aircraft Flight Manuals

The Objective lists the important elements that must be satisfactorily performed to demonstrate competency in a TASK. The Objective includes:

1. specifically what the applicant should be able to do;
2. conditions under which the TASK is to be performed; and
3. acceptable performance standards.

The examiner determines that the applicant meets the TASK Objective through the demonstration of competency in all elements of knowledge and/or skill unless otherwise noted. The Objectives of TASKS in certain AREAS OF OPERATION, such as Fundamentals of Instructing and Technical Subjects, include only knowledge elements. Objectives of TASKS in AREAS OF OPERATION that include elements of skill, as well as knowledge, also include common errors, which the applicant shall be able to describe, recognize, analyze, and correct.

The Objective of a TASK that involves pilot skill consists of four parts. The four parts include determination that the applicant exhibits:

1. instructional knowledge of the elements of a TASK. This is accomplished through descriptions, explanations, and simulated instruction;
2. instructional knowledge of common errors related to a TASK, including their recognition, analysis, and correction;
3. the ability to demonstrate and simultaneously explain the key elements of a TASK. The TASK demonstration must be to the INSTRUMENT PILOT skill level; the teaching techniques and procedures should conform to those set forth in FAA-H-8083-9, Aviation Instructor's Handbook and FAA-H-8083-15, Instrument Flying Handbook; and
4. the ability to analyze and correct common errors related to a TASK.

Use of the Practical Test Standards Book

The FAA requires that all practical tests are conducted in accordance with the appropriate flight instructor practical test standards and the policies set forth in the INTRODUCTION.

All of the procedures and maneuvers in the instrument rating practical test standards have been included in the flight instructor practical test standards; however, to permit completion of the practical test for initial certification within a reasonable time-frame, the examiner shall select one or more TASKS in each AREA OF OPERATION. In certain AREAS OF OPERATION, there are required TASKS which the examiner must select. These required TASKS are identified by a **NOTE** immediately following each AREA OF OPERATION title.

In preparation for each practical test, the examiner shall develop a written "plan of action." The examiner shall vary each "plan of action" to ensure that all TASKS in the appropriate practical test standard are evaluated during a given number of practical tests. Except for required TASKS, the examiner should avoid using the same optional TASKS in order to avoid becoming stereotyped. The "plan of action" for a practical test for initial certification shall include one or more TASKS in each AREA OF OPERATION and shall always include the required TASKS. The "plan of action" for a practical test for the addition of an aircraft category and/or class rating to a flight instructor certificate shall include the required AREAS OF OPERATION as indicated in the Additional Rating Table located on page 12. The required TASKS appropriate to the additional rating(s) sought shall be included. Any TASK selected for evaluation during the practical test shall be evaluated in its entirety.

Applicant shall be expected to perform TASK H in AREA OF OPERATION VI, Recovery from Unusual Attitudes and TASK A in AREA OF OPERATION VIII, Non-precision Instrument Approach using a view-limiting device.

The flight instructor applicant shall be prepared in **all** knowledge and skill areas and demonstrate the ability to instruct effectively in **all** TASKS included in the AREAS OF OPERATION of this practical test standard. Throughout the flight portion of the practical test, the examiner shall evaluate the applicant's ability to demonstrate and simultaneously explain the selected procedures and maneuvers, and to give flight instruction to students at various stages of flight training and levels of experience. The term "instructional knowledge" means that the flight instructor applicant's discussions, explanations, and descriptions should follow the recommended teaching procedures and techniques explained in FAA-H-8083-9, Aviation Instructor's Handbook.

The purpose of including common errors in certain TASKS is to assist the examiner in determining that the flight instructor applicant has the ability to recognize, analyze, and correct such errors. The examiner will not simulate any condition that may jeopardize safe flight or result in possible damage to the aircraft. The common errors listed in the TASKS objective may or may not be found in the TASK References. However, the FAA considers their frequency of occurrence justification for inclusion in the TASK Objectives.

Special Emphasis Areas

The examiner shall place special emphasis on the applicant's demonstrated ability to teach precise aircraft control and sound judgment in aeronautical decision making. Evaluation of the applicant's ability to teach judgment shall be accomplished by asking the applicant to describe the presentation of practical problems that would be used in instructing students in the exercise of sound judgment. The examiner shall also emphasize the evaluation of the applicant's demonstrated ability to teach ***spatial disorientation***, ***wake turbulence*** and ***low level wind shear avoidance, checklist usage, positive exchange of flight controls, runway incursion avoidance, and any other directed special emphasis areas.***

Practical Test Prerequisites

An applicant for a flight instructor—instrument initial certification practical test is required by 14 CFR part 61 to:

1. be at least 18 years of age;
2. be able to read, speak, write, and understand the English language. If there is a doubt, use AC 60-28, English Language Skill Standards;
3. hold either a commercial/instrument pilot or airline transport pilot certificate with an aircraft category rating appropriate to the flight instructor rating sought;
4. have an endorsement from an authorized instructor on the fundamentals of instructing appropriate to the required knowledge test;
5. have passed the appropriate flight instructor knowledge test(s) since the beginning of the 24th month before the month in which he or she takes the practical test;
6. have an endorsement from an authorized instructor certifying that the applicant has been given flight training in the AREAS OF OPERATION listed in 14 CFR part 61, section 61.187 and a written statement from an authorized flight instructor within the preceding 60 days, in accordance with 14 CFR part 61, section 61.39, that instruction was given in preparation for the practical test. The endorsement shall also state that the instructor finds the applicant prepared for the required practical test, and that the applicant has demonstrated satisfactory knowledge of the subject area(s) in which the applicant was deficient on the airman knowledge test.

An applicant holding a flight instructor certificate who applies for an **additional** rating on that certificate is required by 14 CFR to:

1. hold a valid pilot certificate with ratings appropriate to the flight instructor rating sought;
2. have at least 15 hours as pilot-in-command in the category and class aircraft appropriate to the rating sought;
3. have passed the appropriate knowledge test prescribed for the issuance of a flight instructor certificate with the rating sought since the beginning of the 24th month before the month in which he/she takes the practical test; and
4. have an endorsement from an authorized instructor certifying that the applicant has been given flight training in the AREAS OF OPERATION listed in 14 CFR part 61, section 61.187 and a written statement from an authorized flight instructor within the preceding 60 days, in accordance with 14 CFR part 61, section 61.39, that instruction was given in preparation for the practical test. The endorsement shall also state that the instructor finds the applicant prepared for the required practical test, and that the applicant has demonstrated satisfactory knowledge of the subject area(s) in which the applicant was deficient on the airman knowledge test.

Aircraft and Equipment Required for the Practical Test

The flight instructor—instrument applicant is required by 14 CFR part 61, section 61.45, to provide an airworthy, certificated aircraft for use during the practical test. This section further requires that the aircraft must:

1. have fully functioning dual controls, and
2. be capable of performing all AREAS OF OPERATION appropriate for the instructor rating sought and have no operating limitations, which prohibit its use in any of the AREAS OF OPERATION required for the practical test.

Flight Instructor Responsibility

An appropriately rated flight instructor is responsible for training the flight instructor applicant to acceptable standards in **all** subject matter areas, procedures, and maneuvers included in the TASKS within each AREA OF OPERATION in the appropriate flight instructor practical test standard.

Because of the impact of their teaching activities in developing safe, proficient pilots, flight instructors should exhibit a high level of knowledge, skill, and the ability to impart that knowledge and skill to students. The flight instructor shall certify that the applicant is:

1. able to make a practical application of the fundamentals of instructing;
2. competent to teach the subject matter, procedures, and maneuvers included in the standards to students with varying backgrounds and levels of experience and ability;
3. able to perform the procedures and maneuvers included in the standards to the INSTRUMENT PILOT skill level while giving effective flight instruction; and
4. competent to pass the required practical test for the issuance of the flight instructor certificate with the associated category and class ratings or the addition of a category and/or class rating to a flight instructor certificate.

Throughout the applicant's training, the flight instructor is responsible for emphasizing the performance of, and the ability to teach, effective visual scanning and collision avoidance procedures.

Examiner[1] Responsibility

The examiner conducting the practical test is responsible for determining that the applicant meets acceptable standards of teaching ability, knowledge, and skill in the selected TASKS. The examiner makes this determination by accomplishing an Objective that is appropriate to each selected TASK, and includes an evaluation of the applicant's:

1. ability to apply the fundamentals of instructing;
2. knowledge of, and ability to teach, the subject matter, procedures, and maneuvers covered in the TASKS;
3. ability to perform the procedures and maneuvers included in the standards to the INSTRUMENT PILOT skill level while giving effective flight instruction; and
4. ability to analyze and correct common errors related to the procedures and maneuvers covered in the TASKS. It is intended that oral questioning be used at any time during the ground or flight portion of the practical test to determine that the applicant can instruct effectively and has a comprehensive knowledge of the TASKS and their related safety factors.

During the flight portion of the practical test, the examiner shall act as a student during selected maneuvers. This will give the examiner an opportunity to evaluate the flight instructor applicant's ability to analyze and correct simulated common errors related to these maneuvers. The examiner will also evaluate the applicant's use of visual scanning and collision avoidance procedures, and the applicant's ability to teach those procedures.

Examiners should to the greatest extent possible test the applicant's application and correlation skills. When possible scenario based questions should be used.

Satisfactory Performance

The practical test is passed if, in the judgment of the examiner, the applicant demonstrates satisfactory performance with regard to:

1. knowledge of the fundamentals of instructing;
2. knowledge of the technical subject areas;
3. knowledge of the flight instructor's responsibilities concerning the pilot certification process;
4. knowledge of the flight instructor's responsibilities concerning logbook entries and pilot certificate endorsements;
5. ability to demonstrate the procedures and maneuvers selected by the examiner to the INSTRUMENT PILOT skill level while giving effective instruction;
6. competence in teaching the procedures and maneuvers selected by the examiner;
7. competence in describing, recognizing, analyzing, and correcting common errors simulated by the examiner; and
8. knowledge of the development and effective use of a course of training, a syllabus, and a lesson plan.

Unsatisfactory Performance

If, in the judgment of the examiner, the applicant does not meet the standards of performance of any TASK performed, the associated AREA OF OPERATION is failed and therefore, the practical test is failed. The examiner or applicant may discontinue the test at any time when the failure of an AREA OF OPERATION makes the applicant ineligible for the certificate or rating sought. The test may be continued ONLY with the consent of the applicant. If the test is discontinued, the applicant is entitled to credit for only those AREAS OF OPERATION and TASKS satisfactorily performed; however, during the retest and at the discretion of the examiner, any TASK may be re-evaluated, including those previously passed. Specific reasons for disqualification are:

1. failure to perform a procedure or maneuver to the INSTRUMENT PILOT skill level while giving effective flight instruction;
2. failure to provide an effective instructional explanation while demonstrating a procedure or maneuver (explanation during the demonstration must be clear, concise, technically accurate, and complete with no prompting from the examiner);
3. any action or lack of action by the applicant which requires corrective intervention by the examiner to maintain safe flight; and

---

[1] The word "examiner" is used throughout the standards to denote either the FAA inspector or FAA designated pilot examiner who conducts an official practical test.

4. failure to use proper and effective visual scanning techniques to clear the area before and while performing maneuvers. When a notice of disapproval is issued, the examiner shall record the applicant's unsatisfactory performance in terms of AREAS OF OPERATION and TASKS. If the applicant fails the practical test because of a special emphasis area, the Notice of Disapproval shall indicate the associated task. An example would be; AREA OF OPERATION VIII, CIRCLING APPROACH (AIRPLANE), failure to use proper runway incursion avoidance procedures.

Emphasis on Attitude Instrument Flying and Partial Panel Skills

The FAA is concerned about numerous fatal aircraft accidents involving spatial disorientation of instrument rated pilots who have attempted to control and maneuver their aircraft in clouds with inoperative gyroscopic heading and attitude indicators.

Many of the light aircraft operated in instrument meteorological conditions (IMC) are not equipped with dual, independent, gyroscopic heading or attitude indicators. In addition, many are equipped with only a single vacuum source. Therefore, the FAA has stressed that it is imperative for instrument rated pilots to acquire and maintain adequate partial panel skills and that they be cautioned not to be overly reliant upon the gyroscopic instruments.

FAA-S-8081-4, Instrument Rating Practical Test Standards, and FAA-S-8081-9, Flight Instructor—Instrument Practical Test Standards, place increased emphasis on basic attitude instrument flying and require the demonstration of partial panel, non-precision instrument approach procedures. This practical test book emphasizes these areas from an instructional standpoint.

AREA OF OPERATION VI requires the applicant to demonstrate the ability to teach basic instrument flight TASKS under both full panel and partial panel conditions. These TASKS are described in detail in FAA-H-8083-15, Instrument Flying Handbook. The TASKS require the applicant to exhibit instructional knowledge of attitude instrument flying techniques and procedures and to demonstrate the ability to teach basic instrument maneuvers with both full panel and partial panel.

Examiners should determine that the applicant demonstrates and fully understands the PRIMARY AND SUPPORTING or the CONTROL and PERFORMANCE CONCEPT method of attitude instrument flying.

Crew Resource Management (CRM)

CRM refers to the effective use of all available resources; human resources, hardware, and information. Human resources includes all groups routinely working with the cockpit crew or pilot who are involved with decisions that are required to operate a flight safely. These groups include, but are not limited to: dispatchers, cabin crewmembers, maintenance personnel, air traffic controllers, and weather services. CRM is not a single TASK, but a set of competencies that must be evident in all TASKS in this practical test standard as applied to either single pilot or crew operations.

Applicant's Use of Checklists

Throughout the practical test, the applicant is evaluated on the use of an appropriate checklist. Proper use is dependent on the specific TASK being evaluated. The situation may be such that the use of the checklist, while accomplishing elements of an Objective, would be either unsafe or impractical, especially in a single-pilot operation. In this case, a review of the checklist after the elements have been accomplished would be appropriate. Division of attention and proper visual scanning should be considered when using a checklist.

Use of Distractions During Practical Tests

Numerous studies indicate that many accidents have occurred when the pilot has been distracted during critical phases of flight. To evaluate the applicant's ability to utilize proper control technique while dividing attention both inside and outside the cockpit, the examiner shall cause a realistic distraction during the flight portion of the practical test to evaluate the applicant's ability to divide attention while maintaining safe flight.

Positive Exchange of Flight Controls

During flight training, there must always be a clear understanding between students and flight instructors of who has control of the aircraft. Prior to flight, a briefing should be conducted that includes the procedure for the exchange of flight controls. A positive three-step process in the exchange of flight controls between pilots is a proven procedure and one that is strongly recommended.

When the instructor wishes the student to take control of the aircraft, he or she will say, "You have the flight controls." The student acknowledges immediately by saying, "I have the flight controls." The flight instructor again says, "You have the flight controls." When control is returned to the instructor, follow the same procedure. A visual check is recommended to verify that the exchange has occurred. There should never be any doubt as to who is flying the aircraft.

ADDITION OF AN INSTRUMENT RATING TO A FLIGHT INSTRUCTOR CERTIFICATE

AREA OF OPERATION	FLIGHT INSTRUCTOR CERTIFICATE AND RATING HELD			
	AP	RTR	G	IA or H
I	N	N	N	N
II	A&C	A&C	A&C	C
III	B&C	B&C	B&C	C
IV	N	N	N	N
V	Y	Y	Y	N
VI	Y	Y	Y	Y
VII	Y	Y	Y	N
VIII	Y	Y	Y	*A or B
IX	Y	Y	Y	Y
X	Y	Y	Y	Y

LEGEND

AP Airplane
RTR Helicopter/Gyroplane
G Glider
IA or H Instrument Airplane or Helicopter

NOTE: N indicates that the AREA OF OPERATION is not required. Y indicates that the AREA OF OPERATION is to be performed or based on the note in the AREA OF OPERATION. If a TASK (or TASKs) is listed for an AREA OF OPERATION, that TASK (or TASKs) is mandatory.

* Combine with C, D, or E.

RENEWAL OR REINSTATEMENT OF A FLIGHT INSTRUCTOR

REQUIRED AREAS OF OPERATION	NUMBER OF TASKS
II	1
III	1
V	1
VI	2
VII	1
VIII	A OR B COMBINED WITH TASKS C, D, or E
IX	1

The Renewal or reinstatement of one rating on a Flight Instructor Certificate renews or reinstates all privileges existing on the certificate. (14 CFR part 61, sections 61.197 and 61.199)

APPLICANT'S PRACTICAL TEST CHECKLIST

Flight Instructor—Instrument

APPOINTMENT WITH INSPECTOR OR EXAMINER:

NAME ______________________________

DATE/TIME ______________________________

- ☐ View-limiting Device:
- ☐ Aircraft Documents:
  - Airworthiness Certificate
  - Registration Certificate
  - Operating Limitations
- ☐ Aircraft Maintenance Records:
  - Logbook Record of Airworthiness Inspections and AD Compliance
- ☐ Pilot's Operating Handbook, FAA-Approved Flight Manual

PERSONAL EQUIPMENT

- ☐ Practical Test Standards
- ☐ Lesson Plan Library
- ☐ Current Aeronautical Charts
- ☐ Computer and Plotter
- ☐ Flight Plan and Flight Log Forms
- ☐ Current AIM, Airport Facility Directory, and Appropriate Publications

PERSONAL RECORDS

- ☐ Identification - Photo/Signature ID
- ☐ Pilot Certificate
- ☐ Current and Appropriate Medical Certificate
- ☐ Completed FAA Form 8710-1, Airman Certificate and/or Rating Application
- ☐ Airman Knowledge Test Report
- ☐ Pilot Logbook with Appropriate Instructor Endorsements
- ☐ FAA Form 8060-5, Notice of Disapproval (if applicable)
- ☐ Approved School Graduation Certificate (if applicable)
- ☐ Examiner's Fee (if applicable)

EXAMINER'S PRACTICAL TEST CHECKLIST

Flight Instructor—Instrument

APPLICANT'S NAME ______________________________

LOCATION ______________________________

DATE/TIME ______________________________

I. FUNDAMENTALS OF INSTRUCTING

- ☐ **A.** The Learning Process
- ☐ **B.** Human Behavior and Effective Communication
- ☐ **C.** The Teaching Process
- ☐ **D.** Teaching Methods
- ☐ **E.** Critique and Evaluation
- ☐ **F.** Flight Instructor Characteristics and Responsibilities
- ☐ **G.** Planning Instructional Activity

II. TECHNICAL SUBJECT AREAS

- ☐ **A.** Aircraft Flight Instruments and Navigation Equipment
- ☐ **B.** Aeromedical Factors
- ☐ **C.** Regulations and Publications Related to IFR Operations
- ☐ **D.** Logbook Entries Related to Instrument Instruction

III. PREFLIGHT PREPARATION

- ☐ **A.** Weather Information
- ☐ **B.** Cross-Country Flight Planning
- ☐ **C.** Instrument Cockpit Check

IV. PREFLIGHT LESSON ON A MANEUVER TO BE PERFORMED IN FLIGHT

- ☐ Maneuver Lesson

V. AIR TRAFFIC CONTROL CLEARANCES AND PROCEDURES

- ☐ **A.** Air Traffic Control Clearances
- ☐ **B.** Compliance with Departure, Enroute, and Arrival Procedures and Clearances

VI. FLIGHT BY REFERENCE TO INSTRUMENTS

- ☐ **A.** Straight-and-Level Flight
- ☐ **B.** Turns
- ☐ **C.** Change of Airspeed in Straight-and-Level and Turning Flight
- ☐ **D.** Constant Airspeed Climbs and Descents
- ☐ **E.** Constant Rate Climbs and Descents
- ☐ **F.** Timed Turns to Magnetic Compass Headings
- ☐ **G.** Steep Turns
- ☐ **H.** Recovery from Unusual Flight Attitudes

VII. NAVIGATION SYSTEMS

- ☐ **A.** Intercepting and Tracking Navigational Systems and DME Arcs
- ☐ **B.** Holding Procedures

VIII. INSTRUMENT APPROACH PROCEDURES

- ☐ **A.** Non-Precision Instrument Approach
- ☐ **B.** Precision Instrument Approach
- ☐ **C.** Missed Approach
- ☐ **D.** Circling Approach (Airplane)
- ☐ **E.** Landing from a Straight-In Approach

IX. EMERGENCY OPERATIONS

- ☐ **A.** Loss of Communications
- ☐ **B.** Loss of Gyro Attitude and Heading Indicators
- ☐ **C.** Engine Failure During Straight-and-Level Flight and Turns
- ☐ **D.** Instrument Approach—One Engine Inoperative

X. POSTFLIGHT PROCEDURES

- ☐ **A.** Checking Instruments and Equipment

I. AREA OF OPERATION: FUNDAMENTALS OF INSTRUCTING

NOTE: The examiner shall select at least TASK E, F, and G and one other task.

A. TASK: THE LEARNING PROCESS

REFERENCE: FAA-H-8083-9.

Objective. To determine that the applicant exhibits instructional knowledge of the elements of the learning process by describing:

1. Learning theory.
2. Characteristics of learning.
3. Principles of learning.
4. Levels of learning.
5. Learning physical skills.
6. Memory.
7. Transfer of learning.

B. TASK: HUMAN BEHAVIOR AND EFFECTIVE COMMUNICATION

REFERENCE: FAA-H-8083-9.

Objective. To determine that the applicant exhibits instructional knowledge of the elements related to human behavior and effective communication by describing:

1. Human behavior—
   a. control of human behavior.
   b. human needs.
   c. defense mechanisms.
   d. the flight instructor as a practical psychologist.
2. Effective communication
   a. basic elements of communication.
   b. barriers of effective communication.
   c. developing communication skills.

C. TASK: THE TEACHING PROCESS

REFERENCE: FAA-H-8083-9.

Objective. To determine that the applicant exhibits instructional knowledge of the elements of the teaching process by describing:

1. Preparation of a lesson for a ground or flight instructional period.
2. Presentation methods.
3. Application, by the student, of the material or procedure that was presented.
4. Review and evaluation of student performance.

D. TASK: TEACHING METHODS

REFERENCE: FAA-H-8083-9.

Objective. To determine that the applicant exhibits instructional knowledge of the elements of teaching methods by describing:

1. Material organization.
2. The lecture method.
3. The cooperative or group learning method.
4. The guided discussion method.
5. The demonstration-performance method.
6. Computer-based training method.

E. TASK: CRITIQUE AND EVALUATION

REFERENCE: FAA-H-8083-9.

Objective. To determine that the applicant exhibits instructional knowledge of the elements of critique and evaluation by explaining:

1. Critique—
   a. purpose and characteristics of an effective critique.
   b. methods and ground rules for a critique.
2. Evaluation—
   a. characteristics of effective oral questions and what types to avoid.
   b. responses to student questions.
   c. characteristics and development of effective written test.
   d. characteristics and uses of performance tests, specifically, the FAA Practical Test Standards.

F. TASK: FLIGHT INSTRUCTOR CHARACTERISTICS AND RESPONSIBILITIES

REFERENCE: FAA-H-8083-9.

Objective. To determine that the applicant exhibits instructional knowledge of the elements of instructor responsibilities and professionalism by describing:

1. Aviation instructor responsibilities in—
   a. providing adequate instruction.
   b. establishing standards of performance.
   c. emphasizing the positive.
2. Flight instructor responsibilities in—
   a. providing student pilot evaluation and supervision.
   b. preparing practical test recommendations and endorsements.
   c. determining requirements for conducting additional training and endorsement requirements.
3. Professionalism as an instructor by—
   a. explaining important personal characteristics.
   b. describing methods to minimize student frustration.

G. TASK: PLANNING INSTRUCTIONAL ACTIVITY

REFERENCE: FAA-H-8083-9.

Objective. To determine that the applicant exhibits instructional knowledge of the elements of planning instructional activity by describing:

1. Developing objectives and standards for a course of training.
2. Theory of building blocks of learning.
3. Requirements for developing a training syllabus.
4. Purpose and characteristics of a lesson plan.

II. AREA OF OPERATION: TECHNICAL SUBJECT AREAS

NOTE: The examiner shall select TASK A and D and at least one other TASK.

A. TASK: AIRCRAFT FLIGHT INSTRUMENTS AND NAVIGATION EQUIPMENT

REFERENCES: FAA-H-8083-15

Objective. To determine that the applicant exhibits instructional knowledge of aircraft:

1. Flight instrument systems and their operating characteristics to include—
   a. pitot- static system.
   b. attitude indicator.
   c. heading indicator/horizontal situation indicator/remote magnetic indicator.
   d. magnetic compass.
   e. turn-and-slip indicator/turn coordinator.
2. Navigation equipment and their operating characteristics to include—
   a. VHF omnirange (VOR).
   b. distance measuring equipment (DME).
   c. instrument landing system (ILS)
   d. marker beacon receiver/indicators.
   e. automatic direction finder (ADF).
   f. global positioning system (GPS).
3. Antiice/deicing and weather detection equipment and their operating characteristics to include—
   a. airframe.
   b. propeller or rotor.
   c. air intake.
   d. fuel system.
   e. pitot-static system.
   f. radar/lightning detection system.

B. TASK: AEROMEDICAL FACTORS

REFERENCES: AC 61-23; AIM.

Objective. To determine that the applicant exhibits instructional knowledge of the elements related to aeromedical factors by describing the effects, corrective action, and safety considerations of:

1. Hypoxia.
2. Hyperventilation.
3. Middle ear and sinus problems.
4. Spatial disorientation.
5. Motion sickness.
6. Alcohol and drugs.
7. Carbon monoxide poisoning.
8. Evolved gases from scuba diving.
9. Stress and fatigue.

C. TASK: REGULATIONS AND PUBLICATIONS RELATED TO IFR OPERATIONS

REFERENCES: 14 CFR parts 61, 91, 95, and 97; FAA-H-8083-15; AIM.

Objective. To determine that the applicant exhibits instructional knowledge of the elements related to regulations and publications, (related to instrument flight and instrument flight instruction) their purpose, general content, availability, and method of revision by describing:

1. 14 CFR parts 61, 91, 95, and 97.
2. FAA-H-8083-15, Instrument Flying Handbook.
3. Aeronautical Information Manual.
4. Practical Test Standards.
5. Airport Facility Directory.
6. Standard Departures/Terminal Arrivals.
7. En route Charts.
8. Standard Instrument Approach Procedure Charts.

D. TASK: LOGBOOK ENTRIES RELATED TO INSTRUMENT INSTRUCTION

REFERENCES: 14 CFR part 61; AC 61-65; AC 61-98.

Objective. To determine that the applicant exhibits instructional knowledge of logbook entries related to instrument instruction by describing:

1. Logbook entries or training records for instrument flight/instrument flight instruction or ground instruction given.
2. Preparation of a recommendation for an instrument rating practical test, including appropriate logbook entry.
3. Required endorsement of a pilot logbook for satisfactory completion of an instrument proficiency check.
4. Required flight instructor records.

III. AREA OF OPERATION: PREFLIGHT PREPARATION

NOTE: The examiner shall select at least one TASK.

A. TASK: WEATHER INFORMATION

NOTE: Where current weather reports, forecasts, or other pertinent information is not available, this information shall be simulated by the examiner in a manner, which shall adequately measure the applicant's competence.

REFERENCES: AC 00-6, AC 00-45; FAA-S-8081-4; AIM.

Objective. To determine that the applicant exhibits instructional knowledge related to IFR weather information.

1. Sources of weather—
   a. AWOS, ASOS, and ATIS reports.
   b. PATWAS AND TIBS.
   c. TWEB.
2. Weather reports and charts—
   a. METAR, TAF, FA, and radar reports.
   b. inflight weather advisories.
   c. surface analysis, weather depiction, and radar summary charts.
   d. significant weather prognostic charts.
   e. winds and temperatures aloft charts.

B. TASK: CROSS-COUNTRY FLIGHT PLANNING

REFERENCES: 14 CFR part 91; FAA-H-8083-15, FAA-S-8081-4; AIM.

Objective. To determine that the applicant exhibits instructional knowledge of cross-country flight planning by describing the:

1. Regulatory requirements for instrument flight within various types of airspace.
2. Computation of estimated time en route and total fuel requirement for an IFR cross-country flight.
3. Selection and correct interpretation of the current and applicable en route charts, DP's, STAR's, and standard instrument approach procedure charts.
4. Procurement and interpretation of the applicable NOTAM information.
5. Preparation and filing of an actual or simulated IFR flight plan.

C. TASK: INSTRUMENT COCKPIT CHECK

REFERENCES: 14 CFR part 91; FAA-H-8083-15, FAA-S-8081-4.

Objective. To determine that the applicant exhibits instructional knowledge of an instrument cockpit check by describing the reasons for the check and the detection of defects that could affect safe instrument flight. The check shall include:

1. Communications equipment.
2. Navigation equipment.
3. Magnetic compass.
4. Heading indicator/horizontal situation indicator/remote magnetic indicator.
5. Attitude indicator.
6. Altimeter.
7. Turn-and-slip indicator/turn coordinator.
8. Vertical-speed indicator.
9. Airspeed indicator.
10. Outside air temperature.
11. Clock.

IV. AREA OF OPERATION: PREFLIGHT LESSON ON A MANEUVER TO BE PERFORMED IN FLIGHT

NOTE: The examiner shall select at least one maneuver from AREAS OF OPERATION VI through IX and ask the applicant to present a preflight lesson on the selected maneuver as the lesson would be taught to a student. Previously developed lesson plans from the applicant's library may be used.

TASK: MANEUVER LESSON

REFERENCES: FAA-H-8083-9, FAA-H-8083-15; FAA-S-8081-4.

Objective. To determine that the applicant exhibits instructional knowledge of the selected maneuver by:

1. Using a lesson plan that includes all essential items to make an effective and organized presentation.
2. Stating the objective.
3. Giving an accurate, comprehensive oral description of the maneuver, including the elements and associated common errors.
4. Using instructional aids, as appropriate.
5. Describing the recognition, analysis, and correction of common errors.

V. AREA OF OPERATION: AIR TRAFFIC CONTROL CLEARANCES AND PROCEDURES

NOTE: The examiner shall select at least one TASK.

A. TASK: AIR TRAFFIC CONTROL CLEARANCES

REFERENCES: 14 CFR part 91; FAA-H-8083-15; FAA-S-8081-4.

Objective. To determine that the applicant exhibits instructional knowledge of air traffic control clearances by describing:

1. Pilot and controller responsibilities to include tower, en route control, and clearance void times.
2. Correct and timely copying of an ATC clearance.
3. Correct and timely read-back of an ATC clearance, using standard phraseology.
4. Correct interpretation of an ATC clearance and, when necessary, request for clarification, verification, or change.
5. Setting of communication and navigation frequencies in compliance with an ATC clearance.

B. TASK: COMPLIANCE WITH DEPARTURE, EN ROUTE, AND ARRIVAL PROCEDURES AND CLEARANCES

REFERENCES: 14 CFR part 91; FAA-H-8083-15; FAA-S-8081-4; AIM.

Objective. To determine that the applicant exhibits instructional knowledge of the elements related to compliance with departure, en route, and arrival procedures and clearances by describing:

1. Selection and use of current and appropriate navigation publications.
2. Pilot and controller responsibilities with regard to SID's, En Route Low and High Altitude Charts, and STAR's.
3. Selection and use of appropriate communications frequencies.
4. Selection and identification of the navigation aids.
5. Accomplishment of the appropriate checklist items.
6. Pilot's responsibility for compliance with vectors and also altitude, airspeed, climb, descent, and airspace restrictions.
7. Pilot's responsibility for the interception of courses, radials, and bearings appropriate to the procedure, route, or clearance.
8. Procedures to be used in the event of two-way communications failure.

VI. AREA OF OPERATION: FLIGHT BY REFERENCE TO INSTRUMENTS

NOTE: The examiner shall select TASK H and at least one other TASK. The applicant shall select either the primary and supporting or control and performance method for teaching this AREA OF OPERATION.

A. TASK: STRAIGHT-AND-LEVEL FLIGHT

REFERENCES: FAA-H-8083-9, FAA-H-8083-15; FAA-S-8081-4.

Objective. To determine that the applicant:

1. Exhibits instructional knowledge of teaching straight-and-level flight by describing—
   a. the relationship of pitch, bank, and power in straight-and-level flight.
   b. procedure using full panel and partial panel
   c. coordination of controls and trim.
2. Exhibits instructional knowledge of common errors related to straight-and-level flight by describing—
   a. slow or improper cross-check during straight-and-level flight.
   b. improper power control.
   c. failure to make smooth, precise corrections, as required.
   d. uncoordinated use of controls.
   e. improper trim control.
3. Demonstrates and simultaneously explains straight-and-level flight from an instructional standpoint.
4. Analyzes and corrects simulated common errors related to straight-and-level flight.

B. TASK: TURNS

REFERENCES: FAA-H-8083-9, FAA-H-8083-15; FAA-S-8081-4.

Objective. To determine that the applicant:

1. Exhibits instructional knowledge of teaching turns by describing—
   a. the relationship of true airspeed and angle of bank to a standard rate turn.
   b. technique and procedure using full panel and partial panel for entry and recovery of a constant rate turn, including the performance of a half-standard rate turn.
   c. coordination of controls and trim.
2. Exhibits instructional knowledge of common errors related to turns by describing—
   a. improper cross-check procedures.
   b. improper bank control during roll-in and roll-out.
   c. failure to make smooth, precise corrections, as required.
   d. uncoordinated use of controls.
   e. improper trim technique.
3. Demonstrates and simultaneously explains turns from an instructional standpoint.
4. Analyzes and corrects simulated common errors related to turns.

C. TASK: CHANGE OF AIRSPEED IN STRAIGHT-AND-LEVEL AND TURNING FLIGHT

REFERENCES: FAA-H-8083-9, FAA-H-8083-15; FAA-S-8081-4.

Objective. To determine that the applicant:

1. Exhibits instructional knowledge of teaching change of airspeed in straight-and-level flight and turns by describing—
   a. procedure using full panel and partial panel for maintaining altitude and changing airspeed in straight-and-level and turning flight.
   b. coordination of controls and trim technique.
2. Exhibits instructional knowledge of common errors related to changes of airspeed in straight-and-level and turning flight by describing—
   a. slow or improper cross-check during straight-and-level flight and turns.
   b. improper power control.
   c. failure to make smooth, precise corrections, as required.
   d. uncoordinated use of controls.
   e. improper trim technique.
3. Demonstrates and simultaneously explains changes of airspeed in straight-and-level and turning flight from an instructional standpoint.
4. Analyzes and corrects simulated common errors related to changes of airspeed in straight-and-level and turning flight.

D. TASK: CONSTANT AIRSPEED CLIMBS AND DESCENTS

REFERENCES: FAA-H-8083-9, FAA-H-8083-15; FAA-S-8081-4.

Objective. To determine that the applicant:

1. Exhibits instructional knowledge of constant airspeed climbs and descents by describing—
   a. procedure using full panel and partial panel for an entry into a straight climb or climbing turn, from either cruising or climbing airspeed.
   b. a stabilized straight climb or climbing turn.
   c. a level-off from a straight climb or climbing turn, at either cruising or climbing airspeed.
   d. procedure using full panel and partial panel for an entry into a straight descent or descending turn from either cruising or descending airspeed.
   e. a stabilized straight descent or descending turn.
   f. a level-off from a straight descent or descending turn, at either cruising or descending airspeed.
2. Exhibits instructional knowledge of common errors related to constant airspeed climbs and descents by describing—
   a. failure to use a proper power setting and pitch attitude.
   b. improper correction of vertical rate, airspeed, heading, or rate-of-turn errors.
   c. uncoordinated use of controls.
   d. improper trim control.
3. Demonstrates and simultaneously explains a constant airspeed climb and a constant airspeed descent from an instructional standpoint.
4. Analyzes and corrects simulated common errors related to constant airspeed climbs and descents.

E. TASK: CONSTANT RATE CLIMBS AND DESCENTS

REFERENCES: FAA-H-8083-9, FAA-H-8083-15; FAA-S-8081-4.

Objective. To determine that the applicant:

1. Exhibits instructional knowledge of constant rate climbs and descents by describing—
   a. procedure using full panel and partial panel for an entry into a constant rate climb or descent.
   b. a stabilized constant rate straight climb or climbing turn, using the vertical speed indicator.
   c. a level-off from a constant rate straight climb or climbing turn.
   d. an entry into a constant rate straight descent or descending turn.
   e. a stabilized constant rate straight descent or descending turn using the vertical speed indicator.
   f. level-off from a constant rate straight descent or descending turn.

2. Exhibits instructional knowledge of common errors related to constant rate climbs and descents by describing—

   a. failure to use a proper power setting and pitch attitude.
   b. improper correction of vertical rate, airspeed, heading, or rate-of-turn errors.
   c. uncoordinated use of controls.
   d. improper trim control.

3. Demonstrates and simultaneously explains a constant rate climb and a constant rate descent from an instructional standpoint.
4. Analyzes and corrects simulated common errors related to constant rate climbs and descents.

F. TASK: TIMED TURNS TO MAGNETIC COMPASS HEADINGS

REFERENCES: FAA-H-8083-9, FAA-H-8083-15; FAA-S-8081-4.

Objective. To determine that the applicant:

1. Exhibits instructional knowledge of timed turns to magnetic compass headings by describing—

   a. operating characteristics and errors of the magnetic compass.
   b. calibration of the miniature aircraft of the turn coordinator,[2] both right and left, using full panel and the clock.
   c. procedures using full panel and partial panel performing compass turns to a specified heading.

2. Exhibits instructional knowledge of common errors related to timed turns to magnetic compass headings by describing—

   a. incorrect calibration procedures.
   b. improper timing.
   c. uncoordinated use of controls.
   d. improper trim control.

3. Demonstrates and simultaneously explains timed turns to magnetic compass headings from an instructional standpoint.
4. Analyzes and corrects simulated common errors related to timed turns to magnetic compass headings.

G. TASK: STEEP TURNS

REFERENCES: FAA-H-8083-9, FAA-H-8083-15; FAA-S-8081-4.

Objective. To determine that the applicant:

1. Exhibits instructional knowledge of steep turns by describing—

   a. procedure using full panel and partial panel for entry and recovery of a steep turn.
   b. the need for a proper instrument cross-check.
   c. roll-in/roll-out procedure.
   d. coordination of control and trim.

2. Exhibits instructional knowledge of common errors related to steep turns by describing—

   a. failure to recognize and make proper corrections for pitch, bank, or power errors.
   b. failure to compensate for precession of the horizon bar of the attitude indicator.
   c. uncoordinated use of controls.
   d. improper trim technique.

3. Demonstrates and simultaneously explains steep turns from an instructional standpoint.
4. Analyzes and corrects simulated common errors related to steep turns.

H. TASK: RECOVERY FROM UNUSUAL FLIGHT ATTITUDES

REFERENCES: FAA-H-8083-9, FAA-H-8083-15; FAA-S-8081-4.

Objective. To determine that the applicant:

1. Exhibits instructional knowledge of recovery from unusual flight attitudes by describing—

   a. conditions or situations which contribute to the development of unusual flight attitudes.
   b. procedure using full panel and partial panel for recovery from nose-high and nose-low unusual flight attitudes.

2. Exhibits instructional knowledge of common errors related to recovery from unusual flight attitudes by describing—

   a. incorrect interpretation of the flight instruments.
   b. inappropriate application of controls

3. Demonstrates and simultaneously explains recovery from unusual flight attitudes, solely by reference to instruments, from an instructional standpoint.
4. Analyzes and corrects simulated common errors related to recovery from unusual flight attitudes.

VII. AREA OF OPERATION: NAVIGATION SYSTEMS

NOTE: The examiner shall select TASK A and B. If aircraft is not DME equipped, performance of DME arcs shall be tested orally.

A. TASK: INTERCEPTING AND TRACKING NAVIGATIONAL SYSTEMS AND DME ARCS

REFERENCES: 14 CFR part 91; FAA-H-8083-9, FAA-H-8083-15; FAA-S-8081-4; AIM.

Objective. To determine that the applicant:

1. Exhibits instructional knowledge of the elements of intercepting and tracking navigational systems and DME arcs by describing—

   a. tuning and identification of a navigational facility.
   b. setting of a selected course on the navigation selector or the correct identification of a selected bearing on the RMI.
   c. method for determining aircraft position relative to a facility.
   d. procedure for intercepting and maintaining a selected course.
   e. procedure for intercepting and maintaining a DME arc.
   f. procedure for intercepting a course or localizer from a DME arc.
   g. recognition of navigation facility or waypoint passage.
   h. recognition of navigation receiver or facility failure.

2. Exhibits instructional knowledge of common errors related to intercepting and tracking navigational systems and DME arcs by describing—

   a. incorrect tuning and identification procedures.
   b. failure to properly set the navigation selector on the course to be intercepted.
   c. failure to use proper procedures for course or DME arc interception and tracking.
   d. improper procedures for intercepting a course or localizer from a DME arc.

3. Demonstrates and simultaneously explains intercepting and tracking navigational systems and DME arcs from an instructional standpoint.
4. Analyzes and corrects simulated common errors related to intercepting and tracking navigational systems and DME arcs.

B. TASK: HOLDING PROCEDURES

REFERENCES: 14 CFR part 91; FAA-H-8083-9, FAA-H-8083-15; FAA-S-8081-4; AIM.

Objective. To determine that the applicant:

1. Exhibits instructional knowledge of holding procedures by describing—

   a. setting of aircraft navigation equipment.
   b. requirement for establishing the appropriate holding airspeed for the aircraft and altitude.
   c. recognition of arrival at the holding fix and the prompt initiation of entry into the holding pattern.
   d. timing procedure.
   e. correction for wind drift.
   f. use of DME in a holding pattern.
   g. compliance with ATC reporting requirements.

2. Exhibits instructional knowledge of common errors related to holding procedures by describing—

   a. incorrect setting of aircraft navigation equipment.
   b. inappropriate altitude, airspeed, and bank control.
   c. improper timing.
   d. improper wind drift correction.
   e. failure to recognize holding fix passage.
   f. failure to comply with ATC instructions.

---

[2] If the aircraft used for the practical test has a turn needle, substitute turn needle for miniature aircraft of turn coordinator.

3. Demonstrates and simultaneously explains holding procedures from an instructional standpoint.
4. Analyzes and corrects simulated common errors related to holding procedures.

VIII. AREA OF OPERATION: INSTRUMENT APPROACH PROCEDURES

NOTE: The examiner shall select TASKS A and B, to be combined with TASK C, D, or E. At least one non-precision approach procedure shall be accomplished without the use of the gyroscopic heading and attitude indicators under simulated instrument conditions. Circling approaches are not applicable to helicopters.

A. TASK: NON-PRECISION INSTRUMENT APPROACH

REFERENCES: 14 CFR part 91; FAA-H-8083-9, FAA-H-8083-15; FAA-S-8081-4; IAP; AIM.

Objective. To determine that the applicant:

1. Exhibits instructional knowledge of the elements of a nonprecision instrument approach by describing—
   a. selection of the appropriate instrument approach procedure chart.
   b. pertinent information on the selected instrument approach chart.
   c. radio communications with ATC and compliance with ATC clearances, instructions and procedures.
   d. appropriate aircraft configuration, airspeed, and checklist items.
   e. selection, tuning, identification, and determination of operational status of ground and aircraft navigation equipment.
   f. adjustments applied to the published MDA and visibility criteria for the aircraft approach category.
   g. maintenance of altitude, airspeed, and track, where applicable.
   h. establishment and maintenance of an appropriate rate of descent during the final approach segment.
   i. factors that should be considered in determining whether:
      (1) the approach should be continued straight-in to a landing;
      (2) a circling approach to a landing should be made; or
      (3) a missed approach should be performed.
2. Exhibits instructional knowledge of common errors related to a non-precision instrument approach by describing—
   a. failure to have essential knowledge of the information on the instrument approach chart.
   b. incorrect communications procedures or noncompliance with ATC clearances or instructions.
   c. failure to accomplish checklist items.
   d. faulty basic instrument flying technique.
   e. inappropriate descent below the MDA.
3. Demonstrates and simultaneously explains a non-precision instrument approach from an instructional standpoint.
4. Analyzes and corrects simulated common errors related to a non-precision instrument approach.

B. TASK: PRECISION INSTRUMENT APPROACH

REFERENCES: 14 CFR part 91; FAA-H-8083-9, FAA-H-8083-15; FAA-S-8081-4; IAP; AIM.

Objective. To determine that the applicant:

1. Exhibits instructional knowledge of a precision instrument approach by describing—
   a. selection of the appropriate instrument approach chart.
   b. pertinent information on the selected instrument approach chart.
   c. selection, tuning, identification, and determination of operational status of ground and aircraft navigation equipment.
   d. radio communications with ATC and compliance with ATC clearances, instructions and procedures.
   e. appropriate aircraft configuration, airspeed, and checklist items.
   f. adjustments applied to the published DH/DA and visibility criteria for the aircraft approach category.
   g. maintenance of altitude, airspeed, and track, where applicable.
   h. establishment and maintenance of an appropriate rate of descent during the final approach segment.
   i. factors that should be considered in determining whether:
      (1) the approach should be continued straight-in to a landing;
      (2) a circling approach to a landing should be made; or
      (3) a missed approach should be performed.
2. Exhibits instructional knowledge of common errors related to a precision instrument approach by describing—
   a. failure to have essential knowledge of the information on the instrument approach procedure chart.
   b. incorrect communications procedures or noncompliance with ATC clearances.
   c. failure to accomplish checklist items.
   d. faulty basic instrument flying technique.
   e. inappropriate application of DH/DA.
3. Demonstrates and simultaneously explains a precision instrument approach from an instructional standpoint.
4. Analyzes and corrects simulated common errors related to a precision instrument approach.

C. TASK: MISSED APPROACH

REFERENCES: 14 CFR part 91; FAA-H-8083-9, FAA-H-8083-15; FAA-S-8081-4; IAP; AIM.

Objective. To determine that the applicant:

1. Exhibits instructional knowledge of a missed approach procedure by describing—
   a. pertinent information on the selected instrument approach chart.
   b. conditions requiring a missed approach.
   c. initiation of the missed approach, including the prompt application of power, establishment of a climb attitude, and reduction of drag.
   d. required report to ATC.
   e. compliance with the published or alternate missed approach procedure.
   f. notification of ATC if the aircraft is unable to comply with a clearance, instruction, restriction, or climb gradient.
   g. performance of recommended checklist items appropriate to the go-around procedure.
   h. importance of positive aircraft control.
2. Exhibits instructional knowledge of common errors related to a missed approach by describing—
   a. failure to have essential knowledge of the information on the instrument approach chart.
   b. failure to recognize conditions requiring a missed approach.
   c. failure to promptly initiate a missed approach.
   d. failure to make the required report to ATC.
   e. failure to comply with the missed approach procedure.
   f. faulty basic instrument flying technique.
   g. descent below the MDA prior to initiating a missed approach.
3. Demonstrates and simultaneously explains a missed approach from an instructional standpoint.
4. Analyzes and corrects simulated common errors related to a missed approach.

D. TASK: CIRCLING APPROACH (Airplane)

REFERENCES: 14 CFR part 91; FAA-H-8083-9, FAA-H-8083-15; FAA-S-8081-4; IAP; AIM.

Objective. To determine that the applicant:

1. Exhibits instructional knowledge of the elements of a circling approach by describing—
   a. selection of the appropriate circling approach maneuver considering the maneuvering capabilities of the aircraft.
   b. circling approach minimums on the selected instrument approach chart.
   c. compliance with advisories, clearances instructions, and/or restrictions.
   d. importance of flying a circling approach pattern that does not exceed the published visibility criteria.
   e. maintenance of an altitude no lower than the circling MDA until in a position from which a descent to a normal landing can be made.

2. Exhibits instructional knowledge of common errors related to a circling approach by describing—
   a. failure to have essential knowledge of the circling approach information on the instrument approach chart.
   b. failure to adhere to the published MDA and visibility criteria during the circling approach maneuver.
   c. inappropriate pilot technique during transition from the circling maneuver to the landing approach.
3. Demonstrates and simultaneously explains a circling approach from an instructional standpoint.
4. Analyzes and corrects simulated common errors related to a circling approach.

E. TASK: LANDING FROM A STRAIGHT-IN APPROACH

REFERENCES: 14 CFR part 91; FAA-H-8083-9, FAA-H-8083-15; FAA-S-8081-4; IAP; AIM.

Objective. To determine that the applicant:

1. Exhibits instructional knowledge of the elements related to landing from a straight-in approach by describing—
   a. effect of specific environmental, operational, and meteorological factors.
   b. transition to, and maintenance of, a visual flight condition.
   c. adherence to ATC advisories, such as NOTAM's, wind shear, wake turbulence, runway surface, and braking conditions.
   d. completion of appropriate checklist items.
   e. maintenance of positive aircraft control.
2. Exhibits instructional knowledge of common errors related to landing from a straight-in approach by describing—
   a. inappropriate division of attention during the transition from instrument to visual flight conditions.
   b. failure to complete required checklist items.
   c. failure to properly plan and perform the turn to final approach.
   d. improper technique for wind shear, wake turbulence, and crosswind.
   e. failure to maintain positive aircraft control throughout the complete landing maneuver.
3. Demonstrates and simultaneously explains a landing from a straight-in approach from an instructional standpoint.
4. Analyzes and corrects simulated common errors related to landing from a straight-in approach.

IX. AREA OF OPERATION: EMERGENCY OPERATIONS

NOTE: The examiner shall select at least one TASK. The examiner shall omit TASKS C and D unless the applicant furnishes a multiengine airplane for the practical test, then TASK C or D is mandatory.

A. TASK: LOSS OF COMMUNICATIONS

REFERENCES: 14 CFR part 91; FAA-H-8083-9, FAA-H-8083-15; FAA-S-8081-4; IAP; AIM.

Objective. To determine that the applicant exhibits instructional knowledge of the elements related to loss of communications by describing:

1. Recognition of loss of communications.
2. When to continue with flight plan as filed or when to deviate.
3. How to determine the time to begin an approach at destination.

B. TASK: LOSS OF GYRO ATTITUDE AND HEADING INDICATORS

REFERENCES: 14 CFR part 91; FAA-H-8083-9, FAA-H-8083-15; FAA-S-8081-4; IAP; AIM.

Objective. To determine that the applicant:

1. Exhibits instructional knowledge of the elements related to loss of gyro attitude and heading indicators by describing—
   a. recognition of inaccurate or inoperative gyro instruments.
   b. notification of ATC of gyro loss and whether able to continue with flight clearance.
   c. importance of timely transition from full to partial panel condition.
2. Exhibits instructional knowledge of common errors related to loss of gyro attitude and heading indicators by describing—
   a. slow to recognize inaccurate or inoperative gyro instruments.
   b. failure to notify ATC of situation.
   c. failure to adequately transition from full to partial panel condition.
3. Demonstrates and simultaneously explains loss of gyro attitude and heading indicators by conducting a non-precision instrument approach without the use of these instruments. (Use Task A, AREA OPERATION VIII)
4. Analyzes and corrects common errors related to loss of gyro attitude and heading indicators.

C. TASK: ENGINE FAILURE DURING STRAIGHT-AND-LEVEL FLIGHT AND TURNS

REFERENCES: 14 CFR part 91; FAA-H-8083-9; FAA-S-8081-4; FAA-S-8081-12; FAA-S-8081-14; Aircraft Flight Manual.

Objective. To determine that the applicant:

1. Exhibits instructional knowledge of the elements related to engine failure during straight-and-level flight and turns, solely by reference to instruments, by describing—
   a. appropriate methods to be used for identifying and verifying the inoperative engine.
   b. technique for maintaining positive aircraft control by reference to instruments.
   c. importance of accurately assessing the aircraft's performance capability with regard to action that maintains altitude or minimum sink rate considering existing conditions.
2. Exhibits instructional knowledge of common errors related to engine failure during straight-and-level flight and turns, solely by reference to instruments, by describing—
   a. failure to recognize an inoperative engine.
   b. hazards of improperly identifying and verifying the inoperative engine.
   c. failure to properly adjust engine controls and reduce drag.
   d. failure to establish and maintain the best engine inoperative airspeed.
   e. failure to follow the prescribed checklist.
   f. failure to establish and maintain the recommended flight attitude for best performance.
   g. failure to maintain positive aircraft control while maneuvering.
   h. hazards of exceeding the aircraft's operating limitations.
   i. faulty basic instrument flying technique.
3. Demonstrates and simultaneously explains straight-and-level flight and turns after engine failure, solely by reference to instruments, from an instructional standpoint.
4. Analyzes and corrects simulated common errors related to straight-and-level flight and turns after engine failure, solely by reference to instruments.

D. TASK: INSTRUMENT APPROACH—ONE ENGINE INOPERATIVE

REFERENCES: 14 CFR part 91; FAA-H-8083-9; FAA-S-8081-4; FAA-S-8081-12; FAA-S-8081-14; Aircraft Flight Manual.

Objective. To determine that the applicant:

1. Exhibits instructional knowledge of the elements related to an instrument approach with one engine inoperative by describing—
   a. maintenance of altitude, airspeed and track appropriate to the phase of flight or approach segment.
   b. procedure if unable to comply with an ATC clearance or instruction.
   c. application of necessary adjustments to the published MDA and visibility criteria for the aircraft approach category.
   d. establishment and maintenance of an appropriate rate of descent during the final approach segment.
   e. factors that should be considered in determining whether:

      (1) the approach should be continued straight-in to a landing;
      or
      (2) a circling approach to a landing should be performed.
2. Exhibits instructional knowledge of common errors related to an instrument approach with one engine inoperative by describing—
   a. failure to have essential knowledge of the information that appears on the selected instrument approach chart.
   b. failure to use proper communications procedures.
   c. noncompliance with ATC clearances.
   d. incorrect use of navigation equipment.
   e. failure to identify and verify the inoperative engine and to follow the emergency checklist.
   f. inappropriate procedure in the adjustment of engine controls and the reduction of drag.
   g. inappropriate procedure in the establishment and maintenance of the best engine inoperative airspeed.
   h. failure to establish and maintain the proper flight attitude for best performance.
   i. failure to maintain positive aircraft control.
   j. faulty basic instrument flying technique.
   k. inappropriate descent below the MDA or DH.
   l. faulty technique during roundout and touchdown.
3. Demonstrates and simultaneously explains an instrument approach with one engine inoperative from an instructional standpoint.
4. Analyzes and corrects simulated common errors related to an instrument approach with one engine inoperative.

X. AREA OF OPERATION: POSTFLIGHT PROCEDURES

TASK: CHECKING INSTRUMENTS AND EQUIPMENT

REFERENCES: FAA-S-8081-4; Aircraft Flight Manual.

Objective. To determine that the applicant exhibits instructional knowledge of the elements related to checking instruments and equipment by describing:

1. Importance of noting instruments and navigation equipment for improper operation.
2. Reasons for making a written record of improper operation and/or calibration of instruments prior to next IFR flight.

Levels of Simulation Devices
TASK VS. SIMULATION DEVICE CREDIT

Examiners conducting the instrument rating practical tests with flight simulation devices should consult appropriate documentation to ensure that the device has been approved for training, testing, or checking. The documentation for each device should reflect that the following activities have occurred:

1. The device must be evaluated, determined to meet the appropriate standards, and assigned the appropriate qualification level by the National Simulator Program Manager. The device must continue to meet qualification standards through continuing evaluations as outlined in the appropriate advisory circular (AC). For airplane flight training devices (FTD's), AC 120-45 (as amended), Airplane Flight Training Device Qualifications, will be used. For simulators, AC 120-40 (as amended), Airplane Simulator Qualification, will be used.
2. The FAA must approve the device for training, testing, and checking the specific flight TASKS listed in this appendix.
3. The device must continue to support the level of student or applicant performance required by this practical test standard.

NOTE: Users of the following chart are cautioned that use of the chart alone is incomplete. The description and Objective of each TASK as listed in the body of the practical test standard, including all NOTES, must also be incorporated for accurate simulation device use.

USE OF CHART

X Creditable.
A Creditable if appropriate systems are installed and operating.

NOTE: 1. Level 1 FTD's that have been issued a letter authorizing their use by the FAA Administrator, and placed in service on or prior to August 2, 1996, may continue to be used only for those TASKS originally found acceptable. Use of Level 1, 2, or 3 FTD's may not be used for aircraft requiring a type rating.
2. If a FTD's or a flight simulator is used for the practical test, the instrument approach procedures conducted in that FTD or flight simulator are limited to one precision and one non-precision approach procedure.
3. Postflight procedures means, closing flight plans, checking for discrepancies and malfunctions, and noting them on a log or maintenance form.

FLIGHT TASK FLIGHT SIMULATION DEVICE LEVEL

Areas of Operation	1	2	3	4	5	6	7	A	B	C	D
III. Preflight Preparation											
C. Instrument Cockpit Check *	_	A	X	A	A	X	X	X	X	X	X
V. Air Traffic Control Clearances and Procedures											
A. Air Traffic Control Clearances *	_	A	X	A	A	X	X	X	X	X	X
B. Compliance with Departure, En Route, Arrival Procedures and Clearances *	_	_	X	_	_	X	X	X	X	X	X
VI. Flight by Reference to Instruments											
A. Straight-and-Level Flight	_	_	X	_	_	X	X	X	X	X	X
B. Turns	_	_	X	_	_	X	X	X	X	X	X
C. Change of Airspeed in Straight-and-Level and Turning Flight											
D. Constant Airspeed Climbs and Descents	_	_	X	_	_	X	X	X	X	X	X
E. Constant Rate Climbs and Descents	_	_	X	_	_	X	X	X	X	X	X
F. Timed Turns to Magnetic Compass Headings	_	_	X	_	_	X	X	X	X	X	X
G. Steep Turns	_	_	X	_	_	X	X	X	X	X	X
H. Recovery from Unusual Attitudes	_	_	_	_	_	_	X	X	X	X	X

FLIGHT TASK FLIGHT SIMULATION DEVICE LEVEL

Areas of Operation	1	2	3	4	5	6	7	A	B	C	D
VII. Navigation Systems											
A. Intercepting and Tracking Navigational Systems and DME ARC'S	_	A	X	_	A	X	X	X	X	X	X
B. Holding Procedures	_	_	X	_	_	X	X	X	X	X	X
VIII. Instrument Approach Procedures											
A. Non-precision Instrument Approach	_	_	X	_	_	X	X	X	X	X	X
B. Precision Instrument Approach	_	_	X	_	_	X	X	X	X	X	X
C. Missed Approach	_	_	X	_	_	X	X	X	X	X	X
D. Circling Approach (Airplane)	_	_	_	_	_	_	_	_	_	X	X
E. Landing from a Straight-in Approach	_	_	_	_	_	_	_	_	_	X	X
IX. Emergency Operations (ME) & (SE) **											
A. Loss of Communications	_	X	X	_	_	X	X	X	X	X	X
B. Loss of Gyro Attitude and Heading Indicators	_	_	_	_	_	_	X	X	X	X	X
C. Engine Failure During Straight-and-Level Flight and Turns	_	_	X	_	_	X	X	X	X	X	X
D. Instrument Approach—One Engine Inoperative	_	_	_	_	_	_	_	X	X	X	X
X. Postflight Procedures											
Checking Instruments and Equipment	_	A	X	_	A	X	X	X	X	X	X

* Aircraft required for those items that cannot be checked using a flight training device or flight simulator.
** Multiengine and Single-engine

APPENDIX D
INSTRUMENT PROFICIENCY CHECK

The purpose of this appendix is to provide you with information on the requirements of an instrument proficiency check and the appropriate time to take one.

A. **Recent IFR Experience (FAR 61.57)**

1. You may not act as pilot in command under IFR, or in IMC, unless you have performed and logged under simulated or actual instrument conditions, within the past 6 calendar months, in either an airplane, an airplane flight simulator, or an airplane flight training device,
   a. At least six instrument approaches
   b. Holding procedures
   c. Intercepting and tracking courses through the use of navigation systems

B. **Instrument Proficiency Check**

1. If you did not meet the recent IFR experience requirement within the past 12 calendar months, you must pass an instrument proficiency check consisting of at least those tasks required by the instrument rating practical test standards in either an airplane, an airplane flight simulator, or an airplane flight training device, given by an FAA inspector, an examiner, an approved (FAA or military) check pilot, or an authorized instructor (i.e., a CFII, AGI, or IGI).
2. Most pilots will seek a CFII to conduct the instrument proficiency check.
   a. A suggested instrument proficiency check plan and checklist form is presented on pages 400 and 401.
      1) This suggested plan and checklist form should not be considered all-inclusive and is not intended to limit either you or your CFII from selecting appropriate maneuvers and procedures.
         a) The form DOES contain the tasks that are required.
      2) You should have your CFII complete this instrument proficiency plan and checklist so that you have a record of the scope and content of the instrument proficiency check.
         a) This form is also good for CFIIs to keep in their records. Photocopy a copy for your CFII to complete during your proficiency check.
3. For an instrument review, see Gleim's *Instrument Pilot Flight Maneuvers and Practical Test Prep* book.

C. The following tasks must be completed during an instrument proficiency check:

1. Chapter II., Preflight Procedures (all tasks), beginning on page 67
2. Chapter III., Air Traffic Control Clearances and Procedures (all tasks), beginning on page 101
3. Chapter IV., Flight by Reference to Instruments (tasks A, B, C, D, and G), beginning on page 137
4. Chapter V., Navigation Systems, beginning on page 193
5. Chapter VI., Instrument Approach Procedures (all tasks), beginning on page 215
6. Chapter VII., Emergency Operations (all tasks), beginning on page 327
7. Chapter VIII., Postflight Procedures, beginning on page 337

D. Logbook endorsement for a satisfactory completion of an instrument proficiency check: FAR 61.57(d)

I certify that (First name, MI, Last name), (pilot certificate) (certificate number) has satisfactorily completed the instrument proficiency check of Sec. 61.57(d) in a (list make and model of aircraft) on (date).

Date ______ *Signature* ______ *CFI No.* ______ *Expiration Date* ______

INSTRUMENT PROFICIENCY CHECK PLAN AND CHECKLIST

Name ________ Pilot Certificate No. ________

Certificate and Ratings ________

Date of Last Check ________

Class of Medical ________ Date of Medical ________

Total Time ________ Time in Type Aircraft ________

Total Instrument Time: ____ Simulated ____ Actual ____ Flight Simulator/Training Device ____

In Last 180 Days: Simulated ____ Actual ____ Flight Simulator/Training Device ____

Approaches/Last 180 Days: Precision ____ Nonprecision ____

Aircraft to Be Used ________ N # ________

Location of Check ________

I. KNOWLEDGE PORTION OF PROFICIENCY CHECK

A. FAR Part 91 Review
   1. Subpart B (Instrument Flight Rules)
   2. Subpart C (Equipment, Instrument, and Certificate Requirements)
   3. Subpart E (Maintenance)

B. Instrument en route and approach charts, including DPs and STARs

C. Weather analysis and knowledge

D. Preflight planning, including performance data, fuel, alternate, NOTAMs, and appropriate FAA publications

E. Aircraft systems related to IFR operations*

F. Aircraft flight instruments and navigation equipment*

G. Airworthiness status of aircraft and avionics for IFR flight

H. Other areas:

____

____

____

*Required by the Instrument Rating Practical Test Standards

INSTRUMENT PROFICIENCY CHECK PLAN (CONTINUED)

II. REQUIRED SKILL PORTION OF PROFICIENCY CHECK** (Include location.)

A. Instrument cockpit check ____________
B. ATC clearances ____________
C. Compliance with departure, en route, and arrival procedures and clearances ____________
D. Holding procedures ____________
E. Straight-and-level flight ____________
F. Change of airspeed ____________
G. Constant airspeed climbs and descents ____________
H. Rate climbs and descents ____________
I. Recovery from unusual flight attitudes ____________
J. Intercepting and tracking navigational systems and DME arcs ____________
K. Nonprecision instrument approach (type) ____________
L. Nonprecision approach (type) ____________
M. ILS instrument approach ____________
N. Missed approach procedures ____________
O. Circling approach procedures ____________
P. Landing from a straight-in or circling approach ____________
Q. Loss of communications ____________
R. One engine inoperative during straight-and-level flight and turns (multiengine only) ____
S. One engine inoperative -- instrument approach (multiengine only) ____________
T. Loss of gyro attitude and/or heading indicators ____________
U. Checking instruments and equipment ____________

III. OVERALL COMPLETION OF PROFICIENCY CHECK

Remarks: ____________

____

____________ Signature of CFII ____________ Date

____________ Certificate No. ____________ Expiration Date

I have received an instrument proficiency check that consisted of the knowledge review and skill demonstration of the procedures noted.

____________ Signature of the Pilot ____________ Date

**Required by the Instrument Rating Practical Test Standards

APPENDIX E
ORAL EXAM GUIDE

Most flight schools and many CFIs recommend that pilots preparing for their practical test study an "Oral Exam Guide." People facing an exam often become anxious and aspire to be well-prepared (i.e., they have an aversion to failure). With the Gleim system you are well-prepared. Four Gleim books that are applicable to the instrument rating oral exam and the entire FAA practical test are

Aviation Weather and Weather Services (AWWS),
Instrument Pilot FAA Written Exam (IPWE),
Instrument Pilot Flight Maneuvers and Practical Test Prep (IPFM), and
Pilot Handbook (PH)

These books contain all the information you need to do well on your instrument rating practical test. *Pilot Handbook* is part of our Private Pilot Kit and not included in our Instrument Pilot Kit. *Pilot Handbook*, however, is recommended.

Consider this appendix your **ORAL EXAM GUIDE**.

1. Review the requirements to obtain an instrument rating on page 2 of IPFM (this book).
2. Essential reading: "Oral Portion of the Practical Test" on page 40 (IPFM).
3. Relatedly, read the following on pages 34 through 39 (IPFM):

   Airplane and Equipment Requirements
   What to Take to Your Practical Test
   Practical Test Application Form
   Authorization to Take the Practical Test

4. FARs: In *Instrument Pilot FAA Written Exam* (IPWE) read the 7-page outline in Chapter 4 for a condensed review of the most important FARs related to instrument flight.

FAR PART 61

61.3	Requirements for Certificates, Ratings, and Authorizations
61.51	Pilot Logbooks
61.57	Recent Flight Experience: Pilot in Command
61.133	Commercial Pilot Privileges and Limitations

FAR PART 91

91.3	Responsibility and Authority of the Pilot in Command
91.21	Portable Electronic Devices
91.103	Preflight Action
91.109	Flight Instruction; Simulated Instrument Flight and Certain Flight Tests
91.123	Compliance with ATC Clearances and Instructions
91.129	Operations in Class D Airspace
91.131	Operations in Class B Airspace
91.135	Operations in Class A Airspace
91.155	Basic VFR Weather Minimums
91.157	Special VFR Weather Minimums
91.167	Fuel Requirements for Flight in IFR Conditions
91.169	IFR Flight Plan: Information Required
91.171	VOR Equipment Check for IFR Operations
91.173	ATC Clearance and Flight Plan Required
91.177	Minimum Altitudes for IFR Operations
91.205	Powered Civil Aircraft with Standard Category U.S. Airworthiness Certificates: Instrument and Equipment Requirements

91.211	Supplemental Oxygen
91.215	ATC Transponder and Altitude Reporting Equipment and Use
91.411	Altimeter System and Altitude Reporting Equipment Tests and Inspections

NTSB PART 830 Notification and Reporting of Aircraft Accidents or Incidents and Overdue Aircraft, and Preservation of Aircraft Wreckage, Mail, Cargo, and Records

In *Pilot Handbook* (PH), Chapter 4, read the following FARs:

91.7	Civil Aircraft Airworthiness
91.9	Civil Aircraft Flight Manual, Marking, and Placard Requirements
91.126-91.135	Airspace FARs
91.167-91.187	Instrument flight FARs
91.403-91.413	Maintenance and inspection FARs

Note: FARs ARE NOW REFERRED TO AS CFRs: The FAA has recently begun to abbreviate Federal Aviation Regulations as "14 CFR" rather than "FARs." CFR stands for Code of Federal Regulations, and the Federal Aviation Regulations are in Title 14. For example, FAR Part 1 and FAR 61.109 are now referred to as 14 CFR Part 1 and 14 CFR Sec. 61.109, respectively. CFIs and pilots continue to use the acronym FAR.

Examiner Questions:

You will be ready for your practical test. Follow the advice on page 40. The following pages contain questions previously asked by designated examiners on instrument rating practical tests. There are cross references to Gleim books if you wish to research the answers to these questions.

Let us know about your practical test and any unexpected questions you received so we can add them to this appendix. GOOD LUCK!

IFR Flight Planning:

1. When is an instrument rating required? (IPWE, Ch. 4)
2. What limitations are imposed on commercial pilot operations performed by commercial pilots who do not possess an instrument rating? (IPWE, Ch. 4)
3. When is an IFR flight plan required? (IPWE, Ch. 4)
4. What is a composite flight plan? (IPWE, Ch. 5)
5. On a flight with multiple planned altitudes, which altitude should be entered in block 7, cruising altitude, of an IFR flight plan? (IPWE, Ch. 5)
6. On what portion of your flight should you base the time entered in block 10, time en-route, of an IFR flight plan? (IPWE, Ch. 5)
7. Based on which criteria would you select a cruising altitude for an IFR flight? (IPFM, pp. 60 through 62)
8. What are the minimum fuel requirements for flight under IFR? (IPFM, p. 62)
9. How can you determine how much fuel will be used for a given flight? (PH, Ch. 5 and 9)
10. When is an IFR clearance required? (IPWE, Ch. 4; PH Ch. 4)
11. What documents must be aboard an aircraft when it is being operated? (PH, Ch. 4)
12. What are the required instruments and equipment for IFR flight? (PH, Ch. 4; IPFM, p. 6)

13. How often must an aircraft's transponder be inspected if it is to be used in visual and/or instrument meteorological conditions (VMC/IMC)? (PH, Ch. 4)
14. How often must an aircraft inspection be performed if the aircraft is not operated for compensation or hire? If it is operated for compensation or hire (i.e., commercial operations)? (PH, Ch. 4)
15. How often must an aircraft's altimeter and static system be inspected if it is to be flown under IFR? (IPWE, Ch. 4)
16. How often must a VOR accuracy check be performed if an aircraft is to be operated under IFR? What are the different kinds of accuracy checks, and what are their tolerances? (PH, Ch. 4)
17. How can you determine if the required aircraft, systems, and equipment inspections have been performed? (PH, Ch. 4)
18. Who is responsible for determining if the aircraft is in an airworthy condition? (PH, Ch. 4)
19. What recent flight experience is required if you are to act as pilot in command of an aircraft under IFR? Within what time frame must this experience be accumulated? If it is not accumulated within the required time frame, what must be done to regain instrument currency? (IPWE, Ch. 4)
20. With what information must the pilot in command familiarize him/herself before beginning a flight under IFR? (IPWE, Ch. 4)
21. What is the minimum amount of time that must pass between consumption of any alcohol and acting as a required flight crewmember? What is the maximum blood alcohol level that is acceptable for a pilot acting as a required flight crewmember? (PH, Ch. 4)
22. What are the basic VFR minimums for Class G airspace at or below 1,200 ft. AGL? Above 1,200 ft. AGL but below 10,000 ft. MSL? At or above 10,000 ft. MSL? (PH, Ch. 4)
23. What are the basic VFR minimums for Class E airspace below 10,000 ft. MSL? At or above 10,000 ft. MSL? (PH, Ch. 4)
24. What are the basic VFR minimums for Class D airspace? (PH, Ch. 4)
25. What are the basic VFR minimums for Class C airspace? (PH, Ch. 4)
26. What are the basic VFR minimums for Class B airspace? (PH, Ch. 4)
27. May you operate under VFR in Class A airspace? (PH, Ch. 4)
28. What are the special VFR minimums? May you operate under special VFR at night? (PH, Ch. 4)
29. You must select an alternate airport unless what minimum weather conditions are forecast for your destination airport? What is the required time frame for these weather conditions? (IPFM, p. 55)
30. What minimum weather conditions must be forecast at your selected alternate airport, which has a non-precision approach, for it to be acceptable as an alternate? If it has a precision approach? If it has no approved instrument approach? What is the required time frame for these weather conditions? (IPFM, pp. 55 and 56)
31. What is hypoxia? (PH, Ch. 6)
32. Between which altitudes must supplemental oxygen be used by the required flight crew after 30 minutes at those altitudes? (IPWE, Ch. 4)
33. Between which altitudes must supplemental oxygen be used by the required flight crew during the entire time at those altitudes? (IPWE, Ch. 4)

34. Above which altitude must supplemental oxygen be provided for each passenger? Must each passenger use supplemental oxygen above this altitude? (IPWE, Ch. 4)
35. Explain the difference between a standard weather briefing, an abbreviated briefing, and an outlook briefing. When would each be appropriate? (IPFM, p. 48)
36. What is a NOTAM? Name the three types of NOTAMs and describe them. (IPFM, p. 63)
37. How often is the Automatic Terminal Information Service (ATIS) broadcast updated? Is it ever updated at a different interval? (IPFM, p. 55)
38. Name some useful charts, reports, and forecasts for IFR flight planning. What information is contained on each of them? (IPFM, pp. 50 through 54; AWWS, Part III)
39. What is the difference between an AIRMET and a SIGMET? (IPFM, pp. 53 and 54)
40. From which weather reports can you obtain the only reliable information about observed icing conditions and cloud tops? (AWWS, Part III, Ch. 3)
41. What useful information is presented by radar weather maps? By satellite weather maps? Are they different in any way; i.e, is one more useful than the other in some cases? (IPFM, p. 50; AWWS, Part III, Ch. 7 and 18)
42. What is a temperature inversion? What kind of weather can you expect when a temperature inversion exists? (IPWE, Ch. 8; AWWS, Part I, Ch. 2 and 9)
43. If the temperature-dew point spread is small and decreasing, what kind of weather should you expect? (IPWE, Ch. 8)
44. What are some different kinds of fog? How are they formed? (AWWS, Part I, Ch. 12)
45. Ice pellets at the surface are a sign of what phenomenon at a higher altitude? (AWWS, Part I, Ch. 5)
46. What conditions are necessary for the formation of structural ice? (AWWS, Part I, Ch. 10)
47. What are the characteristics of stable air? Of unstable air? (AWWS, Part I, Ch. 8)
48. What conditions are necessary for the formation of thunderstorms? What are the three stages of a thunderstorm, and what are their characteristics? (AWWS, Part I, Ch. 11)
49. What are air mass thunderstorms? What are squall line thunderstorms? Are either associated with a frontal system? (AWWS, Part I, Ch. 11)
50. What is wind shear? Why can it be dangerous? (IPWE, Ch. 8)

Departure:

1. What information will be included in an IFR departure clearance? (IPFM, pp. 102 through 104)
2. How can you obtain your departure clearance at a controlled field? At an uncontrolled field? (IPFM, p. 102)
3. What does it mean when your departure clearance indicates that you are "cleared as filed?" (IPFM, p. 104)
4. What is the significance of a clearance void time? (IPFM, p. 106)
5. Under what conditions may you deviate from an ATC clearance? What action may be required of you if you deviate from a clearance? (PH, Ch. 4)
6. Which transponder code is used by VFR aircraft? By aircraft in distress? By hijacked aircraft? By aircraft experiencing a loss of radio communications? (PH, Ch. 3)

7. Are you required to accept a Departure Procedure (DP) if one is assigned by ATC? How can you avoid being assigned a DP? (IPWE, Ch. 6)
8. What is the purpose of DPs? (IPFM, p. 111)
9. What must you possess in order to accept a DP? (IPFM, p. 113)
10. What is the standard minimum climb gradient for a DP? Is it ever greater? If so, how can you tell? (IPFM, p. 111)
11. Are there takeoff minimums for operations under Part 91? (http://www.gleim.com/Aviation/Updates/books/ipfm) You should open the link titled "Additional Instrument Flight Maneuvers."
12. What is a good self-imposed (i.e., non-regulatory) policy regarding the minimum conditions you should consider acceptable for takeoff? (http://www.gleim.com/Aviation/Updates/books/ipfm) You should open the link titled, "Additional Instrument Flight Maneuvers."
13. Is the presence of frost on the aircraft's wings a significant operational consideration? (IPWE, Ch. 8)
14. What is wake turbulence, and under what conditions is it strongest? (PH, Ch. 3)
15. Where should you plan to lift off when departing behind a large aircraft which has just landed on the same runway? When departing behind a large aircraft that has just taken off on the same runway? (PH, Ch. 3)
16. What is an ILS Critical Area? During ground operations, how can you be sure that you are not inside it? (PH, Ch. 3)
17. Describe runway hold-short lines and signs. (PH, Ch. 3)
18. Describe airport runway and taxiway lighting and signage. What color are runway lights? Do they change color along the length of the runway? What special signs are visible on the runway? What color are taxiway lights? How are taxiways labeled? (PH, Ch. 3)
19. If you are climbing in VFR conditions, who is responsible for maintaining separation between your aircraft and other traffic? (IPWE, Ch. 5)
20. When leveling off from a climb, what is the standard rule to determine by how many feet before the desired altitude you should lead the level-off? (IPFM, p. 160)
21. When climbing, at what rate does ATC expect you to climb, unless otherwise authorized? (IPFM, p. 119)
22. What does it mean when you are cleared to climb "at pilot's discretion?" (IPFM, p. 119)
23. What is the aircraft speed limit below 10,000 ft. MSL? (PH, Ch. 4)
24. May portable electronic devices be used indiscriminately on an aircraft that is being operated under IFR? (PH, Ch. 4)
25. What is the standard temperature lapse rate? (PH, Ch. 7)

En-Route:

1. What is a Minimum En-route Altitude (MEA)? (IPFM, p. 29)
2. What is a Minimum Obstruction Clearance Altitude (MOCA)? (IPFM, p. 29)
3. What is a Minimum Reception Altitude (MRA)? (IPFM, p. 29)
4. What information is found on en-route low altitude charts? (IPFM, p. 113; Chart Training Guide)
5. What is a changeover point? When and how is it depicted on en-route charts? (IPFM, p. 116)
6. If no MEA is published for your route of flight, what is the minimum prescribed altitude for operating an aircraft under IFR in non-mountainous areas? In mountainous areas? (PH, Ch. 4)

7. May you operate below the published MEA prescribed for your route? If so, when? (PH, Ch. 4)
8. What report, if any, must be made to ATC in the event that a navigation or communication radio fails? (PH, Ch. 4)
9. What are the procedures for dealing with a loss of radio communications in VMC? In IMC? (PH, Ch. 4)
10. From which sources can you obtain weather information while en-route? (PH, Ch. 8; IPWE, Ch. 9)
11. What is a Center Weather Advisory? (AWWS, Part III, Ch. 12)
12. What is En-Route Flight Advisory service? On what frequency can it be obtained? (AWWS, Part III, Ch. 1)
13. What are the 5 types of special-use airspace which are charted on aeronautical charts? How does each one relate to IFR operations? (PH, Ch. 3)
14. When you transition from the VFR to the IFR portion of a composite flight plan, when must you cancel the VFR portion and obtain your IFR clearance? (IPWE, Ch. 5)
15. Is a VFR-on-top clearance a VFR clearance or an IFR clearance? Do visual flight rules or instrument flight rules apply? (IPFM, p. 105; IPWE, Ch. 10)
16. When operating with a VFR-on-top clearance, how should you select your cruising altitude? May it be below the minimum IFR altitude for your location? (IPWE, Ch. 10)
17. Are VFR-on-top operations allowed in Class A airspace? (IPWE, Ch. 10)
18. What is a cruise clearance? (IPFM, p. 103)
19. What is the significance of primary vs. supporting instruments in attitude instrument flying? (IPFM, pp. 137 and 138)
20. What is Area Navigation (RNAV)? Name some RNAV systems and their principles of operation. (PH, Ch. 10)
21. What is a waypoint? (PH, Ch. 10)
22. What is a VHF Omnidirectional Range (VOR) station? What is a Non-Directional Beacon (NDB)? (PH, Ch. 10)
23. Explain the indications of a standard VOR indicator. Why does reverse sensing occur? (PH, Ch. 10)
24. Explain how you would intercept a particular VOR radial. (IPFM, pp. 196 and 197)
25. What is bracketing? (PH, Ch. 10)
26. Technically, what is the difference between a VOR, a VOR/DME, and a VORTAC? Operationally, what is the difference? (PH, Ch. 10)
27. What is a Horizontal Situation Indicator (HSI)? Explain its indications. (IPFM, pp. 76 through 78)
28. What is a Radio Magnetic Indicator (RMI)? Explain its indications. (IPFM, p. 81)
29. How do you determine on what radial you are located relative to a given VOR station using a standard VOR indicator? Using an HSI? Using an RMI? (PH, Ch. 10; IPFM, p. 81)
30. How do you identify an intersection of two VOR radials? Of a VOR radial and an NDB bearing? (PH, Ch. 10)
31. Explain how Distance Measuring Equipment (DME) works. What kind of range information does it provide? (IPFM, p. 85)

32. Under what conditions will the groundspeed and Estimated Time En-route (ETE) information presented by a DME receiver be accurate? (IPFM, p. 85)
33. Explain the indications of an Automatic Direction Finder (ADF) indicator. (PH, Ch. 10)
34. What is the difference between homing and tracking? Explain how you would home to an NDB vs. how you would track to it. (PH, Ch. 10)
35. Explain how you would intercept a particular NDB bearing. (IPFM, pp. 203 and 204)
36. How should you verify that the navigational facility you are using is the desired facility and that it is operational? (PH, Ch. 10)
37. How do you identify station passage using a VOR? Using an ADF? (PH, Ch. 10; IPFM, p. 205)
38. What are the three types of structural icing? (AWWS, Part I, Ch. 10)
39. Under what conditions will structural ice accumulate at the greatest rate? (IPWE, Ch. 8; AWWS, Part I, Ch. 10)
40. If you inadvertently penetrate a thunderstorm, what is the recommended procedure? (AWWS, Part I, Ch. 11)
41. What is the most effective way to scan for other aircraft in day VFR conditions? In night VFR conditions? (IPWE, Ch. 7)
42. When may you log instrument flight time? (PH, Ch. 4)

Arrival:

1. When descending, at what rate does ATC expect you to descend, unless otherwise authorized? (IPFM, p. 119)
2. What does it mean when you are cleared to descend "at pilot's discretion?" (IPFM, p. 119)
3. When leveling off from a descent, what is the standard rule to determine by how many feet before the desired altitude you should lead the level-off? (IPFM, p. 160)
4. Describe the components of a holding pattern. How is it typically flown? (IPFM, p. 124)
5. What is a standard holding pattern? A non-standard holding pattern? (IPFM, p. 127)
6. Describe the three types of holding pattern entries. (IPFM, p. 127)
7. How much wind correction should you apply on the outbound leg of a holding pattern if you had five degrees of correction on the inbound leg? (IPFM, pp. 131 and 132)
8. In order to obtain the desired performance on the inbound leg, how should you plan the timing of the outbound leg of the holding pattern. (IPFM, p. 130)
9. What are the three altitude blocks for which holding pattern speed limits have been established? What are those speed limits? (IPWE, Ch. 6)
10. Are you required to accept a Standard Terminal Arrival Route (STAR) if one is assigned by ATC? How can you avoid being assigned a STAR? (IPWE, Ch. 6)
11. What is the purpose of a STAR? (IPFM, p. 114)
12. What must you possess to accept a STAR? (IPFM, p. 114)
13. What is the difference between a visual approach and a contact approach? (IPFM, pp. 229 through 231)
14. Can a contact approach be assigned by ATC? (IPWE, Ch. 6)

15. What are the four segments of an Instrument Approach Procedure (IAP)? (IPFM, pp. 211 and 212)
16. What is a feeder route? (IPFM, p. 212)
17. What is an Initial Approach Fix (IAF)? A Final Approach Fix (FAF)? A Missed Approach Point (MAP)? (IPFM, pp. 211 and 212)
18. What is the difference between a Final Approach Fix (FAF) and a Final Approach Point? (IPFM, p. 212)
19. What is a DME arc? What is its purpose? (IPFM, p. 198)
20. How is a DME arc flown with a standard VOR indicator? With an RMI? (IPFM, pp. 198 through 200)
21. Under what conditions may timed approaches from a holding fix be conducted? (IPFM, p. 223)
22. What is a procedure turn? What is its purpose? (IPFM, p. 220)
23. What is the significance of the symbol, "NoPT" on an IAP chart? (IPFM, p. 220)
24. What is the difference between a Decision Height (DH) and a Minimum Descent Altitude (MDA)? (IPFM, p. 28)
25. What is a Visual Descent Point (VDP)? (IPFM, p. 217)
26. What is a Minimum Safe/Sector Altitude (MSA)? On what ground object is it normally based? What are the normal dimensions of the area in which the MSA is applicable? (IPFM, p. 29)
27. If you have been cleared for an approach and are established on a published segment of the approach, may you descend from your assigned altitude to the MEA of the segment of the published approach on which you are established, even if you are not specifically cleared to do so? (IPWE, Ch. 6)
28. If you must divert to your alternate, which minimums should you use when performing the approach procedure? (IPWE, Ch. 4)
29. Explain the procedures for flying a typical non-precision approach which incorporates a procedure turn. (IPFM, pp. 233 through 241)
30. Explain the procedures for flying a typical precision approach which involves radar vectors to the final approach course. (IPFM, pp. 296 through 305)
31. Name and describe several types of non-precision approaches. (IPFM, p. 233)
32. Name and describe two types of precision approaches. (IPFM, pp. 224 and 296)
33. What is the difference between a Simplified Directional Facility (SDF), a Localizer-type Directional Aid (LDA), and a standard localizer (LOC)? (IPWE, Ch. 6)
34. What methods can be used to define/determine the missed approach point for a non-precision approach? (IPFM, pp. 240, 249, and 262)
35. What is Receiver Autonomous Integrity Monitoring (RAIM)? Why is it important? (IPFM, p. 92)
36. Where is the missed approach point on an Instrument Landing System (ILS) approach? (IPFM, p. 304)
37. If you break out of the clouds at the Minimum Descent Altitude (MDA) for a non-precision approach to a runway with a Visual Approach Slope Indicator (VASI) system and you are below the glide slope, what should you do? (IPWE, Ch. 5)
38. What is the significance of the different aircraft approach categories? On what criterion is an aircraft's approach category based? (IPFM, p. 219)

39. How is an aircraft's approach speed determined? (IPFM, p. 219)
40. If a category A aircraft makes an approach at a category B approach speed, which approach minimums (i.e., category A or B minimums) apply? (IPFM, p. 219)
41. Describe precision instrument runway markings. What is their purpose? (PH, Ch. 3)
42. Where should you plan to touch down when landing behind a large aircraft which has just landed on the same runway? When landing behind a large aircraft which has just taken off on the same runway? (PH, Ch. 3)
43. When can you expect to encounter hazardous wind shear? (IPWE, Ch. 8)
44. What is a microburst? What effect does it have on an airplane which traverses it? (AWWS, Part I, Ch. 11)
45. Under what conditions are microbursts likely? How long do they normally last? (AWWS, Part I, Ch. 11)
46. What is induction (carburetor) icing? What causes it? (AWWS, Part I, Ch. 10)
47. What can be done to prevent the formation of induction ice? (PH, Ch. 2)

ABBREVIATIONS AND ACRONYMS IN INSTRUMENT PILOT FLIGHT MANEUVERS AND PRACTICAL TEST PREP

A/FD *Airport/Facility Directory*
AC Advisory Circular
ADF automatic direction finder
AFM aircraft flight manual
AI attitude indicator
AIM *Aeronautical Information Manual*
AIRMET airman's meteorological information
ALT altimeter
APT WP airport waypoint
ARTCC air route traffic control center
ASI airspeed indicator
ASOS automated surface observing system
ASR airport surveillance radar
ATC air traffic control
ATD along track distance
ATIS automatic terminal information service
AWOS automated weather observing system
BC back course
CDI course deviation indicator
CFII certificated flight instructor-instrument
CFR Code of Federal Regulations
COP changeover point
CTAF common traffic advisory frequency
CVFP charted visual flight procedures
DH decision height
DME distance-measuring equipment
DP instrument departure procedure
EFC expect further clearance
ETA estimated time of arrival
ETE estimated time en route
FAA Federal Aviation Administration
FAF final approach fix
FAR Federal Aviation Regulation
FAWP final approach waypoint
FBO fixed-base operator
FDC Flight Data Center
FSS Flight Service Station
GPS global positioning system
Hg mercury
HI heading indicator
hr. hour(s)
HSI horizontal situation indicator
IAF initial approach fix
IAP instrument approach procedure
IAS indicated airspeed
IF intermediate fix
IFR instrument flight rules
ILS instrument landing system
IM inner marker
IMC instrument meteorological conditions
JEPP Jeppesen Sanderson, Inc.
kt. knots (nautical miles per hour)
LDA localizer-type directional aid
LLWAS low-level wind shear alert system
LOC localizer
LOC BC localizer back course
LOM locator outer marker
LORAN long-range navigation
MAA maximum authorized altitude
MAHWP missed approach holding waypoint
MAP missed approach point
MAWP missed approach waypoint
MB magnetic bearing
MC magnetic compass
MCA minimum crossing altitude
MDA minimum descent altitude
MEA minimum en route altitude
MHA minimum holding altitude
min. minute(s)
MM middle marker
MOA military operations area
MOCA minimum obstruction clearance altitude
MP manifold pressure gauge
MRA minimum reception altitude
MSA minimum safe/sector altitude
MSL mean sea level
MTR military training routes
MVA minimum vectoring altitude
NACO National Aeronautical Charting Office
NAVAID navigational aid
NDB nondirectional beacon
NM nautical mile
NoPT no procedure turn
NOTAM Notice to Airmen
NTAP *Notice to Airmen Publication*
NTZ no transgression zone
NWS National Weather Service
OBS omnibearing selector
OM outer marker
OROCA off-route obstruction clearance altitude
PAR precision approach radar
PCATD personal computer-based aviation training device
POH *Pilot's Operating Handbook*
PRM precision runway monitor
PT procedure turn
PTS practical test standards
RAIM receiver autonomous integrity monitoring
RB relative bearing
RIC remote indicating compass
RMI radio magnetic indicator
RNAV area navigation
RPM tachometer
RWY WP runway waypoint
SDF simplified directional facility
SIGMET significant meteorological information
SM statute mile
SSV standard service volume
STAR standard terminal arrival route
T&SI turn-and-slip indicator
TAA terminal arrival area
TAS true airspeed
TC turn coordinator
TCH threshold crossing height
TDWR terminal Doppler weather radar
TDZE touchdown zone elevation
TEC tower en route control
TERPS U.S. Standard for Terminal Instrument Procedures
VDP visual descent point
VFR visual flight rules
VHF very high frequency
VMC visual meteorological conditions
VOR VHF omnidirectional range
VOT VOR test facility
VSI vertical speed indicator
WSP weather systems processor

AUTHOR'S RECOMMENDATION

The Experimental Aircraft Association, Inc. is a very successful and effective nonprofit organization that represents and serves those of us interested in flying, in general, and in sport aviation, in particular. I personally invite you to enjoy becoming a member:

$35 for a 1-year membership
$20 per year for individuals under 19 years old
Family membership available for $45 per year

Membership includes the monthly magazine *Sport Aviation*.

Write to: Experimental Aircraft Association, Inc.
P.O. Box 3086
Oshkosh, Wisconsin 54903-3086

Or call: (414) 426-4800
(800) 564-6322

The annual EAA Oshkosh AirVenture is an unbelievable aviation spectacular with over 12,000 airplanes at one airport! Virtually everything aviation-oriented you can imagine! Plan to spend at least 1 day (not everything can be seen in a day) in Oshkosh (100 miles northwest of Milwaukee).

Convention dates: 2002 -- July 23 through July 29
2003 -- July 29 through August 14

The annual Sun 'n Fun EAA Fly-In is also highly recommended. It is held at the Lakeland, FL (KLAL) airport (between Orlando and Tampa). Visit the Sun 'n Fun web site at http://www.sun-n-fun.org.

Convention dates: 2002 -- April 7 through April 13
2003 -- April 6 through April 12

BE-A-PILOT: INTRODUCTORY FLIGHT

Be-A-Pilot is an industry-sponsored marketing program designed to inspire people to "Stop dreaming, start flying." Be-A-Pilot has sought flight schools to participate in the program and offers a $49 introductory flight certificate that can be redeemed at a participating flight school.

The goal of this program is to encourage people to experience their dreams of flying through an introductory flight and to begin taking flying lessons.

For more information, you can visit the Be-A-Pilot home page at http://www.beapilot.com or call 1-888-BE-A-PILOT.

CIVIL AIR PATROL: CADET ORIENTATION FLIGHT PROGRAM

The Civil Air Patrol (CAP) Cadet Orientation Flight Program is designed to introduce CAP cadets to general aviation operations. The program is voluntary and primarily motivational, and it is designed to stimulate the cadet's interest in and knowledge of aviation.

Each orientation flight includes at least 30 min. of actual flight time, usually in the local area of the airport. Except for takeoff, landing, and a few other portions of the flight, cadets are encouraged to handle the controls. The Cadet Orientation Flight Program is designed to allow five front-seat and four back-seat flights. But you may be able to fly more.

For more information about the CAP cadet program nearest you, visit the CAP home page at http://www.cap.af.mil or call 1-800-FLY-2338.

LOGBOOK ENDORSEMENT FOR PRACTICAL TEST

The following endorsement must be in your logbook and be presented to your examiner at your practical test.

1. Endorsement for flight proficiency/practical test: FAR 61.39(a)(6) and 61.65(c)

I certify that __(First name, MI, Last name)__ has received the required training of Sec. 61.65(c) and (d). I have determined he/she is prepared for the Instrument-Airplane practical test. He/She has demonstrated satisfactory knowledge of the subject areas found deficient on his/her knowledge test.

Signed	Date	Name	CFI Number	Expiration Date

USE GLEIM'S *FAA TEST PREP* -- A POWERFUL TOOL IN THE GLEIM KNOWLEDGE TRANSFER SYSTEM

Give yourself the competitive edge! Because all of the FAA's "written" tests have been converted to computer testing, Gleim has developed software specifically designed to prepare you for the computerized pilot knowledge test.

- ➥ ***FAATP*** emulates the computer testing vendor of your choice -- CATS, LaserGrade, or AvTEST. You will be completely familiar with the computer testing system you will be using.
- ➥ ***FAATP*** has two interactive modes: "Study" and "Test." Study mode permits you to select questions from specific sources, e.g., Gleim modules, questions that you missed from the last session, etc. You can also determine the order of the questions (Gleim or random), and you can randomize the order of the answer choices for each question.
- ➥ ***FAATP*** precludes you from looking at the answers before you commit to an answer and provides the actual testing environment. This is a major difference from the book.
- ➥ ***FAATP*** contains the well-known Gleim answer explanations which are intuitively appealing and easy to understand.
- ➥ ***FAATP*** maintains a history of your proficiency in each topic. This enables you to focus your study only on topics that need additional study.
- ➥ ***FAATP*** is the most versatile and complete software available.

AN OVERVIEW OF GLEIM'S *FAA TEST PREP* SOFTWARE FOR WINDOWS

Gleim's *FAA Test Prep* for Windows™ contains many of the same features found in earlier versions. However, we have simplified the study process by incorporating the outlines and figures from our books into the new software. Everything you need to study for any of the FAA knowledge tests will be contained in one unique, easy-to-use program. Below are some of the enhancements you will find with our new study software.

Gleim's
FAA Test Prep
for Windows™

32-Bit Version

NEW for Students:

- A complete on-screen library of FAA figures
- The familiar Gleim outlines and questions contained in one convenient program
- Customizable test sessions that emulate the testing vendors (AvTest, CATS, LaserGrade)
- Improved performance analysis charts and graphs to track your study progress

NEW for Instructors:

- More ways to create and customize tests
- Print options that allow you to design and create quizzes for your students
- One comprehensive program to meet all of your students' needs

Visit our web site at www.gleim.com for more information!

NEW!

Gleim Internet Flight Instructor Refresher Course

only $99.95

- FAA-approved
- Fully computerized and easy to use at your own convenience
- 16 one-hour lessons, each using standard lesson format:
  - 30 true/false diagnostic and study questions
  - 10- to 20-page study outline
  - 10 multiple-choice question final quiz

Lesson 1 is FREE, with no obligation

Someone had to be first to offer an Internet FIRC: We are. We offer you a free trial lesson so you can determine for yourself that the Gleim FIRC will save you time, money, and frustration at your next CFI renewal. As with all Gleim products, this FIRC is different AND BETTER.

Why 30 true/false questions before you study the subject matter of each lesson? In addition to being a diagnostic self-evaluation, the true/false questions are an excellent study process. For most lessons, you will answer the true/false questions, print out and study the study outlines, and go immediately to the 10-question test that concludes each lesson. You can even use the outline as a reference and teaching tool to help your students!

Never used the Internet? Great! Here is an opportunity to see how it works. Find a friend who uses the Internet to show you FIRC at www.gleim.com.

Follow the steps on the Gleim web page to access the Gleim Internet Flight Instructor Refresher Course

INDEX

PILOT KNOWLEDGE (WRITTEN EXAM) BOOKS AND SOFTWARE

Before pilots take their FAA pilot knowledge tests, they want to understand the answer to every FAA test question. Gleim's pilot knowledge test books are widely used because they help pilots learn and understand exactly what they need to know to pass. Each chapter opens with an outline of exactly what you need to know to pass the test. Additional information can be found in our reference books and flight maneuver/practical test prep books.

Use ***FAA Test Prep*** software with the appropriate Gleim book to prepare for success on your FAA pilot knowledge test.

PRIVATE PILOT AND RECREATIONAL PILOT FAA WRITTEN EXAM ($15.95)

The test for the private pilot certificate consists of 60 questions out of the 738 questions in our book. Also, the FAA's pilot knowledge test for the recreational pilot certificate consists of 50 questions from this book.

INSTRUMENT PILOT FAA WRITTEN EXAM ($18.95)

The test consists of 60 questions out of the 899 questions in our book. Also, become an instrument-rated flight instructor (CFII) or an instrument ground instructor (IGI) by taking the FAA's pilot knowledge test of 50 questions from this book.

COMMERCIAL PILOT FAA WRITTEN EXAM ($14.95)

The test consists of 100 questions out of the 595 questions in our book.

FUNDAMENTALS OF INSTRUCTING FAA WRITTEN EXAM ($12.95)

The test consists of 50 questions out of the 192 questions in our book. This test is required for any person to become a flight instructor or ground instructor. The test needs to be taken only once. For example, if someone is already a flight instructor and wants to become a ground instructor, taking the FOI test a second time is not required.

FLIGHT/GROUND INSTRUCTOR FAA WRITTEN EXAM ($14.95)

The test consists of 100 questions out of the 855 questions in our book. This book is to be used for the Flight Instructor--Airplane (FIA), Basic Ground Instructor (BGI), and the Advanced Ground Instructor (AGI) knowledge tests.

AIRLINE TRANSPORT PILOT FAA WRITTEN EXAM ($26.95)

The test consists of 80 questions each for the ATP Part 121, ATP Part 135, and the flight dispatcher certificate. Studying for the ATP will now be a learning and understanding experience rather than a memorization marathon -- at a lower cost and with higher test scores and less frustration!!

FLIGHT ENGINEER FAA WRITTEN EXAM ($26.95)

The FAA's flight engineer turbojet and basic knowledge test consists of 80 questions out of the 686 questions in our book. This book is to be used for the turbojet and basic (FEX) and the turbojet-added rating (FEJ) knowledge tests.

REFERENCE AND FLIGHT MANEUVERS/PRACTICAL TEST PREP BOOKS

Our Flight Maneuvers and Practical Test Prep books are designed to simplify and facilitate your flight training and will help prepare pilots for FAA practical tests as much as the Gleim written exam books help prepare pilots for FAA pilot knowledge tests. Each task, objective, concept, requirement, etc., in the FAA's practical test standards is explained, analyzed, illustrated, and interpreted so pilots will gain practical test proficiency as quickly as possible.

Private Pilot Flight Maneuvers and Practical Test Prep	362 pages	($16.95)
Instrument Pilot Flight Maneuvers and Practical Test Prep	426 pages	($18.95)
Commercial Pilot Flight Maneuvers and Practical Test Prep	330 pages	($14.95)
Flight Instructor Flight Maneuvers and Practical Test Prep	538 pages	($17.95)

PILOT HANDBOOK ($13.95)

A complete pilot ground school text in outline format with many diagrams for ease in understanding. This book is used in preparation for private, commercial, and flight instructor certificates and the instrument rating. A complete, detailed index makes it more useful and saves time. It contains a special section on biennial flight reviews.

AVIATION WEATHER AND WEATHER SERVICES ($22.95)

A complete rewrite of the FAA's *Aviation Weather 00-6A* and *Aviation Weather Services 00-45E* into a single easy-to-understand book complete with maps, diagrams, charts, and pictures. Learn and understand the subject matter much more easily and effectively with this book.

FAR/AIM ($15.95)

The purpose of this book is to consolidate the common Federal Aviation Regulations (FAR) parts and the *Aeronautical Information Manual* into one easy-to-use reference book. The Gleim book is better because of bigger type, better presentation, improved indexes, and full-color figures. FAR Parts 1, 43, 61, 67, 71, 73, 91, 97, 103, 105, 119, Appendices I and J of 121, 135, 137, 141, and 142 are included.

GLEIM'S PRIVATE PILOT KIT

Gleim's *Private Pilot FAA Written Exam* book, *Private Pilot Flight Maneuvers and Practical Test Prep*, *Pilot Handbook*, *FAR/AIM*, a combined syllabus/logbook, a flight computer, a navigational plotter, and a versatile, all-purpose flight bag. Our introductory price (substantial savings over purchasing items separately) is far lower than similarly equipped kits found elsewhere. The Gleim Kit retails for **$119.95**. Gleim's *FAR/AIM*, *Private Pilot Syllabus/Logbook*, flight computer, navigational plotter, and flight bag are also available for individual sale. See our order form for details.

Please forward your suggestions, corrections, and comments concerning typographical errors, etc., to **Irvin N. Gleim • c/o Gleim Publications, Inc. • P.O. Box 12848 • University Station • Gainesville, Florida • 32604**. Please include your name and address on the back of this page so we can properly thank you for your interest. Also, please refer to both the page number and the FAA question number for each item.

1. ______________________________

2. ______________________________

3. ______________________________

4. ______________________________

5. ______________________________

6. ______________________________

7. ______________________________

8. ______________________________

9. ______________________________

10. ______________________________

11. ______________________________

12. ______________________________

13. ______________________________

14. ______________________________

15. ______________________________

16. ______________________________

17. ______________________________

Remember for superior service: Mail, e-mail, or fax questions about our books or software.
Telephone questions about orders, prices, shipments, or payments.

Name: ______________________________

Address: ______________________________

City/State/Zip: ______________________________

Telephone: Home: ________ Work: ________ FAX: ________

E-mail ______________________________